Fortran IV
in
Chemistry

Fortran IV
in
Chemistry

An introduction to computer-assisted methods

G. Beech

Senior Lecturer in Inorganic Chemistry
The Polytechnic, Wolverhampton

JOHN WILEY & SONS

London · New York · Sydney · Toronto

Library of Congress Cataloging in Publication Data:
Beech, Graham.
    FORTRAN IV in chemistry: an introduction to
        computer-assisted methods
    Includes bibliographical references and index.
1. Electronic data processing—Chemistry.
2. FORTRAN (Computer program language) I. Title.
QD39.3.E46B43   1976   540′.28′5424   75−2488

ISBN 0 471 06165 4

Typeset by Preface Ltd, Salisbury, Wiltshire and
printed in Great Britain at The Pitman Press, Bath.

Preface

In the last decade we have witnessed a great increase in the use of computer-assisted techniques in science teaching. This is particularly true in the undergraduate chemistry curriculum where great efforts have been made to use the computer as a laboratory calculator, tutor, tester and even as a tool for simulating chemical reactions. I feel that we are now entering a period of consolidation in which we will be able to view the previous exciting developments in perspective and we can begin to make decisions on the best ways to implement computer methods into our own subject.

FORTRAN has been used in the development of almost all of the programs at present in use in our Department. Although this choice was partly accidental, I feel that it was fortunate since it has permitted me to write this book on a computer language that is easily transferable. Transferability is likely to become much more important as we enter an era in which cooperative efforts must replace duplication.

I am pleased to thank all those who helped, in many ways in the preparation of this book. In particular, I thank:

Dr Keith Miller — who contributed Chapter 5.
My colleagues and students — for using the programs.
Our Computer Unit — whose girls first attracted me to computers.
Mrs Joan Thurstans — for typing it all and still being able to smile.

In view of the last acknowledgment, any remaining errors are my responsibility. I therefore welcome comments and opinions from readers of this book.

1975
Wolverhampton   Graham Beech

Contents

x

Chapter 1

FORTRAN, Computers and Chemistry

1.1 Introduction

The purpose of this book is not to teach FORTRAN. Of the 500 or so texts already published,[1,1a] there are several adequate for this purpose.[2] Indeed it can even be argued that FORTRAN is not the most useful language for a chemist who may also have access to other algebraic compilers such as ALGOL, PL-1, APL and, for conversational usage, BASIC, FOCAL and others. The 'best' language depends entirely upon the kind of usage the chemist has in mind. These can be classified[3] as: general purpose calculation; data reduction; statistical analysis; modelling; theoretical calculations; data acquisition and control; and data retrieval and file searching. Both FORTRAN and ALGOL find applications in almost all of these areas but, because of the relative simplicity of the language and its ease of transferability (i.e. machine independence), FORTRAN has remained by far the most popular programming language for chemists.

At a more general level, investigations are in progress at several centres with the object of evaluating the educational advantages of computer-assisted learning. Whatever the outcome there is no doubt that there are already many useful FORTRAN programs currently available to the teacher which will enable him to enrich the learning situation.

It is, however, unfortunate that many published programs are so simple that they would be better suited to a desk calculator, unless they are specifically intended as an introduction, in which case the wider applications should be made apparent. Programs that simulate chemical systems or perform repetitive calculations with great precision are not only more satisfactory, but also will emphasize to a student the power of computer techniques in chemistry. Many examples of programs in these categories will be found in the following chapters.

Throughout this book, the emphasis is on batch-processing of programs. Not only does this reflect the manner in which the programs were developed but, also, many users do not yet have access to interactive processing. Despite the limitations of batch work, we hope to show that many areas of chemistry are amenable to computer assistance.

The main purpose of this book is to provide the core material for a course in which chemistry students study computing as an ancillary subject. By solving problems of chemical significance, chemistry students can perceive the relevance of computer-assisted techniques, and therefore we have used data and systems closely related to the subjects normally studied in most undergraduate courses. There is a general increase in difficulty as one progresses through the book but it is important to remember that the aim has been to use computers as an aid to chemistry; therefore, we have not used strictly graded exercises as would be necessary if we were teaching the FORTRAN language and we assume that users of this book have attended an introductory course of lectures on FORTRAN IV.

1.2 Hardware and Software Used in Program Development

The computer used in the development of the programs in this book was an ICL 1903A with a 48K word core store, used in a batch-processing mode. This belongs to a relatively popular series of machines used in the U.K. We have avoided the use of magnetic tape or disc files so that modifications for other types of computer should be minimal. The majority of the programs described will run on less than 32K of core store and many will run on less than 16K.

One of the greatest problems in writing a book about FORTRAN programs is the extent to which machine independence can be achieved. Up to a few years ago there was a great lack of standardization of the FORTRAN language, in addition to the parallel existence of both FORTRAN II and IV dialects. Now, FORTRAN IV is much more widely used but differences between compilers still persist. A number of organizations have published recommendations for the standardization of FORTRAN IV including the American National Standards Institute (ANSI)[4] and the (U.K.) National Computing Centre (NCC).[5] Also, other bodies have summarized the ANSI recommendations.

Certain differences exist between ANSI FORTRAN, as defined in 1966,[4] and the dialect used on our ICL 1903A computer; the differences which concern this book are listed in Table 1.1. However, if we had not used the additional options offered by the ICL compiler, our programs would have lost some flexibility and ease of use, although this would not, by itself, have been sufficient justification for the adoption of the special 1900 features. However, since the 1966 ANSI report, many of these features have become standard on newer

Table 1.1

Important differences between 1900 and ANS1 FORTRAN IV. (Relevant to this text)

Feature	1900
(i) Length of name	32 characters (ANS1, 6 characters)
(ii) Exponentiation	Integer bases raised to real exponents
(iii) Mixed mode arithmetic	Operators + — * and/combining integer and real elements
(iv) Mixed mode logical expressions	Relational operators may combine integer and real elements
(v) Free format (a) nF0.0	n real (floating point) numbers will be read as input in any floating point form. Each number separated by a space, e.g. 25.2 493.7 0.478
(b) nI0	n integer (fixed point) numbers as input regardless of length of integer, e.g. 253 4972 12 493878
(vi) MASTER	This statement is required to define the beginning of the main program

Table 1.2

Working Party Proposals	Status in 1971
P1 Integer, real and double precision constants, variables and array elements shall be permitted to be mixed within an arithmetic or relational expression	Many compilers permit expressions of mixed type
P2 An arithmetic assignment shall be assignable to an arithmetic variable or array element of any type	Many compilers provide this facility
P11 An alternative form of Hollerith constants and description (e.g. 4HDATA) shall be permitted. The characters comprising the datum shall be preceded and followed an apostrophe (e.g. 'DATA')	Permitted by many compilers although apostrophe is not the universal delimiting character
P14 A formatted READ statement of the form READ $(u, f,$ END = 1)K shall be permitted such that statement 1 is executed when an endfile record is read	Many compilers provide this facility

Working party suggestion

That there should be facilities for 'free-format' (see Table 1.1)

4

compilers and this trend is likely to continue. We are optimistic that this attitude is correct in view of the 1971 report[6] of the Working Party of the British Computer Society's Specialist Group on FORTRAN. In this report, several extensions to ANSI FORTRAN were proposed and the status of the extensions (with regard to their availability) was presented. The relevant sections of this report to this book are summarized in Table 1.2, in which it can be seen that there is a strong trend towards the adoption of the extended FORTRAN features of Table 1.1. Therefore, in view of the current status of compiler development we feel that we are justified in using features (i) to (v) in Table 1.1. Feature (vi) was forced upon us and users will need to change this for their own programs. We were also able to use proposal P11 (Table 1.2) and would recommend its use as a more convenient alternative to Hollerith statements. This feature is referred to as a *literal* by IBM, on whose machines this is also a common feature. We have made occasional use of proposal P14 in the programs in this book and would mention that it is a very convenient feature if a program is to handle consecutive sets of data before transferring control to the END statement of a program. Users of this book having computers with older compilers may need to make minor changes to the programs to allow for the features that we have described, but the amendments should be small.

1.2.1 Program Structure and Conventions Adopted

We have tried to use a 'building block' principle so that fairly complex programs can be constructed from a simple main program and one or more of the subroutines described in Chapter 2. In order to permit this continuity of approach we always use:

Input Device 1 = 80 column card reader.
Output Device 2 = 120 column line printer.
Input Device 3 = 8 track paper tape reader.

For example:

```
C          THE NEXT CARD READS FROM THE LINE PRINTER
           READ(1,50)N
    50     FORMAT(I0)
C          N IS THE NUMBER OF INTEGERS TO BE READ
C          FROM A PAPER TAPE
           READ (3,54) (NUM(I), I = 1,N)
    54     FORMAT (I5)
           WRITE (2,72)N
    72     FORMAT (1H1, 'NUMBER OF POINTS', I4)
```

In common with standard practice, we use the following for control of output in format statements:

        1H1 = new page

either 1H = new line

   or 1X = new line
        1 = line space

For alphanumeric titles the use of enclosing quote marks '   ', as in statement 72 above is far more convenient than the use of a Hollerith field description such as 16HNUMBER OF POINTS (see Table 1.2).

1.3 Examples of Techniques
In this section, two basic techniques will be discussed which have applications in the succeeding chapters. These concern the manipulation of characters and graphical output. Typical programs which are useful in themselves illustrate the discussion.

1.3.1 Character Manipulation
Although FORTRAN programs are mainly concerned with the manipulation of numerical data, there are many occasions on which characters (e.g. letters of the alphabet) need to be input as data or to be output for graphical or aesthetic reasons. For example, one may wish to input chemical formulae and later output them together with numerical results.

    Characters can be stored in any FORTRAN variable but there are strong arguments for the use of integers and we shall only use integer variables for their storage. A DATA statement can be used to store variables in a program; for one group of characters, the form of this statement is

    DATA n_1 /TEXT/

where n_1 is a variable name and TEXT is either a Hollerith string or a character group enclosed by literals. For example, each of the following statements

    DATA JOT/5HCUSO4/
    DATA JOT/'CUSO4'/

would store the characters CUSO4 in the variable JOT. If the variable n_1 is an array, consecutive elements of the array can be used to store successive groups of characters. For example, if we have an array NAME(5), we could write a statement:

    DATA NAME/ONE,TWO,THREE,FOUR,FIVE/

This would store ONE in NAME(1), TWO in NAME(2) and so on. DATA statements are normally placed after the DIMENSION statement which must, of course, contain dimension specifications for any arrays in DATA statements.

Characters can also be stored by an input statement using an A*n* format where *n* must not exceed the number of characters stored per machine word. It is usually safe to have *n* up to 8 although on some smaller machines *n* must not exceed 4. For example, we could store a descriptive title thus:

        READ(1,10) (NAME(I), I = 1,10)
    10 FORMAT(10A8)

This would enable us to store a title up to 80 characters in length. A write statement can be used with a similar A*n* format (plus carraige control) for output. Although the user is unlikely to do so, it should be noted that a fixed list of groups of characters can be assigned to array variables with A*n* read formats rather than with a DATA statement. Further details of character storage are given by Day,[7] who also mentions the problems that can arise when variables which store characters are compared by an IF statement. Dependent on the way in which the IF statement is performed it is possible for *overflow* to occur since variables which store characters are invariably large positive or negative numbers.

Overflow can be avoided by a triple test in which only character values of the same sign are compared. To quote the example given by Day: to test for the equality of the character variables I and J, we could use the following:

        IF (I)1, 2, 2
    1   IF (J)3, 10, 10
    2   IF (J)10, 3, 3
    3   logical or arithmetic IF statement
   10   CONTINUE

It was found necessary to use this type of test on our machine if overflow was to be avoided in such comparisons.

As an example of a program which uses the features described in this section, refer to Table 1.3 for a listing of the program QUAN which calculates formula weights and percentage compositions of any (real or hypothetical) chemical compound. Up to 20 types of atom are allowed and the chemical symbol and number of each type are input in line 155 which makes use of the formatted READ statement allowing consecutive sets of data to be input until an endfile record is reached (proposal P14 in Table 1.2). The chemical symbol is read in A2 format and the number of that type of atom follows in I2 format. The data format is 20(A2, I2) which conveniently fills one 80 column card. In lines 157 to 162 the array ATM(J) is scanned until a blank entry is found which transfers control to statement 205 and assigns NUM to the number of types of atom present in the formula. Notice that the 'triple test' referred to above is used in this section. In the DO loops from lines

Table 1.3
Program QUAN for calculation of formula weights and percentage compositions of
chemical compounds

```
0008                    MASTER QUAN
0009                    INTEGER ATM(20),EL(110),S(20)
0010                    REAL MASS(20)
0011                    DIMENSION N(20),NO(20),PERCNT(20),T(20),U(20),ELMASS(110)
0012        C
0013        C              THIS PROGRAM CALCULATES FORMULA WEIGHTS AND PERCENT COM-
0014        C           POSITIONS.  DATA IS READ FROM UNIT 5, AND WRITTEN ON UNIT 6.
0015        C
0016        C           DATA CONSISTS OF THE ATOMIC SYMBOL AND THE NUMBER OF EACH TYPE
0017        C           OF ATOM.  THE FORMAT IS: 20(A2,I2), THIS ALLOWS FOR ONE CARD
0018        C           PER COMPOUND, AND 20 ATOMS PER COMPOUND, IF THE NUMBER OF
0019        C           ATOMS IS LEFT BLANK, THE PROGRAM ASSUMES THERE IS ONE ATOM,
0020        C           IF A TYPE OF ATOM IS PUT IN SEPERATELY( AS 2 DIFFERENT ATOMS
0021        C           WOULD BE),TWO DIFFERENT COMPUTATIONS ARE PRINTED,  THE COMPU-
0022        C           TATIONS ARE THEN SUMMED.  SAMPLE DATA IS GIVEN FOR
0023        C           CU SO4-5(H2O)
0024        C           ALL THREE DATA CARDS GIVE THE SAME DATA, ALTHOUGH THE FIRST TWO
0025        C           GIVE THE COMPUTATIONS ON THE OXYGEN IN THE WATER SEPERATELY.
0026        C           ONE LETTER ATOMIC SYMBOLS ARE IN THE 2ND COLUMN OF THE A2 FORMAT
0027        C              THE ATOMIC MASSES WERE TAKEN FROM "THE HANDBOOK OF CHEMISTRY
0028        C           AND PHYSICS". WEAST,ED.,(49TH ED.,COLUMBUS: THE CHEMICAL RUBBER
0029        C           CO,)
0030        C           PROGRAM WRITTEN BY:    WILLIAM R. VINCENT, JR.
0031                    DATA EL/' H','HE','LI','BE',' B',' C',' N',' O',' F','NE','NA','MG
0032                   A','AL','SI',' P',' S','CL','AR',' K','CA','SC','TI',' V','CR','MN'
0033                   B,'FE','CO','NI','CU','ZN','GA','GE','AS','SE','BR','KR','RB','SR',
0034                   C' Y','ZR','NB','MO','TC','RU','RH','PD','AG','CD','IN','SN','SB','
0035                   DTE',' I','XE','CS','BA','LA','CE','PR','ND','PM','SM','EU','GD','T
0036                   EB','DY','HO','ER','TM','YB','LU','HF','TA',' W','RE','OS','IR','PT
0037                   F','AU','HG','TL','PB','BI','PO','AT','RN','FR','RA','AC','TH','PA'
0038                   G,' U','NP','PU','AM','CM','BK','CF','ES','FM','MD','NO','LR',' ',
0039                   H' ',' ',' ',' ',' ',' ',' '/
0040                    DATA LBLANK/' '/
0041                    ELMASS( 1)=1.00797
0042                    ELMASS( 2)=4.0026
0043                    ELMASS( 3)=6.939
0044                    ELMASS( 4)=9.0122
0045                    ELMASS( 5)=10.811
0046                    ELMASS( 6)=12.01115
0047                    ELMASS( 7)=14.0067
0048                    ELMASS( 8)=15.9994
0049                    ELMASS( 9)=18.9984
0050                    ELMASS( 10)=20.183
0051                    ELMASS( 11)=22.9898
0052                    ELMASS( 12)=24.312
0053                    ELMASS( 13)=26.9815
0054                    ELMASS( 14)=28.086
0055                    ELMASS( 15)=30.9738
0056                    ELMASS( 16)=32.064
0057                    ELMASS( 17)=35.453
0058                    ELMASS( 18)=39.948
0059                    ELMASS( 19)=39.102
0060                    ELMASS( 20)=40.08
0061                    ELMASS( 21)=44.956
0062                    ELMASS( 22)=47.90
0063                    ELMASS( 23)=50.942
0064                    ELMASS( 24)=51.996
0065                    ELMASS( 25)=54.9380
0066                    ELMASS( 26)=55.847
0067                    ELMASS( 27)=58.9332
0068                    ELMASS( 28)=58.71
0069                    ELMASS( 29)=63.546
0070                    ELMASS( 30)=65.37
0071                    ELMASS( 31)=69.72
0072                    ELMASS( 32)=72.59
0073                    ELMASS( 33)=74.9216
0074                    ELMASS( 34)=78.96
0075                    ELMASS( 35)=79.904
0076                    ELMASS( 36)=83.80
0077                    ELMASS( 37)=85.47
0078                    ELMASS( 38)=87.62
0079                    ELMASS( 39)=88.905
0080                    ELMASS( 40)=91.22
0081                    ELMASS( 41)=92.906
0082                    ELMASS( 42)=95.94
```

```
0083                    ELMASS( 43)=97.0
0084                    ELMASS( 44)=101.07
0085                    ELMASS( 45)=102.905
0086                    ELMASS( 46)=106.4
0087                    ELMASS( 47)=107.868
0088                    ELMASS( 48)=112.40
0089                    ELMASS( 49)=114.82
0090                    ELMASS( 50)=118.69
0091                    ELMASS( 51)=121.75
0092                    ELMASS( 52)=127.60
0093                    ELMASS( 53)=126.9044
0094                    ELMASS( 54)=131.30
0095                    ELMASS( 55)=132.905
0096                    ELMASS( 56)=137.34
0097                    ELMASS( 57)=138.91
0098                    ELMASS( 58)=140.12
0099                    ELMASS( 59)=140.907
0100                    ELMASS( 60)=144.24
0101                    ELMASS( 61)=145.0
0102                    ELMASS( 62)=150.35
0103                    ELMASS( 63)=151.96
0104                    ELMASS( 64)=157.25
0105                    ELMASS( 65)=158.924
0106                    ELMASS( 66)=162.50
0107                    ELMASS( 67)=164.930
0108                    ELMASS( 68)=167.26
0109                    ELMASS( 69)=168.934
0110                    ELMASS( 70)=173.04
0111                    ELMASS( 71)=174.97
0112                    ELMASS( 72)=178.49
0113                    ELMASS( 73)=180.948
0114                    ELMASS( 74)=183.85
0115                    ELMASS( 75)=186.2
0116                    ELMASS( 76)=190.2
0117                    ELMASS( 77)=192.2
0118                    ELMASS( 78)=195.09
0119                    ELMASS( 79)=196.967
0120                    ELMASS( 80)=200.59
0121                    ELMASS( 81)=204.37
0122                    ELMASS( 82)=207.19
0123                    ELMASS( 83)=208.980
0124                    ELMASS( 84)=209.0
0125                    ELMASS( 85)=210.0
0126                    ELMASS( 86)=222.0
0127                    ELMASS( 87)=223.0
0128                    ELMASS( 88)=226.0
0129                    ELMASS( 89)=227.0
0130                    ELMASS( 90)=232.0
0131                    ELMASS( 91)=231.0
0132                    ELMASS( 92)=238.0
0133                    ELMASS( 93)=237.0
0134                    ELMASS( 94)=244.0
0135                    ELMASS( 95)=243.0
0136                    ELMASS( 96)=247.0
0137                    ELMASS( 97)=247.0
0138                    ELMASS( 98)=251.0
0139                    ELMASS( 99)=254.0
0140                    ELMASS(100)=257.0
0141                    ELMASS(101)=256.0
0142                    ELMASS(102)=254.0
0143                    ELMASS(103)=257.0
0144                    ELMASS(104)=0.0
0145                    ELMASS(105)=0.0
0146                    ELMASS(106)=0.0
0147                    ELMASS(107)=0.0
0148                    ELMASS(108)=0.0
0149                    ELMASS(109)=0.0
0150                    ELMASS(110)=0.0
0151              1 READ (1,204,END=99999)(ATM(J),NO(J),J=1,20)
0152                    TMASS=0.0
0153                    DO 200 J=1,20
0154                    IF(ATM(J))11,12,12
0155             11 IF(LBLANK)13,200,200
0156             12 IF(LBLANK)200,13,13
0157             13 IF(ATM(J).EQ.LBLANK)GO TO 205
0158            200 CONTINUE
0159                    IF(ATM(20))4,5,5
0160              4 IF(LBLANK)7,205,205
0161              5 IF(LBLANK)205,7,7
0162              7 IF(ATM(20).NE.LBLANK)J=21
0163            205 NUM=J-1
0164                    DO 1040 J=1,NUM
```

```
0165                    IF(NO(J).EQ.0)NO(J)=1
0166                    DO 110 JOT=1,110
0167                    IF(ATM(J))31,32,32
0168                 31 IF(EL(JOT))33,110,110
0169                 32 IF(EL(JOT))110,33,33
0170                 33 IF(ATM(J).EQ.EL(JOT))GO TO 111
0171                110 CONTINUE
0172                    JOT=110
0173                111 CONTINUE
0174                    MASS(J)=NO(J)*ELMASS(JOT)
0175               1040 TMASS=TMASS+MASS(J)
0176                    WRITE(2,820)(ATM(J),NO(J),J=1,NUM)
0177                    WRITE(2,5490)TMASS
0178                    DO 1041 J=1,NUM
0179                    PERCNT(J)=MASS(J)/TMASS*100
0180               1041 WRITE(2,300)NO(J),ATM(J),MASS(J),PERCNT(J)
0181                    DO 303 J=1,NUM
0182                303 N(J)=0
0183                    DO 304 I=1,NUM
0184                    K=0
0185                    S(I)=NO(I)
0186                    T(I)=MASS(I)
0187                    U(I)=PERCNT(I)
0188                    DO 305 J=I,NUM
0189                    IF(J.EQ.I)GO TO 305
0190                    IF(N(I).NE.0)GO TO 304
0191                    IF(ATM(I))21,22,22
0192                 21 IF(ATM(J))23,305,305
0193                 22 IF(ATM(J))305,23,23
0194                 23 IF(ATM(I).NE.ATM(J))GO TO 305
0195                    K=K+1
0196                    N(J)=I
0197                    IF(K.GT.1)N(J)=25
0198                    S(I)=S(I)+NO(J)
0199                    T(I)=T(I)+MASS(J)
0200                    U(I)=U(I)+PERCNT(J)
0201                305 CONTINUE
0202                304 CONTINUE
0203                    DO 401 I=1,NUM
0204                    IF(N(I).EQ.0)GO TO 401
0205                    IF(N(I).EQ.25)GO TO 401
0206                    J=N(I)
0207                    WRITE(2,522)S(J),ATM(J),T(J),U(J)
0208                401 CONTINUE
0209                    GO TO 1
0210              99999 CONTINUE
0211                204 FORMAT(20(A2,I2))
0212                300 FORMAT(1X,I2,' ATOMS OF ',A2,' HAVE A MASS OF ',F10.5,' AMU., AND
0213                   * ',F8.4,'% OF THE TOTAL FORMULA MASS')
0214                522 FORMAT(1X,/' A TOTAL OF ',I3,' ATOMS OF ',A2,' = ',F12.5,' AMU. ,A
0215                   *ND A TOTAL OF ',F8.4,'% OF THE FORMULA MASS')
0216                820 FORMAT(1H1,////' MOLECULE :'/ 11X,20(A2,I2))
0217               5490 FORMAT(1X,'HAS A FORMULA MASS OF ',F15.5,' AMU.')
0218                    STOP
0219                    END
```

```
MOLECULE :
          CU 1 S 1 O 4 H10 O 5
HAS A FORMULA MASS OF      249.68430 AMU.
  1 ATOMS OF CU HAVE A MASS OF      63.54600 AMU., AND      25.4505% OF THE TOTAL FORMULA MASS
  1 ATOMS OF  S HAVE A MASS OF      32.06400 AMU., AND      12.8418% OF THE TOTAL FORMULA MASS
  4 ATOMS OF  O HAVE A MASS OF      63.99760 AMU., AND      25.6314% OF THE TOTAL FORMULA MASS
 10 ATOMS OF  H HAVE A MASS OF      10.07970 AMU., AND       4.0370% OF THE TOTAL FORMULA MASS
  5 ATOMS OF  O HAVE A MASS OF      79.99700 AMU., AND      32.0393% OF THE TOTAL FORMULA MASS

A TOTAL OF   9 ATOMS OF   O =     143.99460 AMU. ,AND A TOTAL OF   57.6707% OF THE FORMULA MASS




MOLECULE :
          CU 1 S 1 O 9 H10
HAS A FORMULA MASS OF      249.68430 AMU.
  1 ATOMS OF CU HAVE A MASS OF      63.54600 AMU., AND      25.4505% OF THE TOTAL FORMULA MASS
  1 ATOMS OF  S HAVE A MASS OF      32.06400 AMU., AND      12.8418% OF THE TOTAL FORMULA MASS
  9 ATOMS OF  O HAVE A MASS OF     143.99460 AMU., AND      57.6707% OF THE TOTAL FORMULA MASS
 10 ATOMS OF  H HAVE A MASS OF      10.07970 AMU., AND       4.0370% OF THE TOTAL FORMULA MASS
```

168 to 179, the identity of each atom stored in ATM(J) is found by a logical IF statement (line 174) which compares ATM(J) with the list of element symbols EL(JOT) which were assigned in the DATA statement near the commencement of the program. The formula weight is stored in TMASS by summing the individual mass contributions in the array MASS(J) which is computed (lines 178–179) from the product of the number of atoms of type J and their precise atomic weights which were assigned in lines 45–154 using the array ELMASS. Notice that the last seven entries in ELMASS are left vacant to allow for extension of the periodic table! The input data is listed (line 180) using A format together with the total mass. This is followed by a computation which gives the percentage composition on an atom-by-atom basis (line 184). In the final section of the program, the presence of any identical atoms in the input data is sought. If the same types of atom are found as different entries on the input card, then the total number of these atoms is computed with their total mass and their total percentage of the formula mass. Therefore, in the first example in the table, CU01 S01O 04H10 O 05 includes output for 4 atoms, 5 atoms and finally 9 atoms of oxygen. On the other hand, CU01 S01O 09H10 only gives output for the 9 atoms of oxygen. Input for this program is one card per molecule in the format 20(A212). The final card must be an end-of-life record which in the case of ICL1900 series computers is ****, although it may be different on your own machine. If your compiler is not able to read until an end-of-file record you can easily amend the program to read in a set number of data cards.

1.3.2 Graphical Output

Graphical display of data is important to a chemist as a means of quickly examining such features as trends in data, linearity, signal noise and other important properties. The most accurate method is to use a digital plotter but these tend to be slow output devices and are highly machine-orientated both mechanically and in the use of plotting subroutines. For this reason we have avoided, where possible, the use of this type of output device even when it would seem to be desirable (for example, the pH program in Section 3.2.2). In most cases, little labour is required on the part of the programmer to produce graphical output on a plotter. The exception that we make to our use of the line printer is in the plotting of contour diagrams, dealt with in Chapter 5.

The use of the line printer for both numeric and graphical data is not however without advantages, as it is a fast output device and also lends itself well to the plotting of histograms and density functions.

As an example of the use of a line printer for data presentation, let us assume that we have the results of N analytical measurements and we wish to display the magnitude of each result in the form of a simple bar histogram. Our line printer can print up to 120 characters across a page.

Therefore, we may decide that the plot can begin in column 10 and end in column 110. The data must then be scaled to lie between 1 and 100 so that we can print histogram bars of the correct relative lengths. Presuming our data to be in the array DATA(100), we require data scaling statements of the form:

```
      MAX = DATA(1)
      DO  1   I = 2, N
1     IF (DATA(I). GT.MAX) MAX = DATA(I)
      DO  2   I = 1, N
2     DATA(I) = DATA(I) * 100/MAX
```

Let us now output a histogram in horizontal bars consisting of dashed lines. This requires a DATA statement and a simple print statement, the bare essentials being:

```
      DATA INT/1H–/
      {                    }    data scaling statements
      DO 1 I = 1, N
      LENGTH = DATA(I)
1     WRITE (2,2) (INT, J = 1, LENGTH)
3     FORMAT (10X, 100A1)
```

These statements produce a simple histogram of the form:

```
 —   —   —
 —   —   —   —
 —   —   —   —   —   —
 —   —   —   —
 —   —   —   —   —
 —   —
```

One line of characters at a time is printed and the array DATA(100) is therefore called a *line buffer*.

Rather more useful is the device called a *page buffer* which is simply a two-dimensional array used to temporarily store characters before being output onto one page of a line printer. For a printer with a width of 120 characters and 66 lines per page, a convenient size page buffer is 100 x 55 giving 5500 plotting positions. This is a similar situation to the use of a raster in building up a picture on a television screen. Since we will be plotting characters we will use an integer array for this buffer and an example would be the array IPOS(100, 55). The X and Y coordinates can be specified as indices, IX and IY and the position of any character on the page is then defined by IPOS(IX, IY). Let us apply this technique to the problem described above and attempt to produce a *vertical* histogram with x and y as axes suitably annotated. Let the y-axis be in column 5 with tick marks on every fifth line and have a height of 55 lines. The x-axis will therefore begin at line 60 and can conveniently extend for 100 characters, again with ticks marks at every fifth position. We presume that we have up to 100 items of data to display and that these have been scaled (see above) to lie between 1 and 55. We require a slightly more complex data statement than that used previously and the page buffer must initially be blanked out.

```
      DIMENSION KAR(5), IPOS(100,55)
      DATA KAR/1H , 1H+, 1H−, 1HI, 1H*/
      DO 1 I = 1,100
      DO 1 J = 1,55
1     IPOS(I,J) = KAR(1)
```

Firstly store the x-axis:

```
      DO 2 I = 1,100
      IPOS(I,1) = KAR(3)
2     IF (MOD(I,5).EQ.0)IPOS (I,1) = KAR(2)
```

Similarly the y-axis:

```
      DO 3 J = 1,55
      IPOS (1,J) = KAR(4)
3     IF (MOD (J,5).EQ.0) IPOS (1,J) = KAR(2)
```

Note the use of the MOD (remaindering) statements. Finally, we add the N experimental data, stored in the array DATA(100). We count backwards from 56 to ensure the histogram is not presented upside down:

```
      DO 4 I = 1,N
      ITEST = 56
5     ITEST = ITEST−1
      IF (DATA (I) LT.ITEST)GO TO 5
      DO 6 J = 2, ITEST
6     IPOS (I,J) = KAR(5)
4     CONTINUE
```

Having stored all this information, we can output it onto the line printer thus:

```
      WRITE (2,9)
9     FORMAT (/////)
      DO 7 I = 1,55
      J = 56−I
7     WRITE(2,8) (IPOS(K,J),K = 1,100)
8     FORMAT (10X, 100 A1)
```

Annotation of a title in the X-direction is done most easily by using a conventional print statement. The Y-axis can be annotated either by different FORMAT statements or by the use of more extended DATA statements. The histogram could be given clarity by making each bar several characters wide. For example, the loop commencing with DO 6 J = 2, ITEST could be modified to:

```
        K = 5
        DO 6 J = 2, ITEST
        DO 6 J1 = K, K+3
6       IPOS(J1,J) = KAR(5)
        K = K+5
```

A complete subroutine based on the above discussion is presented in Chapter 2 for the graphical analysis of statistics.

Another use of a line printer is to produce *density plots*, best known to chemists from atomic orbital theory. The technique generally used in plotting such diagrams is to use characters of such dimensions and shape that they convey roughly the magnitude of a function. For example X can be used to convey the sense of high density, + or * for medium density and blank or − for low density. The only other criterion for the selection of suitable characters is that they should be highly symmetrical, both in shape and in printing position. The array elements of a page buffer, described above, can be used although it is important to remember that the printing density is not equal in the X and Y directions. For example, the printer described in this section has 120 spaces across the page but only 60 spaces downwards. Unless we take precautions, a function of circular symmetry will appear to be oblate. One way around this problem is as follows: let us presume we wish to display the magnitude of a function at various x, y distances from the origin $(x = 0, y = 0)$. Firstly, we set up a page buffer (two-dimensional integer array) say NSYM(M,N) such that N characters in the horizontal direction are as closely as possible equal to the vertical spacing of M new lines. Therefore, if we were to output NSYM(M,N) by the statements:

```
        DO 1 I = 1,M
1       WRITE (2,3) (NSYM(I,J),J = 1,N)
3       FORMAT (100 A1)
```

then the result would be enclosed by a square, with sides of approximately equal length, filled by the contents of the array. The *paper* coordinates, I and J, clearly have circular symmetry so that these can be selected by our program and then converted into real spatial coordinates. For example, suppose that we wish to represent the magnitude of the function:

$$f = xy$$

for values of x and y between ±1. What we must do is to use the indices of the grid described above in order to generate real x and y spatial

14

coordinates. Let us presume that the size of the array is 83 x 51, i.e. 83 character spaces are roughly equal in length to 51 line spaces. We calculate the value of f at each position on the grid, compare it with prescribed limits and then store the appropriate character in the page buffer. The following statements would generate a density map using five characters (including a blank space) for the range x, y equal to ±1:

```
        DIMENSION KTOR(6), F(51,83), NSYM(51,83)
        DATA KTOR/1H , 1H—, 1H+, 1H*/
        FMAX = 0
        DO 4 J = 1,51
        DO 4 K = 1,83
        X = (J—26)/25
        Y = (K—42)/41
        F(J,K) = X * Y
    4   IF(F(J K).GT FMAX)FMAX = F(J,K)
        DO 6 J = 1,51
        DO 5 K = 1,83
        JK = (4.9 * F(J,K)/FMAX) + 1
    5   NSYM(J,K) = KTOR(JK)
    6   WRITE (2,7) (NSYM(J,K), K = 1,83)
    7   FORMAT (18X, 83A1)
```

Note that, in the statement which generates the index JK, the factor 4.9 is used rather than 5.0. Because of the truncation which occurs in the integer assignment, this ensures that only five values of JK can be generated and by the addition of unity, ensures that they are non-zero. A factor of 4.0 is not used since this would only allow JK to equal 5 for the maximum value of F(J,K).

Contour plotting of functions is more accurately done on a graph plotter than on a line printer. Since we make use of contour plotting in Chapter 5, a listing of a typical package is given in Appendix A. As can be seen, it is highly machine-oriented although given the definitions of the plotting routines, it should be convertible to different machines.

References

1. Stock, M., and Stock, K. F., *Bibliography of Programming Languages*, Verlag-Dokumentation, Pullach/Munchen, 1973.
1a. Martin, F., *Computing*, p. 9 (July 5 1973).
2. Examples are (a) Court, R., *Fortran for Beginners*, Holmes McDougall, Edinburgh, 1970; (b) McCracken, D. D., *A Guide to Fortran IV Programming*, 2nd Edn., Wiley-Interscience, New York, 1972; (c) Organick, E. I., *A Fortran IV Primer*, Addison-Wesley, Reading, Mass., 1966.
3. Anderson, R. E., *J. Chromatog. Science*, 10, 8 (1972).
4. American National Standards Institute, *American National*

Standard for FORTRAN, Document X3.9— 1966, New York; ANSI 1966.

5. *Standard FORTRAN Programming Manual*, National Computing Centre, Manchester, 1970.

6. 'The next standard FORTRAN', *The Computer Bulletin*, 15 (1971).

7. Day, A. C., *FORTRAN Techniques with Special Reference to non-Numerical Applications*, Cambridge University Press, Cambridge, 1972.

Chapter 2

Numerical Methods

2.1 Introduction

The subject area of numerical methods that we shall consider includes statistics, some aspects of calculus and the fitting of experimental data to theoretical functions. In order to make the best use of this chapter it would be useful for you to have a knowledge of:

(i) Elementary statistics.[1]
(ii) Differentiation and integration of simple functions.
(iii) The properties of matrices and determinants.[2]

References 1 and 2 will be useful for readers wishing to learn about or to revise the above subjects.

At the end of this chapter, having correctly completed the worked examples and problems, you will be able to write FORTRAN programs to:

(a) Calculate standard deviations, means and probabilities.
(b) Plot histograms for experimental data.
(c) Solve simultaneous equations.
(d) Fit experimental data to theoretical models.
(e) Solve eigenvalue problems.
(f) Perform differentiation and integration.

An important objective of this chapter is to develop useful programs and subroutines that can be applied in later sections of this book and in many routine applications in chemistry. The reader is therefore recommended to work through all of this chapter.

2.2 Statistics

2.2.1 Introduction

Chemical experiments and measurements are always subject to error owing to instrumental limitations and/or the limitations of the experimenter. For example, it is common practice to repeat a titration in order to reduce the random errors associated with the detection of an end point. Random errors are, therefore, thought of as differences between experimental results and the 'true value'. The latter implies that our results are distributed statistically about the true value — which is unlikely to be strictly true unless we make a very large number (an infinite number, to be correct) of experimental observations. To circumvent this problem it is usual to calculate the mean (average) result and to associate with it an estimated uncertainty: the so-called 'standard deviation'; thus we might record an analytical result as $30.5 \pm 0.4\%$. In order to calculate the uncertainty attached to a mean, it is necessary to consider the way in which we expect our data to be distributed about the mean. This distribution is defined by $p(x)$ — the probability density function of the variable, x. It can be thought of as defining the probability of x taking a value between $x \pm dx$, where dx is a small range of x. If we knew the form of $p(x)$, we could define the mean (μ_x) and variance (σ_x^2) of our data:

$$\mu_x = \int_{-\infty}^{\infty} x p(x) \cdot dx \approx \sum_{-\infty}^{\infty} x_i p(x_i) \Delta x$$

$$\sigma_x^2 = \int_{-\infty}^{\infty} (x - \mu_x)^2 p(x) \cdot dx \approx \sum_{-\infty}^{\infty} (x_i - \mu_x)^2 p(x) \Delta x$$

The summations replace integrations in all practical situations with Δx being a constant interval chosen to uniformly divide the range of x as shown in Figure 2.1. In any practical situation, only a finite number of x-values are recorded. This number is a sample of the population for

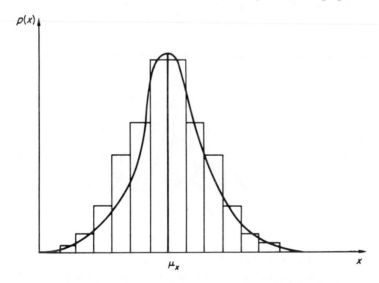

Figure 2.1   Illustration of the form of a Gaussian distribution

which acceptable estimates of μ_x and σ_x^2 are given by the sample mean, $\bar{x}$ and sample variance, s^2 :

$$\bar{x} = \frac{1}{N} \sum_{i=1}^{N} x_i \tag{2.1}$$

$$s^2 = \frac{1}{N-1} \sum_{i=1}^{N} (x_i - \bar{x})^2 \tag{2.2}$$

The denominator $(N - 1)$ in equation (2.2) is preferred to N if an unbiased estimator is to be obtained.[3]

The great majority of experimental data closely follow a Gaussian or normal distribution for which

$$p(x) = [\sigma_x \sqrt{(2\pi)}]^{-1} \exp \left[\frac{(x - \mu_x)^2}{2\sigma_x^2} \right] \tag{2.3}$$

In equation (2.3), σ_x is the standard deviation of x and is equal to the positive square root of the variance. The form of $p(x)$ for various values of σ_x is illustrated in Figure 2.2. The form of $p(x)$ can be modified by using a normalized variable, z,

$$x = (x - \mu_x)/\sigma_x \tag{2.4}$$

so that in terms of z, we have a zero mean and a standard deviation of unity. With this new variable, we obtain

$$p(z) = [\sqrt{(2\pi)}]^{-1} \exp (-z^2/2)$$

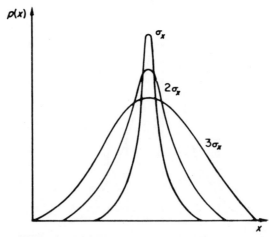

Figure 2.2   Dependence of the shape of a Gaussian curve on the value of σ_x, the standard deviation

Integration of $p(z)$ between limits will tell us the probability of z lying between those limits. For example, with $-1 < z < 1$, approximately 68% of the area of the curve in Figure 2.2 is enclosed; this rises to about 95% for $-2 < z < 2$ and to 99.7% for $-3 < z < 3$. These percentages are equal to the probability of z occurring within the prescribed ranges. So, for example, if the mean is 300.0 and the standard deviation is 15.0 we would expect about 68% of our results (if we sampled for a very long time!) to lie between 285.0 and 315.0. To test this, we can use the program START (Table 2.1) to process a number of experimental values and obtain their mean, standard deviation and the individual differences from the sample mean. The program is a simple one; firstly a title card (10A8) is read and then in line 13 an integer, J, is read in IO format; this represents the number of data points. The J data points are then read by an implied DO (line 15') into the array NUM(1000). The following arithmetic statements calculate the mean, the deviations, the squares of the deviations and the standard deviation.

In Table 2.1 we list the output from START for fifty determinations of the radioactivity of a compound obtained from a Geiger–Muller counter. The mean value is seen to be 413 and the standard deviation is 11. Suppose we are interested to know the number which lie between $413 \pm (11 \times n)$, where n is an integer. By simply counting the number of results within the specified limits, we obtain:

	$n = 1$	$n = 2$	$n = 3$
% Results within limits	66	96	100
'Theoretical'	68	95	99.7

Clearly the results follow the expected distribution quite closely. Some other important properties are based on using a sample mean $\bar{x}$ as an estimator of a population mean. A standard statistical argument shows that if $\bar{x}$ is repeatedly determined then it is always normally distributed about the population mean, μ_x. From this we can calculate the standard deviation of the mean, σ_x:

$$\sigma_{\bar{x}} = \sigma_x / \sqrt{N} \qquad (2.5)$$

where σ_x is the sample standard deviation (estimated by s) of N data points. We also can apply confidence intervals of the type

$$(\bar{x} + z\sigma_{\bar{x}}) \quad \text{to} \quad (\bar{x} - z\sigma_{\bar{x}})$$

where $z = (\bar{x} - \mu_x)/\sigma_x$

The range $z = +1.96$ to $z = -1.96$ contains 95% of the normal curve. Thus with z equal to 1.96, the confidence interval is 95% certain to contain the true mean, σ_x. Similarly with z equal to 2.75, our confidence interval is 99.7% certain to contain μ_x. Strictly speaking the above argument is only true for very large numbers of samples; for

Table 2.1

Program START for the statistical analysis of sets of integers. (The comment line refers to the large free format statement used intentionally in this program.)

```
0008                MASTER START
0009                REAL MEAN
0010                DIMENSION DEV(1000),DEVSQ(1000),H(10),NUM(1000)
0011                READ(1,100)(H(I),I=1,10)
0012      100       FORMAT(10A8)
0013                READ(1,10)J
0014      10        FORMAT(IU)
0015                READ(1,11)(NUM(I),I=1,J)
0016      11        FORMAT(1000I0)
COMMENT 178   IS THIS LARGE A REPEAT COUNT INTENDED AT ABOUT COLUMN 16, LINE 0016

0017                DO 12 I=1,J
0018      12        TOT=TOT+NUM(I)
0019                MEAN=TOT/J
0020                TOT=0
0021                DO 13 I=1,J
0022                DEV(I)=NUM(I)-MEAN
0023                DEVSQ(I)=ABS(DEV(I))**2
0024      13        TOT=TOT+DEVSQ(I)
0025                VAR=TOT/(J-1)
0026                STD=SQRT(VAR)
0027                WRITE(2,14)(H(I),I=1,10)
0028      14        FORMAT(1H1,9X,10A8//10X,'NUMBER',10X,'DEVIATION',10X,'DEVN,SQRD'/1
0029                C0X,6(1H-),2(10X,9(1H-)))
0030                WRITE(2,15)((I,NUM(I),DEV(I),DEVSQ(I)),I=1,J)
0031      15        FORMAT(1 X,I4,8X,I4,12X,F7.2,12X,F7.2)
0032      99        WRITE(2,18)MEAN,VAR,STD
0033      18        FORMAT(1H0,10X,'MEAN = ',F7.2//10X,'VARIANCE = ',F7.2//10X,'STANDA
0034                CRD DEVIATION = ',F7.2)
0035                STOP
0036                END
```

TEST

	NUMBER	DEVIATION	DEVN,SQRD
1	1434	1,11	1,23
2	1387	-45,89	2105,89
3	1430	-2,89	8,35
4	1435	2,11	4,45
5	1403	-29,89	893,41
6	1442	9,11	82,99
7	1418	-14,89	221,71
8	1451	18,11	327,97
9	1448	15,11	228,31
10	1426	-6,89	47,47
11	1433	0,11	0,01
12	1417	-15,89	252,49
13	1440	7,11	50,55
14	1422	-10,89	118,59
15	1385	-47,89	2293,45
16	1463	50,11	906,61
17	1412	-20,89	436,39
18	1420	-12,89	166,15
19	1362	-70,89	5025,39
20	1441	8,11	65,77
21	1399	-53,89	1148,53
22	1371	-61,89	3830,37
23	1397	-35,89	1288,09
24	1497	64,11	4110,09
25	1413	-19,89	395,61
26	1426	-6,89	47,47
27	1447	14,11	199,09
28	1468	35,11	1232,71
29	1404	-28,89	834,63
30	1485	52,11	2715,45
31	1429	-3,89	15,13
32	1487	54,11	2927,89
33	1466	33,11	1096,27
34	1461	28,11	790,17
35	1416	-16,89	285,27
36	1439	6,11	37,33
37	1383	-49,89	2489,01
38	1464	31,11	967,83
39	1384	-48,89	2390,23
40	1435	2,11	4,45
41	1432	-0,89	0,79

	TEST		
---	NUMBER	DEVIATION	DEVN.SQRD
42	1416	-16,89	285,27
43	1437	4,11	16,89
44	1417	-15,89	252,49
45	1406	-26,89	723,07
46	1465	32,11	1031,05
47	1454	21,11	445,63
48	1433	0,11	0,01
49	1411	-21,89	479,17
50	1435	2,11	4,45
51	1468	35,11	1232,71
52	1439	6,11	37,33
53	1405	-27,89	777,85
54	1392	-40,89	1671,99
55	1461	28,11	790,17
56	1463	30,11	906,61
57	1471	38,11	1452,37
58	1416	-16,89	285,27
59	1463	30,11	906,61
60	1448	15,11	228,31
61	1488	55,11	3037,11
62	1409	-23,89	570,73
63	1482	49,11	2411,79
64	1464	31,11	967,83
65	1397	-35,89	1288,09
66	1473	40,11	1608,81
67	1418	-14,89	221,71
68	1446	13,11	171,87
69	1385	-47,89	2293,45
70	1406	-26,89	723,07
71	1446	13,11	171,87
72	1355	-77,89	6066,85
73	1440	7,11	50,55
74	1472	39,11	1529,59
75	1422	-10,89	118,59
76	1434	1,11	1,23
77	1391	-41,89	1754,77
78	1442	9,11	82,99
79	1492	59,11	3493,99
80	1445	12,11	146,65
81	1388	-44,89	2015,11
82	1425	-7,89	62,25
83	1452	19,11	365,19
84	1470	37,11	1377,15
85	1492	59,11	3493,99
86	1416	-16,89	285,27
87	1421	-11,89	141,37
88	1455	22,11	488,85
89	1433	0,11	0,01
90	1462	29,11	847,39
91	1404	-28,89	834,63
92	1447	14,11	199,09
93	1428	-4,89	23,91
94	1397	-35,89	1288,09
95	1451	18,11	327,97
96	1422	-10,89	118,59
97	1420	-12,89	166,15
98	1472	39,11	1529,59
99	1445	12,11	146,65
100	1410	-22,89	523,95

MEAN = 1432,89

VARIANCE = 934,52

STANDARD DEVIATION = 30,57

smaller numbers, z should be increased to obtain the same degree of confidence that μ is within the stated limits. It is usual to assume z to be 2 and 3 for the 95% and 99.7% levels respectively.

2.2.2 Graphical Studies of Statistics
Program START discussed earlier in this section, clearly has some value but it is lacking in visual impact. An improvement would be to

manually plot a histogram but, for a large amount of data, this is tedious and can be more easily achieved by the use of a subroutine such as HTGRAM (Table 2.2). This uses the line printer to produce a simple bar histogram as shown in the table and it is quite a useful example of the techniques described in Chapter 1 concerning graphical output. The subroutine makes use of a page buffer (Section 1.3) which is blanked out in lines 31−33. A MOD statement is used to annotate the x- and y-axes in lines 36−42.|(Remember that the result of MOD(I, J) is the value of the remainder obtained on dividing I by J.) In order to calculate the bar heights we record the number of data lying within each interval and then scale the values (line 59) from zero to 55 since we use 55 lines of the line printer page.

The page buffer is filled in lines 62−73 by using ITEST as a counter which is decreased from 55 to 0. When the bar height equals ITEST, asterisks (KAR (5)) are assigned to the whole bar area from zero to its maximum value. The subroutine is called by the statement.

CALL   HTGRAM(NP,NIN,IW,IL,IU)

in which

NP = number of data points
NIN = number of intervals
IW = width of each interval
IL = lower limit
IU = upper limit

Note that the difference, (IU − IL) should equal NIN*IW. An example of the use of this routine is shown in Table 2.2 for 1000 readings from a Geiger Counter. These are also listed although a smaller number can be used for testing the program. An example of a MASTER program to call the subroutine is also shown in Table 2.2.

2.3 Solution of Simultaneous Equations

Many numerical problems in chemistry require the solution of simultaneous equations of the form (2.6):

$$a_{11}x_1 + a_{12}x_2 \cdots a_{1n}x_n = y_1$$
$$a_{21}x_2 + a_{22}x_2 \cdots a_{2n}x_n = y_2$$

$$\tag{2.6}$$

$$a_{n1}x_1 + a_{n2}x_2 \cdots a_{nn}x_n = y_n$$

This is a problem of n equations in n unknowns. The simplest problem is that of two equations in two unknowns:

$$a_{11}x_1 + a_{12}x_2 = y_1$$

$$a_{21}x_1 + a_{22}x_2 = y_2$$

for which the solution is easily found to be:

$$x_1 = \frac{y_1 a_{22} - y_2 a_{12}}{\Delta} \; ; \quad x_2 \frac{y_2 a_{11} - y_1 a_{21}}{\Delta}$$

where $\Delta = a_{11} a_{22} - a_{12} a_{21}$.

A similar result is obtained for three equations in three unknowns:

$$a_{11}x_1 + a_{12}x_2 + a_{13}x_3 = y_1$$

$$a_{21}x_1 + a_{22}x_2 + a_{23}x_3 = y_2$$

$$a_{31}x_1 + a_{32}x_2 + a_{33}x_3 = y_3$$

$$x_1 = [y_1 (a_{22}a_{33} - a_{32}a_{23})$$
$$+ y_2 (a_{32}a_{13} - a_{12}a_{33}) + y_3 (a_{12}a_{23} - a_{13}a_{22})]/\Delta \quad (2.7)$$

$$x_2 = [y_1 (a_{23}a_{31} - a_{21}a_{33})$$
$$+ y_2 (a_{33}a_{11} - a_{13}a_{31}) + y_3 (a_{13}a_{21} - a_{23}a_{11})]/\Delta \quad (2.8)$$

$$x_3 = [y_1 (a_{21}a_{32} - a_{31}a_{22}) + y_2 (a_{31}a_{12} - a_{32}a_{11})$$
$$+ y_3 (a_{11}a_{22} - a_{21}a_{12})]/\Delta \quad (2.9)$$

where

$$\Delta = a_{11} (a_{22}a_{33} - a_{23}a_{32}) + a_{12} (a_{23}a_{31} - a_{21}a_{33})$$
$$+ a_{13} (a_{21}a_{32} - a_{22}a_{31}) \quad (2.10)$$

The analytical solution of equations in more than three unknowns becomes increasingly more complex and, for this reason, other more general methods are used.

An example of a general method known as recursion is provided by the subroutine S1ML[4] a listing of which is provided in Table 2.3. Its method of operation is actually very simple in that it eliminates parameters successively until only one unknown remains. The other unknowns are then determined by back-substitution. This is a fairly standard method, described in standard texts[2] and also by Simone.[4] There are many similar methods, some of greater efficiency, which are often supplied as standard software. On reading through S1ML, it will be observed that the coefficients a_{ij} of the parameters are stored in the two-dimensional array X(N,N) for a set of N equations. The dependent variables y are stored in the array elements X(I, (N+1)) which, for economy of storage space are also the final locations of the parameters (i.e. x_1 is in X(1, (N+1)), x_2 in X(2, (N+1)) etc.) that we are trying to calculate.

A built-in safeguard in the subroutine is that if a diagonal element X(I,I) is very small (<0.001), rows and columns are interchanged to

Table 2.2
Program TEST – illustrating the use of the histogram plotting subroutine HTGRAM

```
0008          MASTER TEST
0009          COMMON/A/IDATA(1000)
0010          READ(1,1)NP,NIN,IW,IL,IU
0011        1 FORMAT(5I0)
0012          READ(1,3)(IDATA(I),I=1,NP)
0013        3 FORMAT(100I0)
COMMENT 1/8 IS THIS LARGE A REPEAT COUNT INTENDED AT ABOUT COLUMN 16, LINE 0013
0014          CALL HTGRAM(NP,NIN,IW,IL,IU)
0015          STOP
0016          END


0017          SUBROUTINE HTGRAM(NP,NIN,IW,IL,IU)
0018          COMMON/A/IDATA(1000)
0019          DIMENSION KAR(5),IPOS(100,56),IHT(500)
0020          DATA KAR /1H ,1H+,1H-,1HI,1H*/
0021          WRITE(2,92)NP,NIN,IW,IL,IU
0022       92 FORMAT(1H1,40X,'HISTOGRAM PLOTTING ROUTINE',//,1X,'NUMBER OF POINTS
0023         1=',I5,' NUMBER OF INTERVALS=',I3,' WIDTH OF INTERVAL=',I3,5X,'
0024         2LOWER LIMIT=',I5,' UPPER LIMIT=',I5)
0025          WRITE(2,95)
0026       95 FORMAT(//,40X,'ORIGINAL DATA',//)
0027          WRITE(2,94)(IDATA(I),I=1,NP)
0028       94 FORMAT(2016)
0029          K1=0
0030          K2=0
0031          DO 1 I=1,100
0032          DO 1 J=1,56
0033        1 IPOS(I,J)=KAR(1)
0034    C     NOW ADD THE X-AXIS(VARIABLE TICK SPACING)
0035          IT=100/NIN
0036          DO 2 I=1,100
0037          IPOS(I,1)=KAR(3)
0038        2 IF(MOD(I,IT).EQ.0)IPOS(I,1)=KAR(2)
0039    C     AND THE Y-AXIS(FIXED TICK SPACING)
0040          DO 3 J=1,56
0041          IPOS(1,J)=KAR(4)
0042        3 IF(MOD(J-1,5).EQ.0)IPOS(1,J)=KAR(2)
0043    C     CALCULATE BAR HEIGHTS AND RECORD NUMBER OUTSIDE BOUNDS(IL AND IU)
0044       43 MAX=0
```

```
0045        DO 50 I=1,NIN
0046        IHT(I)=0
0047        DO 50 J=1,NP
0048        IF(I.GT.1)GO TO 22
0049        IF(IDATA(J).GE,IL)GO TO 21
0050        K1=K1+1
0051        GO TO 50
0052     21 IF(IDATA(J).LE,IU)GO TO 22
0053        K2=K2+1
0054        GO TO 50
0055     22 IF(IDATA(J).GT.(IL+(I-1)*IW).AND.IDATA(J).LE.(IL+I*IW))IHT(I)=IHT(
0056        1I)+1
0057     50 IF(IHT(I).GT.MAX)MAX=IHT(I)
0058        DO 51 I=1,NIN
0059     51 IHT(I)=55*IHT(I)/MAX
0060     C  ASSIGN CHARACTERS(**) TO THE BARS
0061        IN=100/NIN
0062        DO 4 I=1,NIN
0063        IF(IHT(I).EQ.0)GO TO 4
0064        ITEST=56
0065      5 ITEST=ITEST-1
0066        IF(IHT(I).LT.ITEST)GO TO 5
0067        L1=(I-1)*IN
0068        IF(L1.EQ.0)L1=2
0069        L2=I*IN
0070        DO 6 J=2,ITEST+1
0071      6 K=L1+1,L2-1
0072        IPOS(K,J)=KAR(5)
0073      4 CONTINUE
0074     C  OUTPUT THE HISTOGRAM
0075        WRITE(2,9)
0076      9 FORMAT(1H1,30X,'HISTOGRAM FROM DATA SUPPLIED',//)
0077        WRITE(2,12)MAX,(IPOS(K,56),K=1,100)
0078        DO 7 I=2,55
0079        J=57-I
0080      7 WRITE(2,8)(IPOS(K,J),K=1,100)
0081     12 FORMAT(2X,I8,100A1)
0082        WRITE(2,13)(IPOS(K,1),K=1,100)
0083     13 FORMAT(8X,'0',100A1)
0084        WRITE(2,14)IL,IU
0085     14 FORMAT(6X,I6,94X,I6)
0086      8 FORMAT(10X,100A1)
0087        WRITE(2,10)K1,K2
0088     10 FORMAT(1H1,' NUMBER OF DATA BELOW AND ABOVE ALLOWED RANGES WERE '
0089        1,2I5,' RESPECTIVELY')
0090        RETURN
0091        END
```

ORIGINAL DATA

2898	2895	2817	2831	2857	2806	2756	2919	2768	2924	2636	2867	2861	2854	2813	2960	2827	2804	2841	2840
2810	2903	2837	2893	2891	2890	2835	2846	2945	2934	2857	2816	2868	2322	2872	2866	2830	2897	2863	2883
2949	2869	2886	2780	2780	2812	2882	2925	2879	2820	2803	2818	2856	2837	2801	2867	2902	2890	2827	2820
2954	2895	2851	2860	2890	2918	2873	2882	2834	2857	2635	2914	2849	2905	2883	2924	2873	2828	2863	2956
2889	2914	2933	2894	2876	2883	2905	2908	2879	2931	2821	2861	2882	2933	2862	2924	2849	2850	2878	2865
2927	2901	2945	2889	2883	2870	2905	2913	2831	2842	2800	2448	2892	2872	2835	2460	2877	2850	3007	2843
2901	2851	2851	2900	2860	2878	2878	2909	2809	2918	2832	2784	2892	2859	2910	2942	2861	2799	2883	2911
2854	2841	2924	2906	2911	2823	2902	2909	2893	2781	2884	2837	2905	2972	2935	2389	2912	2762	2844	2845
2827	2874	2858	2911	2849	2897	2809	2891	2839	2903	2911	2810	2821	2913	2877	2815	2817	2888	2798	2951
2918	2906	2863	2874	2841	2822	2853	2977	2913	2856	2878	2847	2795	2331	2880	2840	2795	2836	2827	2832
2904	2856	2796	2856	2849	2749	2809	2918	2906	2903	2911	2864	2751	2799	2847	2901	2862	2783	2804	2870
2866	2840	2840	2892	2845	2899	2853	2888	2839	2880	2878	2877	2795	2887	2880	2331	2854	2818	2843	2851
2872	2623	2965	2885	2845	2804	2857	2892	2771	2923	2649	2462	2891	2904	2858	2693	2911	2856	2842	2886
2790	2850	2781	2796	2859	2781	2794	2850	2928	2802	2751	2891	2837	2904	2890	2836	2890	2891	2833	2886
2966	2839	2854	2866	2902	2800	2893	2925	2880	2885	2795	2281	2929	2881	2854	2890	2899	2801	2913	2880
2823	2843	2916	2878	2838	2759	2893	2809	2844	2918	2826	2781	2837	2895	3000	2944	2868	2897	2862	2819
2883	2843	2837	2854	2800	2867	2891	2846	2900	2887	2851	2762	2861	2644	2695	2958	2831	2831	2864	2931
2868	2865	2898	2872	2894	2894	2872	2841	2833	2887	2859	2648	2848	2841	2965	2947	2941	2862	2878	2859
2851	2877	2897	2792	2950	2867	2877	2721	2841	2887	2903	2774	2907	2903	2932	2615	2860	2930	2926	2844
2893	2845	2845	2845	2859	2872	2829	2867	2762	2981	2898	2774	2847	2864	2864	2932	2871	2903	2923	2826
2869	2810	2945	2787	2659	2880	2878	2847	2847	2812	2971	2940	2959	2959	2916	2819	2867	2902	2880	2800
2648	2813	3002	2868	2913	2862	2927	2877	2819	2835	2849	2848	2885	2842	2909	2857	2918	2926	2932	2911
2873	2907	2859	2898	2898	2835	2859	2839	2819	2865	2865	2871	2871	2871	2412	2933	2921	2873	2831	2840
2616	2827	2633	2942	2836	2833	2817	2846	2894	2919	2867	2614	2785	2790	2820	2377	2921	2926	2841	2934
2796	2844	2859	2917	2917	2835	2828	2841	2894	2828	2884	2667	2656	2792	2813	2289	2860	2946	2839	2835
2829	2885	2817	2924	2705	2892	2876	2948	2912	2894	2893	2656	2668	2906	2926	2277	2910	2917	2815	2826
2830	2866	2810	2895	2845	2979	2876	2898	2777	2932	2665	2888	2937	2937	2651	2951	2774	2840	2916	2881
2886	2851	2870	2894	2903	2835	3004	2879	2779	2847	2665	2904	2500	2931	2654	2758	2859	2809	2905	2895
2891	2842	2877	2867	2918	2915	2917	3000	2950	2868	2818	2627	2879	2579	2867	2818	2935	2898	2806	2824
2835	2868	2926	2901	2934	2864	2906	2946	2936	2861	2871	2754	2783	2749	2873	2919	2487	2856	2806	2912
2879	2868	2919	2901	2963	2884	2917	2878	2899	2899	2934	2894	2834	2834	2242	2828	2450	2827	2932	2893
2867	2992	2903	2931	2866	2884	2906	2876	2936	2862	2860	2795	2378	2474	2460	2633	2888	2884	2804	2907
2665	2894	2939	2866	2861	2942	2873	2873	2924	2910	2890	2798	2474	2846	2847	2847	2473	2815	2923	2767
2777	2882	2901	2825	2825	2603	2928	2843	2929	2807	2957	2852	2851	2790	2860	2822	2852	2897	2829	2895
2892	2835	2870	2927	2653	2882	2797	2867	2952	2845	2850	2880	2901	2705	2207	2871	2902	2852	2851	2900
2937	2815	2857	2997	2918	2692	2836	2834	2447	2845	2635	2894	2686	2788	2705	2465	2422	2859	2851	2811
2904	2906	2866	2843	2776	2852	2905	2876	2843	2877	2625	2870	2686	2623	2715	2847	2405	2949	2863	2961
2897	2897	2859	2874	2874	2862	2853	2853	2859	2878	2860	2826	2912	2856	2183	2847	2822	2922	2875	2887
2655	2894	2920	2902	2850	2875	2824	2927	2859	2864	2930	2822	2893	2917	2917	2824	2841	2942	2877	2937
2933	2948	2856	2863	2863	2905	2877	2932	2819	2862	2654	2890	2855	2307	2843	3004	2945	2967	2863	2910
2717	2830	2837	2983	2961	2921	2877	2944	2819	2862	2654	2941	2807	2882	2675	2859	2983	2828	2891	2898
2899	2844	2912	2841	2929	2872	2829	2825	2870	2820	2854	2673	2852	2789	2675	2919	2955	2830	2819	2912
2917	2878	2916	2916	2929	2843	2858	2843	2843	2840	2891	2852	2853	2830	2822	2897	2840	2930	2875	2861
2918	2819	2800	2837	2929	2807	2944	2861	2915	2833	2834	2491	2869	2814	2269	2865	2863	2797	2880	2866
2909	2951	2903	2935	2935	2941	2882	2906	2931	2877	2954	2749	2964	2909	2668	2953	2931	2951	2863	2926
2642	2857	2805	2906	2764	2866	2884	2884	2930	2853	2802	2653	2922	2913	2918	2888	2866	2869	2878	2932
2801	2805	2945	2906	2854	2861	2913	2774	2774	2805	2805	2850	2805	2847	2960	2857	2853	2951	2929	2906
2820	2912	2862	2878	2924	2862	2905	2901	2930	2918	2854	2931	2918	2852	2431	2827	2952	2909	2985	2916
2755	2846	2867	2893	2830	2830	2896	2967	2967	2990	2932	2861	2961	2653	2935	2841	2652	2969	2884	2959

HISTOGRAM FROM DATA SUPPLIED

NUMBER OF DATA BELOW AND ABOVE ALLOWED RANGES WERE   4   6 RESPECTIVELY

Table 2.3
Subroutine S1ML — a subroutine for solving simultaneous equations

```
        SUBROUTINE SIML
C       SUBROUTINE FOR SIMULTANEOUS EQUATIONS
        DIMENSIONLROW(20),LAB(20)
        COMMON/XVAR/N,X(20,20)
        M=N+1
        DO 2 I=1,M
        LROW(I)=I
     2  LAB(I)=I
        M1=N-1
        DO 13 I=1,M1
        IF(ABS (X(I,I))-.001)3,3,11
     3  DO 4 JJ=I,N
        DO 4 III=I,N
        IF(ABS (X(III,JJ))-.001)4,4,5
     4  CONTINUE
        WRITE(2,25)
        GO TO 22
    25  FORMAT(1X,'NO SOLUTION')
        GO TO 21
     5  IF(JJ-I)8,8,6
     6  LAB(M)=LAB(I)
        LAB(I)=LAB(JJ)
        LAB(JJ)=LAB(M)
        DO 7 JJJ=1,N
        X(JJJ,M+1)=X(JJJ,I)
        X(JJJ,I)=X(JJJ,JJ)
     7  X(JJJ,JJ)=X(JJJ,M+1)
     8  IF(III-I)11,11,9
     9  DO 10 IK=I,M
        X(N+1,IK)=X(I,IK)
        X(I,IK)=X(III,IK)
    10  X(III,IK)=X(N+1,IK)
        LROW(M)=LROW(I)
        LROW(I)=LROW(III)
        LROW(III)=LROW(M)
    11  CONTINUE
        A=X(I,I)
        L1=I+1
        DO 13 J=L1,N
        B=X(J,I)
        IF(ABS(B)-.001)13,15,12
    12  DO 13 K=I,M
        X(J,K)=X(J,K)-(B*X(I,K)/A)
    13  CONTINUE
        IF(ABS(X(N,N))-.0001)14,14,15
    14  WRITE(2,25)
        GO TO 22
    15  X(N,M)=X(N,M)/X(N,N)
        L=M
        DO 17 J=1,M1
        L=L-1
        L1=L-1
        SX=0
        DO 16 I=L,N
    16  SX=SX+X(L1,I)*X(I,M)
    17  X(L1,M)=(X(L1,M)-SX)/X(L1,L1)
        DO 21 L=1,N
        IF(LAB(L)-L)18,21,18
    18  DO 20 J=1,N
        IF(LAB(J)-L)20,19,20
    19  X(M+1,M)=X(L,M)
        X(L,M)=X(J,M)
        X(J,M)=X(M+1,M)
        LAB(M)=LAB(L)
        LAB(L)=LAB(J)
        LAB(J)=LAB(M)
        GO TO 22
    20  CONTINUE
    21  CONTINUE
    22  CONTINUE
        RETURN
        END
```

place a different element on the diagonal. This technique is called 'pivoting' and avoids the accumulation of large errors which can form on division by a very small number.

The subroutine is called by the simple statement

CALL S1ML

No dummy arguments are used, but instead a common block must be present in any program which calls S1ML:

COMMON/XVAR/M,X(20, 20)

M is the number of unknowns (equal to the number of simultaneous equations) and the array X is that which stores the values of the dependent and independent variables, as described above.

For the reader concerned with the solution of very large sets of simultaneous equations, the conventional methods can become inefficient. An interesting procedure for these cases is described by Wilson et al.[4a] who also give a complete FORTRAN IV listing of the subroutine. Examples are quoted in which up to 8036 simultaneous equations are solved!

As a simple example, let us consider the following:

The % yield, Y, of an organic reaction was thought to be given by a linear equation:

$$Y = ap + bT + cL$$

where p was the applied pressure (atmospheres), T was the temperature (K) and L was the reaction time (hours). We find a, b and c from the data:

$$4.5 = 1.0\,a + 1.0\,b + 1.0\,c$$
$$9.3 = 1.0\,a + 2.0\,b + 3.0\,c$$
$$24.6 = 4.0\,a + 8.0\,b + 5.0\,c$$

Using equations (2.7)–(2.10), we would obtain

$$a = 1.5$$
$$b = 1.2$$
$$c = 1.8$$

We can also solve the equations by the program TEST (which uses S1ML) listed in Table 2.4, giving identical results to the manual method.

This method of testing our subroutine exemplifies an important principle in program development: always test the program or subroutine with known data to which you know the answer.

A further point is that, having obtained a solution to a set of simultaneous equations, it is always useful to check the correctness of

Table 2.4

Illustrative use of S1ML for a problem in three unknowns

```
0008                    MASTER TEST
0009            C       PROGRAM TO SOLVE 3X3 SET OF EQTNS
0010                    COMMON/XVAR/N,X(20,20)
0011                    N=3
0012                    DO 4 I=1,N
0013            4   READ(1,1) (X(I,J),J=1,(N+1))
0014            1   FORMAT(10F7.3)
0015                    WRITE(2,6)
0016            6   FORMAT(1H1,' 3X3 SET OF EQUATIONS',//)
0017                    DO 5 I=1,N
0018            5   WRITE(2,5) (X(I,J),J=1,(N+1))
0019            5   FORMAT(10F7.3)
0020                    CALL S1ML
0021                    WRITE(2,2)(X(I,(N+1)),I=1,N)
0022            2   FORMAT(1H ,/,' VALUES OF A,B,AND C ARE  ',3F7.2)
0023                    STOP
0024                    END


        3X3 SET OF EQUATIONS

        1.000   1.000   1.000   4.500
        1.000   2.000   3.000   9.500
        4.000   8.000   5.000  24.600

VALUES OF A,B,AND C ARE     1.50   1.20   1.80
```

the results either by hand or by using a goodness-of-fit test (Section 2.4.3) in the case of curve-fitting problems.

As we shall be using S1ML quite a lot, we should be quite clear about its mode of use: for a set of N x N equations we place the dependent variables in array positions $X(1, (N+1))$ to $X(N, (N+1))$. Coefficients of dependent variables are placed in the array $X(I, J)$ where I and J run from 1 to N. The output values of the parameters that we are trying to determine are placed in the array elements $X(1, N+1)$ to $X(N, N+1)$, thereby overwriting the original contents (which were the dependent variables).

2.4 Least Squares Curve Fitting

Having analysed our data by the methods of Section 2.1 we are often interested in the relationship (if any) between a measured variable and some other variable, such as temperature, that can be adjusted by the experimenter. These variables are called dependent and independent respectively. For example, in equation (2.11), y is the dependent variable and x the independent variable:

$$y_i = a + bx_i \tag{2.11}$$

Clearly y_i is linearly dependent on the variable x_i and also on the parameters a and b. Equation (2.12) shows a linear dependence on the parameters whereas (2.13) is a non-linear equation:

$$y_i = a + bx_i + cx_i^2 + \log x_i \tag{2.12}$$

$$y_i = \sqrt{(ax_i + b)} + cx_i \tag{2.13}$$

The experimenter may know a theoretical y–x relationship but, in some cases, such knowledge is lacking. For example, a calibration curve of concentration *vs.* absorbance can be non-linear with no explicit relationship.

In this section we are concerned with the problem of estimating the values of the parameters in such equations as (2.11)–(2.13). The most popular method is to minimize the sum of squares, S_N :

$$S_N = \sum_{i=1}^{N} (y_i - \hat{y}_i)^2$$

where y_i and $\hat{y}_i$ are the observed and estimated y-values respectively.

2.4.1 Linear Parameter Dependence (Two Parameters)

Relationships which depend linearly on the parameters are very common. Furthermore, seemingly non-linear relationships can sometimes be cast in a linear form. For example,

$$K_i/K_0 = \exp\left[\frac{-\Delta H}{R} \left(\frac{1}{T_i} - \frac{1}{T_0} \right) \right]$$

is equivalent to

$$\log_e K_i = \log_e K_0 + \frac{\Delta H}{RT_0} - \frac{\Delta H}{RT_i}$$

This is of the form

$$y_i = a + bx_i$$

Transformations of this type can be very useful in experimental data fitting. Some other transformations are shown in Table 2.5.

As an example of the least squares approach, let us assume that our y–x data should follow equation (2.11). The least squares criterion asserts that we must minimize the sum, S:

$$S = \sum_i (y_i - \hat{a} - \hat{b}x_i)^2 \tag{2.14}$$

where the hat symbols ($\hat{}$) indicate that the parameter values are estimates. Differentiation of (2.14) with respect to $\hat{a}$ and then $\hat{b}$, then setting each derivative to zero, yields simultaneous equations (2.15)

$$\sum y_i = \hat{a}N + \hat{b}\sum x_i$$
$$\sum x_i y_i = \hat{a}\sum x_i + \hat{b}\sum x_i^2 \tag{2.15}$$

often called the *normal* equations, which have solutions (2.16) and (2.17).

$$\hat{b} = (N\sum x_i y_i - \sum y_i \sum x_i)/\{N\sum x_i^2 - (\sum x_i)^2\} \tag{2.16}$$
$$\hat{a} = (\sum y_i - \hat{b}\sum x_i)/N \tag{2.17}$$

Table 2.5
Transformations of various functions into the general form $y' = a + bx'$.
(Logarithms to base e will ensure that $\log e = 1$)[a]

Function	Required transformation		Values of a and b	
	$y' =$	$x' =$	$a = :$	$b =$
$y = cx^n$	$\log y$	$\log x$	$\log c$	n
$y = cn^n + d$	y	x^n	d	c
or	$\log(y - c)$	$\log x$	$\log c$	n
$y = ce^{nx}$	$\log x$	x	$\log c$	$n \log e$
$y = c(1 - e^{-nx})$	$\log(c - y)$	x	$\log c$	$-n \log e$
$y = c/x + d$	y	$1/x$	d	c
$y = c/x^n + d$	$\log(y - c)$	$\log x$	$\log c$	n
or	y	$1/x^n$	d	c
$y = x/(cx + d)$	x/y	x	d	c

[a] Based on Figure 9.12 in *Problem Solving with Computers* by P. Calter, McGraw-Hill, New York, 1973.

It can be shown that the variances b and a are given by

$$\sigma_{\hat{b}}^2 = N\sigma_y^2 / \{N\Sigma x_i^2 - (\Sigma x_i)^2\} \tag{2.18}$$

$$\sigma_{\hat{a}}^2 = \sigma_y^2 \Sigma x_i^2 / \{N\Sigma x_i - (\Sigma x_i)^2\} \tag{2.19}$$

The positive square roots, $\sigma_{\hat{b}}$ and $\sigma_{\hat{a}}$ being the corresponding standard deviations. An unbiased estimator of σ_y^2 is given by

$$s_y^2 = \hat{\sigma}_y^2 = \frac{\Sigma(y_i - \hat{a} - \hat{b}x_i)^2}{N - 1} \tag{2.20}$$

Therefore we can easily estimate the values of a and b and their variances if we wish to do so. Are we sure, however, that this linear relationship is correct? A good way of investigating the linear dependence of two variables is to calculate the *correlation coefficient* defined by (2.21):

$$r = \Sigma(x_i - \bar{x})(y_i - \bar{y}) / \sqrt{\{[\Sigma(x_i - \bar{x})^2][\Sigma(y_i - \bar{y})^2]\}} \tag{2.21}$$

r can lie between ± 1. If r tends to these limits, the x, y data is said to be well correlated in a positive or negative sense.

As a simple example of the least squares technique, let us consider the data in Table 2.6 which shows the readings of an isoteniscope as a function of temperature. The atmospheric pressure was equivalent to 76 cm of mercury. The vapour pressure is simply the atmospheric pressure *minus* the isoteniscope reading. Theory suggests that the linear relationship (2.22) should hold for the vapour pressure, P, of a substance

Table 2.6
Readings of an isoteniscope (pressure
differences) as a function of temperature

Temperature/K	Pressure difference (/cm of mercury)
329.9	0.0
326.8	5.2
323.5	12.8
321.0	18.8
317.9	25.0
314.5	31.0
309.5	39.2

$$\ln P = \frac{-\Delta H}{RT} + \text{const.} \tag{2.22}$$

R is the gas constant (8.314 J K^{-1} mol) and ΔH is the enthalpy of vaporization (J mol^{-1}) of the substance. Accordingly we write the program VAPOUR (Table 2.7). Input to the program is:

Card 1: Title (10A8).
Card 2: Number of data points, N; atmospheric pressure (I2, F4.1).
Cards 3 to 2+N: Isoteniscope, temperature readings (F4.1, F5.1).

From the input data we set up the normal equations for the problem (lines 35–38) and compute ΔH from the coefficient of $(1/T)$ in line 40, having used equation (2.16). The correlation coefficient is calculated in lines 50–56. Clearly, from the results in Table 2.7 the data have a high negative correlation.

2.4.2 Linear Dependence on More than Two Parameters
It is fortuitous for our data to follow equation (2.11) and a more common situation is when a linear relationship exists between the y_i and several parameters. Some examples are:

Polynomials   E.g.

$$C_p = a + bT + cT^2 \qquad \text{Empirical heat capacity } (C_p),$$
$$C_p = a + bT + cT^{-\frac{1}{2}} \qquad \text{temperature } (T) \text{ relations.} \tag{2.23}$$

Logarithmic Transformations   E.g.

$$dx/dt = A \exp(-E/RT) \cdot (1 - \alpha)^n \tag{2.24}$$

34

Table 2.7
Program VAPOUR – calculation of enthalpy of vaporization

```
0008                    MASTER VAPOUR
0009                    DIMENSION P(20),T(20),H(10),PAT(20),PPAT(20),TE(20)
0010                    R=8.514
0011                    READ(1,100)(H(I),I=1,10)
0012          100       FORMAT(10A8)
0013                    READ(1,101) N,PA
0014          101       FORMAT(I2,F4.1)
0015                    DO 1 J=1,N
0016          1         READ(1,102)P(J),T(J)
0017          102       FORMAT(F4.1,F5.1)
0018                    TH=0.0
0019                    SX=0.0
0020                    SY=0.0
0021                    SXY=0.0
0022                    SX2=0.0
0023                    S1=0.0
0024                    S2=0.0
0025                    S3=0.0
0026                    TL=999
0027                    DO 2 J=1,N
0028                    IF(T(J).GT.TH)TH=T(J)
0029                    IF(T(J).LT.TL)TL=T(J)
0030                    X=1./T(J)
0031                    TE(J)=X*10.**3
0032                    PAT(J)=PA-P(J)
0033                    Y=ALOG10(PAT(J))
0034                    PPAT(J)=Y
0035                    SX=SX+X
0036                    SY=SY+Y
0037                    SXY=SXY+X*Y
0038          2         SX2=SX2+X*X
0039                    EM=((N*SXY)-(SX*SY))/((N*SX2)-(SX*SX))
0040                    EL=-2.305*R*EM
0041                    WRITE(2,200)(H(I),I=1,10),TL,TH
0042          200       FORMAT(1H1,10A8///1H0,'TEMPERATURE RANGE',F8.1,' TO',F8.1//)
0043                    WRITE(2,105)
0044          105       FORMAT(1H ,5X,'TEMP/K   PRESS.DIFF   PAT-P=Y   LOG(Y)      10**3/T')
0045                    DO 3 I=1,N
0046          3         WRITE(2,202) T(I),P(I),PAT(I),PPAT(I),TE(I)
0047          202       FORMAT(1H ,3F10.1,2F10.3)
0048                    WRITE(2,201) EL
0049          201       FORMAT(1H0,'HEAT OF VAPOURISATION =',F9.0,' JMOL-1')
0050                    XB=SX/N
0051                    YB=SY/N
0052                    DO 5 I=1,N
0053                    S1=S1+((TE(I)/1000.)-XB)*(PPAT(I)-YB)
0054                    S2=S2+((TE(I)/1000.)-XB)**2
0055          5         S3=S3+(PPAT(I)-YB)**2
0056                    AR=S1/SQRT(S2*S3)
0057                    WRITE(2,6)AR
0058          6         FORMAT(1H ,'CORRELATION COEFFICIENT = ',F5.3)
0059          7         STOP
0060                    END
```

EXAMPLE

TEMPERATURE RANGE   309.5 TO   329.9

TEMP/K	PRESS.DIFF	PAT-P=Y	LOG(Y)	10**3/T
329.9	0.0	76.0	1.881	3.031
326.8	5.2	70.8	1.850	3.060
323.5	12.8	63.2	1.801	3.091
321.0	18.8	57.2	1.757	3.115
317.9	25.0	51.0	1.708	3.146
314.5	31.0	45.0	1.653	3.180
309.5	39.2	36.8	1.566	3.231

HEAT OF VAPOURISATION =   50762. JMOL-1
CORRELATION COEFFICIENT = -.999

rearranged to

$$\log_e (dx/dt) = \log_e A - E/RT + n \log_e (1 - \alpha)$$

(This is one possible equation for the rate of decomposition dx/dt of a pure substance.)

Miscellaneous transformations

$$y = (px + r)/(1 + qx)$$

an equation of a hyperbola, rearranged to

$$y = r + px - qxy \qquad (2.25)$$

Therefore, we may be provided with a theoretical equation such as (2.23) or with empirical relationships such as (2.24) or (2.25). The choice of equation is often a subjective matter but the important thing is that all such equations are *linear in the unknown parameters* even though they may be distinctly non-linear in the independent variables. Fortunately, the method of solution of such equations is perfectly general. For example, let us assume that we have a polynomial:

$$y_i = a_0 + a_1 x + a_2 x^2 + \ldots + a_n x^n$$

An extension of the least squares approach employed in Section 2.4.1 yields the following normal equations:

$$\Sigma y_i = a_0 N + a_1 \Sigma x_i \ldots + a_n \Sigma x_i^n$$
$$\Sigma x_i y_i = a_0 \Sigma x_i + a_1 \Sigma x_i^2 \ldots + a_n \Sigma x_i^{n+1} \qquad (2.26)$$

$$\cdot \qquad \cdot \qquad \cdot \qquad \cdot$$
$$\cdot \qquad \cdot \qquad \cdot \qquad \cdot$$
$$\cdot \qquad \cdot \qquad \cdot \qquad \cdot$$

$$\Sigma x_i^n y_i = a_0 \Sigma x_i^n + a_1 \Sigma x_i^{n+1} \ldots + a_n \Sigma x_i^{2n}$$

(Note the symmetry about the diagonal elements of the quantities on the right-hand side.)

Similarly from equation (2,25) we would obtain

$$\Sigma y_i = rN + p\Sigma x_i - q\Sigma x_i y_i$$
$$\Sigma x_i y_i = r\Sigma x_i + p\Sigma x_i^2 - q\Sigma x_i^2 y_i$$
$$\Sigma x_i y^2 = -r\Sigma x_i y_i - p\Sigma x_i^2 y_i + q\Sigma x_i^2 y_i^2$$

The reader should be able to obtain the normal equations for any such linear relationship.

The solution of these equations is not as simple as for the two-parameter case and for three or more unknown parameters (three sets of simultaneous equations) it is more convenient to use a subroutine such as S1ML (listed in Table 2.3) to calculate the unknown parameters from the simultaneous equations. If S1ML is to be used, note that you would use the summations $(\Sigma y_i; \Sigma x_i; \Sigma x_i y_i$ etc.) in place of the dependent variables and the coefficients of the parameters in the array elements $X(I, N+1)$ and $X(I, I)$ respectively.

As an example we find that it has been proposed that[5 a,b] the hyperbolic relationship (2.25) should be useful in the construction of calibration tables in spectrophotometry. We shall use the program LSHY (Table 2.8) which in turn, uses the subroutine HYPB to perform the least squares fit to the equation:

$$y_i = r + px_i - qx_i y_i \qquad (2.27)$$

The values of r, p, q are to be evaluated with subroutine SIML, as described earlier.

In Table 2.8 we see that two common blocks are used:

COMMON/XVAR/N, X(20, 20)
COMMON/A/A(100), NP, YCALC(100)

The purpose of the first block is to provide communication between subroutine HYPB and S1ML. In common block A, the arrays and variable have the following significance:

A(100) = array containing the independent variables
Y(100) = array containing the dependent variables
NP = number of data points
YCALC(100) = calculated values (from equation (2.27)) of the dependent variables

In subroutine HYPB the normal equations are set-up from (2.27). The values of p, q and r are calculated (array $X(J, 4)$) and substituted into the theoretical equation. Specimen results are included in Table 2.8 for some spectroscopic data. Input to the MASTER program is a title (5A8), the number of points NP(I0 format) followed by NP data cards (each 2F0.0). Each of these contains an independent variable (concentration in the example) and a dependent variable (absorbance) in 2F0.0 format.

Another useful subroutine is POLY (Table 2.9). The purpose of this is to fit a list of measured (dependent) variables to a polynomial up to the 20th degree.

$$y_i = a_0 + a_1 x_i + a_2 x_i^2 \ldots + a_n x_i^n$$

Equations (2.26) are used for the least-squares fitting and, again, S1ML is used for the solution of the equations. The same common blocks XVAR and A are needed as were used in the preceding example. The user must specify the degree of the polynomial required by the dummy

Table 2.8
LSHY — a program to fit experimental x, y data to the hyperbolic relationship
$$y_i = r + pq_i - qx_iy_i$$

```
0006          MASTER LSHY
0007          DIMENSIONTITLE(5)
0008          COMMON/A/A(100),Y(100),NP,YCALC(100)
0009          COMMON/XVAR/N,X(20,20)
0010          READ(1,6)(TITLE(I),I=1,5)
0011          READ(1,1)NP
0012        1 FORMAT(I0)
0013        6 FORMAT(5A8)
0014          READ(1,2)(A(I),Y(I),I=1,NP)
0015        2 FORMAT(2F0.0)
0016          WRITE(2,3)(TITLE(I),I=1,5)
0017        3 FORMAT(1H1,5A8)
0018          CALL HYPB
0019          WRITE(2,4)
0020        4 FORMAT(///,1H ,'    X        YOBS        YCALC')
0021          WRITE(2,5)(A(I),Y(I),YCALC(I),I=1,NP)
0022        5 FORMAT(1H ,F7.3,2X,F8.4,2X,F8.4)
0023          STOP
0024          END
```

```
0056          SUBROUTINE HYPB
0057          COMMON/A/A(100),Y(100),NP,YCALC(100)
0058          COMMON/XVAR/N,X(20,20)
0059          N=3
0060          DO 1 I=1,4
0061          DO 1 J=1,4
0062        1 X(I,J)=0.0
0063          X(1,3)=NP
0064          DO 2 I=1,NP
0065          X(1,1)=X(1,1)+A(I)
0066          X(1,2)=X(1,2)-A(I)*Y(I)
0067          X(1,4)=X(1,4)+Y(I)
0068          X(2,1)=X(2,1)+A(I)*A(I)
0069          X(2,2)=X(2,2)-A(I)*A(I)*Y(I)
0070          X(3,2)=X(3,2)-A(I)*A(I)*Y(I)*Y(I)
0071        2 X(3,4)=X(3,4)+A(I)*Y(I)*Y(I)
0072          X(2,3)=X(1,1)
0073          X(2,4)=-X(1,2)
0074          X(3,1)=-X(2,2)
0075          X(3,3)=X(2,4)
0076          CALL SIML
0077          WRITE(2,3)(X(J,4),J=1,3)
0078        3 FORMAT(1H ,' P,Q,R IN THIS HYPERBOLIC CURVE FITTING  WERE ',3(4X,E
0079       110.4))
0080          DO 4 I=1,NP
0081        4 YCALC(I)=(X(1,4)*A(I)+X(3,4))/(1.+X(2,4)*A(I))
0082          RETURN
0083          END
```

```
   EXAMPLE CALCULATION
P,Q,R IN THIS HYPERBOLIC CURVE FITTING  WERE       0.1617E 02     0.6681E 00     0.1530E 01


    X       YOBS       YCALC
 0.052     2.2700     2.2917
 0.080     2.7000     2.6809
 0.116     3.1700     3.1614
 0.156     3.6800     3.6708
 0.187     4.0400     4.0489
 0.278     5.0700     5.0825
 0.374     6.0700     6.0639
```

Table 2.9
Subroutine POLY — for fitting x, y data to a polynomial
$$y_i = b_0 + b_1 x_i + b_2 x_i^2 + \ldots + b_n x_i^n$$

```
      SUBROUTINE POLY
      COMMON/A/A(100),Y(100),NP,YCALC(100)
      COMMON/XVAR/N,X(20,20)
C     FITS A POLYNOMIAL UP TO 20 TH  DEGREE
      NVAR=N
      DO 8 I=1,NVAR+1
      DO 8 J=1,NVAR+1
    8 X(I,J)=0
      DO 2 I=1,NVAR
      DO 2 J=1,NVAR
      IF((I+J),EQ.2)GO TO 2
      DO 6 K=1,NP
    6 X(I,J)=X(I,J)+(A(K))**(I+J-2)
    2 CONTINUE
      X(1,1)=NP
      DO 3 I=2,NVAR
      DO 3 K=1,NP
    3 X(I,NVAR+1)=X(I,NVAR+1)+(Y(K))*(A(K)**(I-1))
      DO 4 K=1,NP
    4 X(1,NVAR+1)=X(1,NVAR+1)+Y(K)
      CALL SIML
      DO 5 J=1,NP
      YCALC(J)=0
      DO 5 I=1,NVAR
    5 YCALC(J)=YCALC(J)+X(I,(NVAR+1))*((A(J))**(I-1))
      RETURN
      END
```

argument N in common block XVAR. For many purposes a quadratic fit (N = 2) is suitable although higher orders are sometimes necessary.

2.4.3 Testing Goodness-of-Fit

Having worked through the examples above we would naturally like to know which subroutine is better for fitting the data (in this case there is very little data to worry about but the problem is a general one). One method of testing the goodness-of-fit is to use the Chi-squared (χ^2) test.[6] This is defined by the statistic

$$\chi^2 = (n - 1)s^2/\sigma^2 \qquad (2.28)$$

where s^2 is the sample variance and σ^2 is the population variance. If repeated samples of size n are taken, χ^2 follows a distribution similar to that in Figure 2.3. The significance of $\chi^2(\alpha, n)$ is that with n degrees of freedom (n being the number of data points in the sample) there is α percent probability of observing a larger value of χ^2. For our purposes, we note that the statistic

$$\chi^2 = \Sigma \frac{[y_i(\text{observed}) - y_i(\text{calculated})]^2}{y_i(\text{calculated})} \qquad (2.29)$$

follows a distribution similar to (2.28). Some values of $\chi^2(\alpha, n)$ are tabulated in Table 2.10. Clearly, for a given n, the best fit of observed and experimental data is to be associated with the lowest possible value of χ^2. Given two methods of data treatment, that which gives the smaller χ^2 is the preferred method.

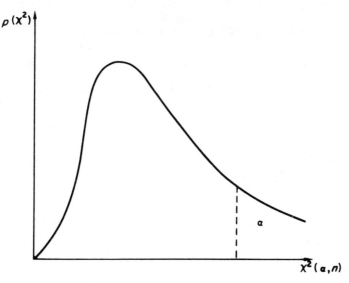

Figure 2.3　The $\chi^2 (\alpha,n)$ distribution

Table 2.10
Examples of percentage points of the $\chi^2_{\alpha, n}$
distribution[a]

α	0.99	0.95	0.50	0.10
ν				
1	0.0002	0.0039	0.45	2.71
2	0.020	0.103	1.39	4.61
3	0.115	0.352	2.37	6.25
4	0.30	0.71	3.36	7.78
5	0.55	1.15	4.35	9.24

[a]ν is the number of degrees of freedom, equal
to $(n-1)$.

As an example of a suitable subroutine for performing a χ^2 test we
may consider CHISQ (SQ) in Table 2.11. This very simple subroutine is
almost self-explanatory. The same common block A as used previously
must be present in a program that uses this subroutine. The value of χ^2
is stored in the variable SQ for return to the main program.

2.4.4 General Comments on the Linear Least Squares Method
Ill conditioning　When using the least squares method for linear
problems, the set of simultaneous equations may become ill-
conditioned in practical problems. This means that one or more of the

Table 2.11

Subroutine CHISQ — approximate calculation of χ^2

```
0059          SUBROUTINE CHISQ(SQ)
0060          COMMON/A/A(100),Y(100),NP,YCALC(100)
0061          SQ=0
0062          DO 1 I=1,NP
0063        1 SQ=SQ+((Y(I)-YCALC(I))**2)/YCALC(I)
0064          RETURN
0065          END
```

equations is a multiple (or very nearly so) of some other in the set. For example, if we try to solve (2.30),

$$1.000\ a - 0.738\ b = 0.976$$
$$0.801\ a - 0.590\ b = 0.781$$

$$(2.30)$$

apparent values could be assigned to a and b; however, the equations are almost exact multiples of each other and the values of a and b are, as it were, being determined by one equation — a physically impossible situation.

We can test for ill-conditioning (which often arises when the wrong equations were chosen in the first place) by calculating the value of the determinant, D, corresponding to the matrix of coefficients of x_i. For example, in the case given above,

$$D = \begin{vmatrix} 1.000 & -0.738 \\ 0.801 & -0.590 \end{vmatrix} = 0.00114$$

A zero, or near-zero, value of D may indicate an ill-conditioned problem and is, generally, a signal to re-examine the equations.

Unequally Weighted Data Points   So far, we have assumed that all of the data points that we require to fit have equal importance. It may be, however, that the standard deviation, σ_i, to be attached to each point is not identical. If this is so, a weighted fit is preferable. For example, if a polynomial function is to be fitted we would minimize:

$$D = \sum_i \frac{1}{\sigma_i^2} \left(y_i - \sum_{j=0}^{n} a_j x_i^j \right)^2$$

$$(2.31)$$

Of course, the weakness in this is that σ_i, the standard deviation of each y_i, has to be known — thereby necessitating many measurements at a single value of x_i. Fortunately, this can be avoided by a method known as *relative deviations*.[7] This makes the reasonable supposition that the absolute error in the dependent variable is proportional to the errors in the independent variables. For example, in spectrophotometry, the absolute error in the solution absorbance is greater at high solute concentrations. Therefore, we define a relative deviation, D', between experimental and calculated data and use this to develop our normal equations.

If a polynomial fit is anticipated, we obtain:

$$D' = \sum_i \left[\left(y_i - \sum_{j=0}^{n} a_j x_i^j \right) / y_i \right]^2$$

This is subjected to an identical treatment to that used previously (differentiation with respect to each of the a_j in turn) so that the final equations are:

$$a_0 \sum_{i=1}^{N} 1/y_i + a_1 \Sigma x_i/y_i + \ldots + a_n \Sigma x_i^n/y_i = N$$

$$a_0 \Sigma x_i/y_i + a_1 \Sigma x_i^2/y_i + \ldots + a_n \Sigma x_i^{n+1}/y_i = \Sigma 1/y_i$$

$$\begin{array}{cccc} \cdot & \cdot & \cdot & \cdot \\ \cdot & \cdot & \cdot & \cdot \\ \cdot & \cdot & \cdot & \cdot \end{array}$$

$$a_0 \Sigma x_i^n/y_i + a_1 \Sigma x_i^{n+1}/y_i \ldots + a_n \Sigma x_i^{2n} = \Sigma x_i^n/y_i$$

This approach is particularly useful when wide ranges of x and y are expected.

2.5 Non-linear Parameter Dependence

Many equations which arise in more-advanced areas of study are non-linear and cannot be recast into a linear form. Examples of situations which give rise to non-linearity include:

(i) The temperature response of a thermistor.
(ii) The concentration dependence of the optical density of a solution containing complexes in equilibria.
(iii) The viscosity of a solution.

Owing to the frequent occurrence of non-linear relations, there is a profusion of suitable programs having greater or lesser efficiency. We shall only examine two of the simpler approaches and the reader is referred to the specialist literature if he wishes to use the more powerful algorithms.

The two methods to be discussed are called:

(i) Direct grid search.
(ii) Steepest descent.

It should be remembered that many programs appear under these generic names and can have quite different characteristics.

The one thing that is common to all of the approaches is that they are *iterative* in nature — unlike the non-iterative methods described so far. At each stage of an iterative program the 'true' answer is approached more and more closely.

42

2.5.1 Direct Grid Search[8]

This is the simpler and more dependable method but it is inefficient for the solution of equations with large numbers of variables. The least squares criterion, (2.32), is again applied

$$S_j = \sum_i [F_i(\beta_j;x) - y_i]^2 \tag{2.32}$$

where $F_i(\beta_j; x)$ is the value of a function with parameters β_j at the jth iteration. The summation is over all data points i.

The possible behaviour of S in the vicinity of a minimum for a two-dimensional problem is illustrated in Figure 2.4. The contours represent constant values of S_j for different values of β_1 and β_2.

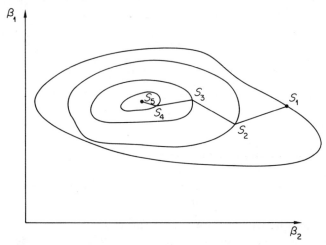

Figure 2.4   Zig-zag path of a direct grid search procedure from an initial guess to a minimum of a function of two parameters

We start with initial guesses of the β_i, arbitrary step sizes, $\Delta\beta_i$, and a condition for termination of the analysis. This is either that S_j is no longer changing or that the β_i parameters have been determined to an acceptable accuracy. There is usually provision for step expansion or contraction when the estimated parameters are far from, or near to, those of the analytical minimum.

A possible flow diagram is shown in Figure 2.5.

In Table 2.12 we present a program and subroutine for direct grid search minimization based on the above discussion. The subroutine is called from the main program (discussed later) by:

CALL BE10(NVAR, AMIN, NP, AL)

as shown in Table 2.12 (line 20) in which NVAR is the number of parameters to be optimized, AMIN is the final value of the minimum,

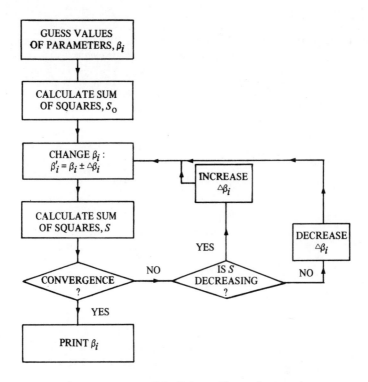

Figure 2.5   A possible direct grid search procedure

NP is the number of points to be fitted to the theoretical expression and AL is the value of any linear parameter. In the common blocks, labelled A and B, the parameters have the following significances:

$XXX(10) =$ the array which stores the input and final values of the parameters

$X(10) =$ the array containing the current values of the parameters

$DX(10) =$ the magnitudes of the step sizes

$DELTA(10) =$ the minimum values of the step size

$YFIN(100) =$ the array containing the final computed values of the dependent variables

$YOBS(100) =$ input (observed) values of the dependent vaiables

$YCALC(100) =$ current calculated values of the dependent variables

$XD(100) =$ array of the independent variables

Of these the user must specify the input values of XXX, DX, DELTA, YOBS and XD.

The current value of the minimum is computed by the use of a function routine called ERROR, which is used in the subroutine by

Table 2.12
A program using the direct grid search subroutine BEIO for fitting experimental viscosity data to a theoretical relationship

```
0008              MASTER TEST
0009              COMMON/A/XXX(10),X(10),DX(10),DELTA(10),YFIN(100)
0010              COMMON/B/YOBS(100),YCALC(100),XD(100)
0011              READ(1,1)NVAR,NP
0012            1 FORMAT(2I0)
0013              READ(1,29)(XD(I),YOBS(I),I=1,NP)
0014           29 FORMAT(2F9.0)
0015            2 FORMAT(3E10.4)
0016              READ(1,2)(XXX(I),DX(I),DELTA(I),I=1,NVAR)
0017              WRITE(2,2004)
0018         2004 FORMAT(1H1,'INPUT   ORIGIN X        STEP SIZE        LIMIT        ')
0019              WRITE(2,222)(XXX(I),DX(I),DELTA(I),I=1,NVAR)
0020          222 FORMAT(1H ,E15.8,3X,E15.8,3X,E15.8)
0021              CALL BEIO(NVAR,AMIN,NP,AL)
0022              WRITE(2,5)
0023            5 FORMAT(1H ,' FINAL  VALUES')
0024              WRITE(2,6)AMIN
0025              WRITE(2,9)AL
0026            9 FORMAT(1H ,' LINEAR PARAMETER ',F10.4,/,'  NON LINEAR',/)
0027            6 FORMAT(1H ,' AMIN =',F11.5)
0028              WRITE(2,8)(XXX(I),DX(I),I=1,NVAR)
0029            8 FORMAT(1H ,F10.4,5X,E10.4)
0030              WRITE(2,51)
0031           51 FORMAT(1H1,'    TEMP        OBS         CALC')
0032              WRITE(2,50)(XD(I),YOBS(I),YFIN(I),I=1,NP)
0033           50 FORMAT(F9.2,5X,F9.4,5X,F9.4)
0034              STOP
0035              END



0036              SUBROUTINE BEIO(NVAR,AMIN,NP,AL)
0037              DIMENSION(10),XX(10)
0038              COMMON/A/XXX(10),X(10),DX(10),DELTA(10),YFIN(100)
0039              COMMON/B/YOBS(100),YCALC(100),XD(100)
0040          302 FORMAT(39H JOB ABANDONED,MORE THAN 100 ITERATIONS)
0041            8 FORMAT(1H ,6H AMIN=,E15.8)
0042          101 FORMAT(1H ,25H NUMBER OF ITERATIONS WAS,I4)
0043           11 FORMAT(1H ,10X,4HSTEP,I2,2H =,E10.4)
0044          900 FORMAT(1H1,'DIRECT GRID SEARCH')
0045          555 FORMAT(1H ,20A4)
0046              NC=0
0047              DO 4 I=1,NVAR
0048              X(I)=XXX(I)
0049              XX(I)=XXX(I)
0050            4 CONTINUE
0051              M1=1
0052              AMIN=ERROR(NP,AL)
0053              NC=0
0054          308 NC=NC+1
0055              WRITE(2,8)AMIN
0056              IF(NC.LT.100)GO TO 301
0057              WRITE(2,302)
0058              GO TO 313
0059          301 CONTINUE
0060              ND=0
0061              NL=0
0062    C         MINIMISATION LOOPS
0063              GO TO(30,31,32,33,34,35,36,37,38,39)NVAR
0064           39 DO 29 J10=1,3
0065              X(10)=XXX(10)+DX(10)*(J10-2)
0066           38 DO 28 J9=1,3
0067              X(9)=XXX(9)+DX(9)*(J9-2)
0068           37 DO 27 J8=1,3
0069              X(8)=XXX(8)+DX(8)*(J8-2)
0070           36 DO 26 J7=1,3
0071              X(7)=XXX(7)+DX(7)*(J7-2)
0072           35 DO 25 J6=1,3
0073              X(6)=XXX(6)+DX(6)*(J6-2)
0074           34 DO 24 J5=1,3
0075              X(5)=XXX(5)+DX(5)*(J5-2)
0076           33 DO 23 J4=1,3
0077              X(4)=XXX(4)+DX(4)*(J4-2)
0078           32 DO 22 J3=1,3
0079              X(3)=XXX(3)+DX(3)*(J3-2)
0080           31 DO 21 J2=1,3
```

```
0081                    X(2)=XXX(2)+DX(2)*(J2-2)
0082            30 DO 20 J1=1,5
0083                    X(1)=XXX(1)+DX(1)*(J1-2)
0084           120 F=ERROR(NP,AL)
0085                    IF(AMIN.LT.F)GO TO 20
0086                    AMIN=F
0087                    DO 66 JK=1,NVAR
0088            66 XX(JK)=X(JK)
0089                    DO 67 K=1,NP
0090            67 YFIN(K)=YCALC(K)
0091            20 CONTINUE
0092                    IF(NVAR.EQ.1)GO TO 40
0093            21 CONTINUE
0094                    IF(NVAR.EQ.2)GO TO 40
0095            22 CONTINUE
0096                    IF(NVAR.EQ.3)GO TO 40
0097            23 CONTINUE
0098                    IF(NVAR.EQ.4)GO TO 40
0099            24 CONTINUE
0100                    IF(NVAR.EQ.5)GO TO 40
0101            25 CONTINUE
0102                    IF(NVAR.EQ.6)GO TO 40
0103            26 CONTINUE
0104                    IF(NVAR.EQ.7)GO TO 40
0105            27 CONTINUE
0106                    IF(NVAR.EQ.8)GO TO 40
0107            28 CONTINUE
0108                    IF(NVAR.EQ.9)GO TO 40
0109            29 CONTINUE
0110            40 CONTINUE
0111                    DO 377 I=1,NVAR
0112       C        STORE X FOR FINAL O/P
0113                    IF((ABS(XXX(I)-XX(I))).GE.DX(I))GO TO 306
0114                    IF(DX(I).GT.DELTA(I))GO TO 800
0115                    GO TO 377
0116           800 DX(I)=DX(I)/2.
0117                    NL=NL+1
0118           602 CONTINUE
0119                    GO TO 377
0120           306 DX(I)=2.*DX(I)
0121                    ND=ND+1
0122           377 XXX(I)=XX(I)
0123                    IF(NL.GT.0.OR.NL.GT.0)GO TO 308
0124           313 CONTINUE
0125                    WRITE(2,101)NC
0126                    RETURN
0127                    END


0128                    FUNCTION ERROR(NP,AL)
0129                    DIMENSIONF(100)
0130                    COMMON/A/XXX(10),X(10),DX(10),DELTA(10),YFIN(100)
0131                    COMMON/B/YOBS(100),YCALC(100),XD(100)
0132                    FSUM=0
0133                    YSUM=0
0134                    R=2.000
0135                    ERROR=0
0136                    DO 2 I=1,NP
0137                    F(I)=EXP((1./R)*(X(1)/XD(I)+X(2)*XD(I)+X(3)*(XD(I)**2)))
0138                    FSUM=FSUM+F(I)
0139             2 YSUM=YSUM+YOBS(I)
0140                    AL=YSUM/FSUM
0141                    DO 1 I=1,NP
0142                    YCALC(I)=AL*F(I)
0143             1 ERROR=ERROR+(YOBS(I)-YCALC(I))**2
0144                    RETURN
0145                    END
```

```
INPUT    ORIGIN X       STEP SIZE       LIMIT
  0.50000000E 04   0.11000000E 04   0.50000000E 02
  0.80000000E-01   0.23000000E-01   0.50000000E-03
 -0.20000000E-08   0.13000000E-09   0.50000000E-10
  AMIN= 0.12076145E 02
  AMIN= 0.30280174E 01
  AMIN= 0.22997293E-02
  AMIN= 0.22995478E-02
  AMIN= 0.22994570E-02
```

```
AMIN= 0.22992755F-02
AMIN= 0.22991848F-02
AMIN= 0.22991594F-02
AMIN= 0.16129528F-02
AMIN= 0.16129258F-02
AMIN= 0.16128718F-02
AMIN= 0.14857902F-02
AMIN= 0.14254154F-02
AMIN= 0.14254068F-02
AMIN= 0.13690354F-02
AMIN= 0.13690270E-02
AMIN= 0.15166551F-02
AMIN= 0.15166468F-02
AMIN= 0.12682706F-02
AMIN= 0.12682625E-02
AMIN= 0.12238781F-02
AMIN= 0.12238703F-02
AMIN= 0.11834738F-02
AMIN= 0.11834661F-02
AMIN= 0.11470539F-02
AMIN= 0.11470464F-02
AMIN= 0.11146144F-02
AMIN= 0.11146071F-02
AMIN= 0.10861516F-02
AMIN= 0.10861446F-02
AMIN= 0.10616686F-02
AMIN= 0.10616617E-02
AMIN= 0.10411542F-02
AMIN= 0.10411476F-02
AMIN= 0.10246047F-02
AMIN= 0.10245982E-02
AMIN= 0.10120161E-02
AMIN= 0.10120099E-02
AMIN= 0.10053848F-02
AMIN= 0.10053788E-02
AMIN= 0.99870693F-03
AMIN= 0.99870113F-03
AMIN= 0.99797869F-03
AMIN= 0.99797307F-03
NUMBER OF ITERATIONS WAS   44
FINAL  VALUES
AMIN =    0.00100
  LINEAR PARAMETER 0.7160E-08
  NUN LINEAR

0.7612E 04      0.3437E 02
0.3867E-01      0.3594E-03
-.4075E-09      0.3250E-10
```

TEMP	OBS	CALC
273.15	1.7870	1.7738
278.15	1.5160	1.5209
283.15	1.3060	1.3157
288.15	1.1380	1.1477
291.15	1.0530	1.0615
293.15	1.0020	1.0092
298.15	0.8903	0.8941
303.15	0.7975	0.7979
308.15	0.7194	0.7169
311.15	0.6783	0.6744
313.15	0.6531	0.6483
318.15	0.5963	0.5899
323.15	0.5467	0.5400
328.15	0.5044	0.4971
333.15	0.4666	0.4601
338.15	0.4342	0.4280
343.15	0.4049	0.4001
348.15	0.3798	0.3758
353.15	0.3554	0.3546
358.15	0.3345	0.3361
363.15	0.3156	0.3198
368.15	0.2985	0.3055
373.15	0.2829	0.2930

such statements as

AMIN = ERROR (NP, AL)

The use of a function routine allows considerable flexibility since the user has only to alter the form of this small routine, thus leaving the subroutine unaltered.

The trial value of each parameter is calculated by statements of the type

X(I) = XXX(I) + DX(I) * (J − 2)

where J is 1, 2, or 3, so that we 'step' to either side of XXX(I) in order to calculate a test value of X. This is accomplished in a series of nested DO loops, the point of entry being determined by a computed GO TO (line 63). In line 85−90 we test to see if a new minimum has been found and, if so, we store the values of the parameters and the corresponding values of Y(I). In line 114 we test to see if the present value of X(I) is at a boundary (i.e. XXX(I) ± DX(I). If so, the step size is doubled (line 120) and the search process is repeated. Similarly, if DX(I) is greater than the limiting value, DELTA(I), it is halved (line 110) and again the process is repeated.

As an example of the use of this subroutine we find that Becsey, Berke and Callan[8] cited the use of a direct grid search method in fitting viscosity data to the equation:

$$\eta = \text{const. exp} \ (\epsilon_0/RT + \epsilon_1 \cdot T/R + \epsilon_2 * T^2/R) \tag{2.33}$$

where η is the viscosity at temperature T.

The program BEST (Table 2.12) calculates the values of ϵ_0, ϵ_1, ϵ_2 and the constant linear term in (2.33). This equation is used in the function routine ERROR (NP, AL) in line 137. Note that we have used 2.0 cal K^{-1} mol^{-1} for the gas constant, R, in accordance with the original reference.[3] The linear term is calculated as a mean value in line 199. The values of ϵ_0, ϵ_1 and ϵ_2 correspond to the array elements XXX(1), XXX(2) and XXX(3) respectively. The progress of the minimization is clear from Table 2.12, in which the value of the computed minimum is printed at each iteration. The number of iterations is limited to 100 in order to avoid excessive use of computer time. At the end of the iterations, the final value of the minimum and mean value of any linear term can be printed. In Table 2.12 we also print the non-linear terms, together with the final step sizes and, finally, list the obseved and calculated values of the viscosity, calculated from equation (2.33).

2.5.2 Steepest Descent

The efficiency of the preceding method can be improved upon if we make use of the rate of change of the value of a function, S, with the change in value of a parameter, β_i. For example, a possible iterative method is

$$\beta_i' = \beta_i - \lambda(\partial S/\partial \beta_i) \qquad (2.34)$$

where λ is a quantity chosen such that β_i lies towards a minimum. A better iterative method is

$$\beta_i' = \beta_i - \Delta\beta_i \cdot G_i \qquad (2.35)$$

where $\Delta\beta_i$ is the parameter step-size (as in the previous section) and G_i is a gradient defined by (2.36).

$$G_i = (\partial s/\partial \beta_i) \Big/ \sqrt{\left[\sum_{j=1}^{n} (\partial s/\partial \beta_i)^2\right]} \qquad (2.36)$$

The G_i are, therefore, relative values normalized to unity. Sensible choices of the $\Delta\beta_i$ ensures a rapid descent to a lower value of S — this is one version of the 'Steepest Descent' method.

As the minimum is approached, this technique becomes slow to converge so, in the program to be described, the following approach is used.[9]

(i) Evaluate the gradient (equation (2.36)) at the chosen x_i values.

(ii) Search along the gradient in a stepwise manner until an apparent minimum is found.

(iii) From the last three values of the function, S_1, S_2, S_3, which straddle the minimum, find the improved values of β_i by parabolic interpolation;

$$\left. \begin{array}{l} \beta_i' = \beta_i - f\,\Delta\beta_i G_i \\ \\ \beta_i'' = \beta_i + (f-1)\,\Delta\beta_i G_i \end{array} \right\} \qquad (2.37)$$

or

where $f = \dfrac{S_3 - S_2}{S_3 - 2S_2 + S_1} + 0.5$

The set of values β_i' or β_i'' is chosen which gives the lower value of the minimum.

(iv) If further iterations are desired, we return to (i). Iteration is terminated when the change in S, the error-square sum, is acceptably small.

A subroutine based on this approach is listed in Table 2.13.

For ease of use, we retain the same common blocks in this subroutine as were used in BE10. It is called by:

CALL GRSD(NVAR, AMIN, PC, NP, AL)

Table 2.13
Subroutine GRSD — minimization by a steepest descent method

```
      SUBROUTINE GRSD(NVAR,AMIN,PC,NP,AL)
      DIMENSIONG(10)
      COMMON/A/XXX(10),X(10),DX(10),DELTA(10),YFIN(100)
      COMMON/B/YOBS(100),YCALC(100),XD(100)
      NC=0
      WRITE(2,6)PC
    6 FORMAT(//,1H ,' TERMINATE AT ',F7.3,' PER CENT')
      PC=PC/100
   99 DO 300 I=1,NVAR
  300 X(I)=XXX(I)
      BEGIN=ERROR(NP,AL)
      START=BEGIN
      IF(START.LT.1.0E-6)GO TO 100
C     CALCULATE GRADIENT
      NC=NC+1
      IF(NC.LT.100)GO TO 90
      WRITE(2,91)
   91 FORMAT(1H ,'MORE THAN 100 ITERATIONS')
      GO TO 100
   90 SUMG=0
  220 DO 1 I=1,NVAR
      X(I)=XXX(I)+0.1*DX(I)
      G(I)=BEGIN-ERROR(NP,AL)
      X(I)=XXX(I)
    1 SUMG=SUMG+G(I)**2
C     NOTE THAT G IS THE NEGATIVE OF THE TRUE GRADIENT.
      DO 2 I=1,NVAR
    2 G(I)=G(I)/SQRT(SUMG)
  120 DO 3 I=1,NVAR
    3 X(I)=X(I)+G(I)*DX(I)
      AMID=ERROR(NP,AL)
C     CHECK STEP SIZE
      IF(AMID.LT.BEGIN)GO TO 4
      DO 5 I=1,NVAR
      X(I)=XXX(I)
    5 G(I)=G(I)/2.0
      GO TO 120
C     INCREMENT X ALONG G
    4 DO 7 I=1,NVAR
    7 X(I)=X(I)+ G(I)*DX(I)
      END=ERROR(NP,AL)
      IF(END-AMID)8,9,9
    8 BEGIN=AMID
      AMID=END
      GO TO 4
C     FIND MINIMUM BY PARABOLIC INTERPOLATION
      WRITE(2,20)BEGIN,AMID,END
   20 FORMAT(1H ,3F10.8)
    9 AF=1./(1.+(BEGIN-AMID)/(END-AMID))+0.5
      DO 10 I=1,NVAR
      X(I)=X(I)-AF*DX(I)*G(I)
   10 XXX(I)=X(I)
      AMIN=ERROR(NP,AL)
      IF(AMID-AMIN)102,110,110
  102 DO 103 I=1,NVAR
      X(I)=X(I)+(AF-1.)*DX(I)*G(I)
  103 XXX(I)=X(I)
      AMIN=ERROR(NP,AL)
  110 WRITE(2,11)AMIN
   11 FORMAT(1H ,50X,F11.5)
      IF(((START-AMIN)/(START)).GT.PC)GO TO 99
  100 WRITE(2,12)NC,AMIN
   12 FORMAT(1H ,10X,I3,' ITERATIONS , MINIMUM AT',F11.5)
      DO 400 I=1,NP
  400 YFIN(I)=YCALC(I)
      RETURN
      END
      FINISH
```

NVAR, AMIN NP and AL have their previous significance; PC is the limiting percentage change in the value of the computed minimum upon which control is transferred to the calling program. The gradient of the function is calculated and the values of the parameters are changed until an approximate minimum is found. Then, in accordance with equations (2.37), the minimum is refined by a parabolic interpolation. When this routine was applied to the example in the previous section, 49 iterations were required but the computer time required was less than half that used by BE10. Furthermore the value of the computed minimum was almost as good (0.00341)

It is worth remembering that gradient methods, such as this, are not without limitations, nor are they so universally useful as the direct grid search methods. The latter, though inefficient, are reliable whereas under certain circumstances, the gradient method can be troublesome and can lead to erroneous results. An example of such a circumstance arises when the gradient of the function is very small. This can happen when one is a long way away from a minimum and can lead to non-detection of the true minimum.

As a general point, it should also be remembered that any minimization problem may have many minima. The 'correct' answer is normally accepted as that which results in the lowest minimum but, if the initial guesses of the parameters are far from their correct values, it is possible that an erroneous *local* minimum may be located – at which point many procedures will stop and presume that a correct set of values has been located.

2.6 Eigenvalue Problems
So-called 'eigenvalue' problems arise in many physical systems and prior to discussing these we shall present the problem in a formal, but general, manner.

If the matrix product $A \cdot X$ can be set equal to λX, where λ is a scalar quantity, the matrix equation (2.38) results:

$$A \cdot X = \lambda X \tag{2.38}$$

Equation (2.38) may be rewritten as (2.39)

$$(A - \lambda I) X = 0 \tag{2.39}$$

where I is the unit matrix having the diagonal elements a_{ii} equal to unity and all other a_{ij} ($i \neq j$) equal to zero.

One solution to (2.39) is $X = 0$, this being a trivial solution; non-trivial solutions, $X \neq 0$, obtain when:

$$| A - \lambda I | = 0$$

This may be written more fully as

$$\begin{vmatrix} a_{11} - \lambda & a_{12} \ldots a_{1n} \\ a_{21} & a_{22} - \lambda \ldots a_{2n} \\ \cdot & \cdot \quad \cdot \\ \cdot & \cdot \quad \cdot \\ \cdot & \cdot \quad \cdot \\ a_{n1} & a_{n2} \ldots a_{nn} - \lambda \end{vmatrix} = 0$$

Expansion of this determinant would give an nth-degree polynomial in λ. The solutions are called *eigenvalues*. Substitution of each of the n values of λ into (2.39) yields, on solution of each set of equations, n column vectors, X, called *eigenvectors*. Thus, for each eigenvalue there is a corresponding eigenvector.

The eigenvalue problem is of widespread physical significance; in chemistry, the most significant application arises in the wave-mechanical treatment of atoms and molecules. As will be shown in Chapter 5, the eigenvalues in a molecular orbital calculation are the energies of the molecular orbitals which are, in turn, obtained from the eigenvectors. We shall discuss the solution of the eigenvalue problem in this chapter and use a subroutine based on the results that will now be obtained.

The analytical solution of the eigenvalue problem for a matrix greater that 3 x 3 is one of some complexity, involving the solution of an nth-degree polynomial. For this reason, approximation methods have been devised and many of these involve *matrix diagonalization*. A diagonal matrix is one which has all off-diagonal elements equal to zero. It can be easily shown that, if A is a matrix for which there exist n eigenvectors, then there is a matrix Q for which

$$Q^{-1} A Q = \text{diag} (\lambda) = D$$

where D is a diagonal matrix, the diagonal elements of which are the eigenvalues of A. Also, the n columns of Q are, in fact, the corresponding eigenvectors so that, by properly choosing Q, the eigenvalue problem is completely solved.

It is the choice of Q which is the difficult step, since generally, one has no *a priori* knowledge of the form of this matrix. Therefore, as in the previous section, iterative procedures are used — the most common of these being the Jacobi rotation method. This is geometric in character and makes use of the transformation properties which relate a frame of reference x, y and a new frame x', y' obtained by rotating the old axes by an angle θ (Figure 2.6).

$$\begin{pmatrix} x \\ y \end{pmatrix} = \begin{pmatrix} \cos \theta & \sin \theta \\ -\sin \theta & \cos \theta \end{pmatrix} \begin{pmatrix} x' \\ y' \end{pmatrix} = Q \begin{pmatrix} x' \\ y' \end{pmatrix}$$

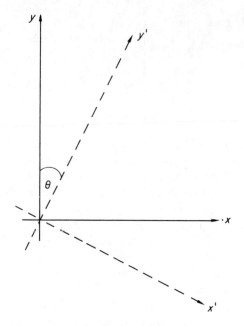

Figure 2.6 Transformation from one set of coordinates (x, y) to a second set (x', y')

In this case, which corresponds to an orthogonal transformation, we observe that

$$Q^{-1} = Q^t, \quad \text{i.e. } Q^{-1} Q = Q^t Q = I$$

and also, that if we have a symmetric matrix, B, then with the correct angle θ

$$Q^t B Q = D$$

where B is a 2 x 2 matrix and D is the corresponding diagonal matrix. This is most easily shown by reference to an example:

Example Find the eigenvalues of A, where

$$A = \begin{pmatrix} 6 & 1 \\ 1 & 2 \end{pmatrix}$$

By expansion of the determinant $| A - \lambda I |$,

$$| A - \lambda I | = \begin{vmatrix} 6 - \lambda & 1 \\ 1 & 2 - \lambda \end{vmatrix} = \lambda^2 - 8\lambda + 11 = 0$$

from which, $\lambda = 6.236$ or 1.764.

Alternatively, using the Jacobi method,

$$\begin{pmatrix} \cos\theta & -\sin\theta \\ \sin\theta & \cos\theta \end{pmatrix} \begin{pmatrix} 6 & 1 \\ 1 & 2 \end{pmatrix} \begin{pmatrix} \cos\theta & \sin\theta \\ -\sin\theta & \cos\theta \end{pmatrix} = \begin{pmatrix} \lambda_1 & 0 \\ 0 & \lambda_2 \end{pmatrix}$$

and expanding

$$\begin{pmatrix} 6\cos^2\theta + 2\sin^2\theta - 2\sin\theta\cos\theta & 4\sin\theta\cos\theta + \cos^2\theta - \sin^2\theta \\ 4\sin\theta\cos\theta + \cos^2\theta - \sin^2\theta & 6\sin^2\theta + 2\cos^2\cos^2\theta + 2\sin\theta\cos\theta \end{pmatrix} = \begin{pmatrix} \lambda_1 & 0 \\ 0 & \lambda_2 \end{pmatrix}$$

This requires

$$4\sin\theta\cos\theta = \sin^2\theta - \cos^2\theta$$

or

$$\tan 2\theta = -\tfrac{1}{2} \quad \text{and} \quad \theta = 76° \, 44'$$

Substitution into the diagonal elements yields

$$\lambda = 6\cos^2\theta + 2\sin^2\theta - 2\sin\theta\cos^2\theta = 1.764$$

$$\lambda = 6\sin^2\theta + 2\cos^2\theta + 2\sin\theta\cos\theta = 6.236$$

This result can be generalised for any symmetric matrix.

$$A = \begin{pmatrix} p & r \\ r & q \end{pmatrix}$$

$$\tan 2\theta = 2r/(q - p)$$

and the eigenvalues are obtained in the same manner as in the above worked example.

As we have previously intimated, the power of the Jacobi rotation method is not in the solution of 2 x 2 matrices! Consider a larger symmetric matrix A of order n and let us isolate the elements a_{ii}, a_{ij}, a_{ji} and a_{jj} in columns i and j, and rows i and j. We subject these elements, which comprise a submatrix of the parent matrix, to a Jacobi rotation such that a_{ij} and a_{ji} are eliminated. This requires a rotation of θ, given by

$$\tan 2\theta = 2a_{ij}/(a_{jj} - a_{ii})$$

and a transformation

$$Q^t A Q$$

in which the transforming matrices Q^t and Q clearly only operate on the elements at the intersection of rows i, j and columns i, j and leave all other elements unchanged. As a point of strategy, it is usual to locate the largest off-diagonal elements and to repeat the procedure successively until each of the diagonal elements is reduced to an acceptably low (pre-set) value. In matrix notation, this means that we begin with:

$$Q_1^t \, A \, Q_1 \,\hat{=}\, \text{Diag} \,(\lambda)$$

as the first step in an iteration.

Subroutine JACB (Table 2.14) makes use of the concepts described above (it is based on the program described in Reference 11). We have

Table 2.14

Program EIGEN for the solution of an eigenvalue problem using subroutine JACB

```
0008                    MASTER EIGEN
0009                    COMMON/JAC01/N,A(40,40),S(40,40)
0010                    READ(1,1)N
0011                    DO 9 I=1,N
0012                  9 READ(1,2) (A(I,J),J=1,N)
0013                    WRITE(2,7)
0014                  7 FORMAT(1H1,' INPUT MATRIX')
0015                    DO 10 I=1,N
0016                 10 WRITE(2,8) (A(I,J),J=1,N)
0017                  1 FORMAT(I2)
0018                  2 FORMAT(10F7,3)
0019                    CALL JACB
0020                    WRITE(2,4)
0021                    DO 3 I=1,N
0022                    WRITE(2,5)A(I,I)
0023                  3 WRITE(2,6)(S(J,I),J=1,N)
0024                  4 FORMAT(1H1,' SOLUTIONS FOR THIS EIGENVALUE PROBLEM',//)
0025                  5 FORMAT(1H ,' EIGENVALUE =',F9,3,//,' EIGENVECTOR IS')
0026                  6 FORMAT(1H ,30X,F9,3)
0027                  8 FORMAT(1H ,20F6,2)
0028                    STOP
0029                    END


0030                    SUBROUTINEJACB
0031                    COMMON/JAC01/N,A(40,40),S(40,40)
0032          C         SET INDICATOR WHICH CHECKS OFF-DIAGONAL ELEMENTS
0033          C         INDIC=0
0034          C         INPUT ARRAY TO BE SOLVED
0035          C         COMPUTE INITIAL NORM
0036                151 VI=0,0
0037                    DO 106 I=1,N
0038                    DO 106 J=1,N
0039                    IF(I-J)107,206,107
0040                107 VI=VI+A(I,J)**2
0041          C         GENERATE S IDENTITY MATRIX
0042                    S(I,J)=0
0043                    GO TO 106
0044                206 S(I,J)=1,0
0045                106 CONTINUE
0046                    VI=SQRT(VI)
0047          C         COMPUTE FINAL NORM
0048                    VF=VI*0,1E-07
0049          C         COMPUTE THRESHOLD NORM
0050                    AN=N
0051                128 VI=VI/AN
0052          C         SET UP SYSTEMATIC SEARCH
0053                137 IQ=1
0054                124 IQ=IQ+1
0055                    IP=0
0056                121 IP=IP+1
0057                    IF(A(IP,IQ))108,120,109
0058                108 IF(-A(IP,IQ)-VI)120,112,112
0059                109 IF(A(IP,IQ)-VI)120,112,112
0060                112 INDIC=1
0061          C         COMPUTE AND COSINE OF MATRIX ROTATION ANGLE
0062                    ALAM=-A(IP,IQ)
```

```
0063              AMU=0.5*(A(IP,IP)-A(IQ,IQ))
0064              IF(AMU)113,114,114
0065          113 SGN=-1.0
0066              GO TO 115
0067          114 SGN=+1.0
0068          115 OMEGA=SGN*ALAM/SQRT(ALAM**2+AMU**2)
0069              STHT=OMEGA/SQRT(2.0+2.0*SQRT(1.0-OMEGA**2))
0070              CTHT=SQRT(1.0-STHT**2)
0071        C     TRANSFORM ELEMENTS OF THE PTH AND QTH COLUMN
0072              DO 116 I=1,N
0073              IF(I-IP)117,118,117
0074          117 IF(I-IQ)119,118,119
0075        C     ROTATE THE SECULAR MATRIX
0076          119 AIPI=A(IP,I)*CTHT-A(IQ,I)*STHT
0077              AIQI=A(IP,I)*STHT+A(IQ,I)*CTHT
0078              A(IP,I)=AIPI
0079              A(IQ,I)=AIQI
0080        C     ROTATE THE MATRIX OF COEFFICIENTS
0081          118 AIPI=S(I,IP)*CTHT-S(I,IQ)*STHT
0082              AIQI=S(I,IP)*STHT+S(I,IQ)*CTHT
0083              S(I,IP)=AIPI
0084          116 S(I,IQ)=AIQI
0085              AIPI=A(IP,IP)*CTHT**2+A(IQ,IQ)*STHT**2-2.*A(IP,IQ)*STHT*CTHT
0086              AIQI=A(IP,IP)*STHT**2+A(IQ,IQ)*CTHT**2
0087             1+2.0*A(IP,IQ)*STHT*CTHT
0088              AIPIQ=(A(IP,IP)-A(IQ,IQ))*CTHT*STHT
0089             1+A(IP,IQ)*(CTHT**2-STHT**2)
0090              A(IP,IP)=AIPI
0091              A(IQ,IQ)=AIQI
0092              A(IP,IQ)=AIPIQ
0093              A(IQ,IP)=A(IP,IQ)
0094        C     TRANSFORM MATRIX
0095              DO 123 I=1,N
0096              A(I,IP)=A(IP,I)
0097          123 A(I,IQ)=A(IQ,I)
0098          120 IF(IP-IQ+1)121,122,122
0099          122 IF(IQ-N)124,125,125
0100          125 IF(INDIC)126,127,126
0101          126 INDIC=0
0102              GO TO 137
0103          127 IF(VI-VF)129,129,128
0104          129 RETURN
0105              END
```

```
INPUT MATRIX
 5.00  2.00  0.00  1.00
 2.00  5.00  1.00  0.00
 0.00  1.00  5.00  2.00
 1.00  0.00  2.00  5.00
```

SOLUTIONS FOR THIS EIGENVALUE PROBLEM

EIGENVALUE =   8.000

EIGENVECTOR IS

```
                              0.500
                              0.500
                              0.500
                              0.500
```

EIGENVALUE =   4.000

EIGENVECTOR IS

```
                             -0.500
                              0.500
                              0.500
                             -0.500
```

EIGENVALUE =   6.000

EIGENVECTOR IS

```
                             -0.500
                             -0.500
                              0.500
                              0.500
```

EIGENVALUE =   2.000

EIGENVECTOR IS

```
                             -0.500
                              0.500
                             -0.500
                              0.500
```

included it in a complete program EIGEN so that an example can be discussed. Examination of the subroutine itself, shows that the sine and cosine of the matrix rotation angles are calculated in lines 69 and 70. These are then used in the matrix rotation and the first approximation to the diagonal matrix is generated. The eigenvalues are stored as the diagonal elements in the array A(40,40) and the eigenvectors in the array S(40,40). The iteration is terminated by the value of VI, the so-called 'norm', computed initially in lines 37—40 and in each cycle in line 51.

The subroutine is called by:

CALL JACB

A common block of the form

COMMON/JAC01/N,A(40,40), S(40,40)

is required, in which N is the order of the eigenvalue problem. As an example of the use of this subroutine, let us verify that the eigenvalues, λ_i, and eigenvectors, q_i, of the matrix (2.40) are:

$$\lambda_1 = 8; q_1 = (0.5, 0.5, 0.5, 0.5)*$$
$$\lambda_2 = 4; q_2 = (-0.5, 0.5, 0.5, -0.5)*$$
$$\lambda_3 = 6; q_3 = (-0.5, -0.5, 0.5, 0.5)*$$
$$\lambda_4 = 2; q_4 = (-0.5, 0.5, -0.5, 0.5)*$$

The matrix in question being

$$A = \begin{pmatrix} 5 & 2 & 0 & 1 \\ 2 & 5 & 1 & 0 \\ 0 & 1 & 5 & 2 \\ 1 & 0 & 2 & 5 \end{pmatrix}$$

The output from the program EIGEN is listed in Table 2.14. The results are precisely those which would be obtained by a manual method. We recommend that the reader convinces himself that the solutions are corrects.

2.7 Differentiation and Integration
There are two methods of differentiation of importance when using a computer, dependent on whether or not an analytical form of a derivative is available. If, for example, we have a function:

$$y = a + bx + cx^2$$

then the first and second derivatives are easily evaluated:

$$dy/dx = b + 2cx$$
$$d^2y/dx^2 = 2c$$

If, however, no differentiable function is available then a convenient method is to calculate the function at the desired values of the parameters $a_0, \ldots, a_n$ and the independent variables, x,

$$y = f(a_0, a_1, \ldots, a_n; x)$$

and then to increment x by a small amount and recalculate y,

$$x = x + \Delta x$$

where Δ is a fraction (say 10^{-3}) so small that the change in y divided by the change in each x approximates to dy/dx.

Differentiation techniques are used in Chapter 4 when we discuss end-point location. They are also used in Chapter 6 when we discuss peak detection in spectroscopy.

Integrations are again easily performed when the function can be inegrated analytically. When this is not possible an approximation formula can be applied. For example, in Figure 2.7 a parabola-like curve joins the 3 points y_1, y_2 and y_3. Simpsons rule[12,13] tells us that the area under the curve is

$$A = \frac{\Delta x}{3}(y_1 + 4y_2 + y_3)$$

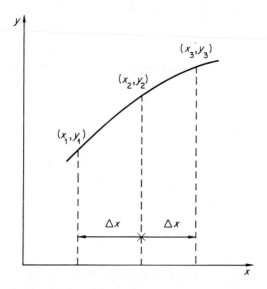

Figure 2.7　A portion of a curve approximated as a parabola

The smaller the value of Δx, the more precise is the calculation of A. A larger area can be divided into smaller units so that an integral of the form

$$I = \int_{x_1}^{x_n} y \, dx$$

is approximated by:

$$I = \frac{\Delta x}{3} [y_0 + 4y_1 + 2y_2 + 4y_3 + \ldots + 4y_{n-1} + y_n]$$

or, as an alternative:

$$I = \frac{\Delta x}{3} [C_1 + C_2 + C_3] \tag{2.41}$$

where

$$C_1 = y_1 + y_n$$
$$C_2 = 4(y_1 + y_3 + y_5 + \ldots + y_{n-1})$$
$$C_3 = 2(y_2 + y_4 + \ldots + y_{n-2})$$

This method is useful in such cases as

(a) The use of heat capacity, C_p, equations to calculate heats of reaction at various temperatures.

(b) Calculation of ΔS from ΔC_p over a temperature range.

Such applications are easily handled merely by changing the form of a function routine (see Problem 2.8.7).

2.8 Problems

2.8.1   Modify the program START (Table 2.1) so that it will list the number of data within the limit $\mu \pm n\sigma$, where $n = 1, 2$ or 3. Also, convert the program into a subroutine.

2.8.2   Rewrite HTGRAM (Table 2.2) to handle non-integer numbers.

2.8.3   Write a subroutine for the least-squares analysis of a 2-parameter relationship:

$$y_i = a + bx_i$$

The subroutine should calculate a, b and their standard deviations and the correlation coefficient between y_i and x_i.

2.8.4   Use the subroutine CHISQ (Table 2.11) to determine whether a polynomial equation or a hyperbolic relationship gives a better fit to the data in Table 2.6.

2.8.5   Use one of the non-linear data analysis subroutines (Section 2.5) to confirm that the local minimum of the function

$$f(x,y) = \exp(x^2 - 4) - \exp(4 - y^2)$$

is approximately -55 with x and y each equal to zero. Use starting values of

$$f(5,5) = 1.3 \times 10^9$$

2.8.6   Use the subroutine JACB (Table 2.14) to confirm that the eigenvalues of A, where

$$A = \begin{pmatrix} 15 & \sqrt{6} & 7 \\ \sqrt{6} & 18 & \sqrt{6} \\ 7 & \sqrt{6} & 15 \end{pmatrix}$$

are 8, 16 and 24 (or multiples thereof).

2.8.7   Write a program to evaluate

$$\int_{-1}^{+1} (x + 1)^2 \, dx$$

to an accuracy of $\pm 0.1\%$ by Simpson's rule.

References
1.   Chatfield, C., *Statistics for Technology*, Penguin, Harmonds-worth, 1970.
2.   Bickley, W. G., and Thompson, R. S. H. G., *Matrices, Their Meaning and Manipulation*, English Universities Press, London, 1964.
3.   Bendat, J. S., and Piersol, A. G., *Random Data: Analysis and Measurement Procedures*, Wiley-Interscience, New York, 1971, p. 101.
4.   Simone, A. J., *The Mathematics Teacher*, 452 (1967).
4a. Wilson, E. L., Bathe, K. J., and Doherty, W. P., *Computers and Structures*, 4, 363 (1974).
5a. Hohmann, E. C., and Lockhart, F. J., *Chem. Tech.* 614 (1972).
5b. A similar approach was used by H. D. McGeachin, Albright and Wilson Ltd., Oldbury Division (Unpublished) from whom the data in Table 2.8 was obtained.
6.   Reference 1, p. 148.
7.   Anderson, K. P., and Snow, R. L., *J. Chem. Ed.*, 44, 756 (1967).
8.   Becsey, J. C., Berke, L., and Callan, J. R., *J. Chem. Ed.*, 45, 728 (1968).
9.   Bevington, P. R., *Data Reduction in the Physical Sciences*, McGraw-Hill, New York, 1969, p. 215.

10. Williams, I. P., *Matrices for Scientists*, Hutchinson University Press, London, 1972, p. 81.
11. Dickson, T. R., *The Computer and Chemistry*, W. H. Freeman, San Francisco, 1968.
12. Reference 9, p. 268.
13. Isenhour, T. L., and Jurs, P. C., *Introduction to Computer Programming for Chemists*, Allyn and Bacon, Boston, 1972.

Chapter 3

Introductory Applications in Practical Chemistry

The applications in this chapter are taken from various areas of practical chemistry. It is not the intention to present a comprehensive list — partly due to space limitations and, also, because many applications have similar underlying principles. We have gathered together an assortment of sharply contrasting experiments — the computer aspects of which exemplify the techniques described in Chapter 2.

3.1 Linear Least Squares Methods in the Treatment of Kinetic and Thermal Data

In Chapter 2, the specific example of the Clausius–Clapeyron equation was discussed. Many computer applications in practical physical chemistry involve a similar method of data analysis so that we now present just a few specific examples. An indication of how to tackle more general problems is given in Section 3.2.

3.1.1 Analysis of First-order Kinetic Data

This can be approached in general terms (see Problem 3.4.4) but we shall consider as an example a widely used reaction in introductory kinetics — that between persulphate and iodide ions:

$$S_2O_8^{2-} + 2I^- \longrightarrow 2SO_4^{2-} + I_2 \qquad (3.1)$$

It is usual to withdraw aliquots from the reaction mixture from time to time and to titrate the free iodine with thiosulphate:

$$2H_2O + I_2 + S_2O_3^{2-} \longrightarrow 2I^- + 2SO_4^{2-} + 4H^+ \qquad (3.2)$$

Therefore, the titration figure is proportional to the extent of reaction. If the persulphate, $S_2O_8^{2-}$, in equation (3.1) is in a large excess, the reaction appears to proceed according to first-order kinetics such that the rate is given by:

$$\text{rate} = \frac{-d[L^-]}{dt} = k[I^-]$$

Or, in integrated form:

$$\ln(a - x)_t = -kt + C \tag{3.3}$$

where

$$a = \text{initial } I^- \text{ concentration}$$
$$(a - x)_t = I^- \text{ concentration at time } t$$
$$k = \text{rate constant}$$
$$C = \text{a constant}$$

The usual graphical determination of k expresses (3.3) in a slightly different form:

$$2.303 \log_{10}(T_\infty - T_t) = -kt + C \tag{3.4}$$

T_∞ and T_t are the thiosulphate titrations at 'infinite' time and at time t, respectively. A graph of $\log_{10}(T_\infty - T_t)$ against t has a gradient of $(-k/2.303)$ from which k may be obtained. Equation (3.4) may also be solved with program KNTT (Table 3.1) which analyses the equation by the linear least squares method (Section 2.4). This can be advantageous, in comparison with the graphical method, when significant scatter is present in the data. The program processes data for 298 K and then reads in further data for a temperature of 288 K. From the two rate constants, the activation energy E^* for this temperature range may be calculated:

$$E^* = R \ln(k_{288}/k_{298})/(1/298 - 1/288)$$

In program KNTT you will observe that, firstly, a title card is read in 10A8 format. This is stored in the array H(10) and then printed in the same format on a new page. The number of data points, N, is input in I2 format and then the time, titre values are read as successive pairs (F5.0,F5.1) for a temperature of 298 K. Finally, an 'infinity' value of the titration is input (F5.1). In succeeding lines the normal equations are set up and in lines 32 and 33 the rate constant is calculated. A similar procedure for the data at 288 K follows. The activation energy and pre-exponential factor are computed in lines 46 and 47. In Table 3.1 you will find test data for this program. Note that the 'infinity' values were 42.0 and 49.7 cm^3 respectively.

3.1.2 Potentiometric Determination of Activity Coefficients

Many numeric aspects of electrochemistry are amenable to computer assistance because of the occurrence of fairly complicated equations.

Table 3.1
Program KNTT for calculating kinetic parameters of the persulphate—iodide reaction

```
0008                    MASTER KNTT
0009                    DIMENSION TIM(20),T(20),H(10),AX(20),AXL(20)
0010                    DIMENSIONXK(2)
0011                    READ(1,100)(H(I),I=1,10)
0012                    NCOUNT=0
0013        100         FORMAT(10A8)
0014                    WRITE(2,200)(H(I),I=1,10)
0015        200         FORMAT(1H1,10A8//)
0016         10 READ(1,101)N
0017        101         FORMAT(I2)
0018                    DO 1 J=1,N
0019          1         READ(1,102)TIM(J),T(J)
0020                    NCOUNT=NCOUNT+1
0021        102         FORMAT(F5.0,F5.1)
0022                    READ(1,103)A
0023        103         FORMAT(F5.1)
0024                    S1,S2,S3,S4=0.0
0025                    DO 2 J=1,N
0026                    X=T(J)
0027                    S1=S1+ALOG10(A-X)
0028                    S2=S2+TIM(J)
0029                    S3=S3+TIM(J)*ALOG10(A-X)
0030                    S4=S4+(TIM(J))**2
0031          2 CONTINUE
0032                    EM=(N*S3-S1*S2)/(N*S4-S2*S2)
0033                    XK(NCOUNT)=(-2.303)*EM
0034                    WRITE(2,500)
0035        500         FORMAT(1H ,' TIME(SECS) TITRE(CM3)')
0036                    DO 3 I=1,N
0037          3         WRITE(2,202)TIM(I),T(I)
0038        202         FORMAT(1H ,F10.0,2F10.1,F10.3)
0039                    IF(NCOUNT.EQ.1)WRITE(2,300)
0040        300         FORMAT(1H0,'RATE CONSTANT AT 298 K')
0041                    IF(NCOUNT.EQ.2)WRITE(2,301)
0042        301         FORMAT(1H0,'RATE CONSTANT AT 288K')
0043                    WRITE(2,201)XK(NCOUNT)
0044        201         FORMAT(1H0,'RATE CONSTANT =',F10.6)
0045                    IF(NCOUNT.EQ.1)GO TO 10
0046                    EACT=(8.314*ALOG(XK(2)/XK(1))/((1./298.)-(1./288.)))
0047                    PREEX=XK(1)/(EXP(-EACT/(8.314*298)))
0048                    WRITE(2,104)EACT
0049        104         FORMAT(1H0,'ACTIVATION ENERGY=',F9.3,'J MOL-1')
0050                    WRITE(2,105)PREEX
0051        105         FORMAT(1H0,'PRE-EXPONENTIAL FACTOR =',E10.3)
0052                    STOP
0053                    END
```

```
TEST RUN   INFINITY TITRES ARE 42.0 AND 49.7

TIME(SECS) TITRE(CM3)
      0,        0.0
    180,       12.6
    600,       21.8
    960,       27.5
   1260,       31.9
   1800,       37.5
   2400,       39.5
   2940,       40.1
   3540,       40.5

RATE CONSTANT AT 298 K

RATE CONSTANT = 0.000981
TIME(SECS) TITRE(CM3)
    600,       15.1
   1200,       25.2
   1800,       32.5
   3000,       40.8
   6000,       48.0

RATE CONSTANT AT 288K

RATE CONSTANT = 0.000557

ACTIVATION ENERGY=40383.159J MOL-1

PRE-EXPONENTIAL FACTOR = 0.118E 05
```

We exemplify this with an example in potentiometry — the determination of activity coefficients from potentiometric data.[1]

The e.m.f., E, of the cell

$$\text{Pt,H}_2\,(1\text{ atm.}) \mid \text{HCl(m)} \mid \text{AgCl(s),Ag}$$

for which the cell reaction is

$$\text{AgCl(s)} + \tfrac{1}{2}\text{H}_2\,(\text{g}) \longrightarrow \text{H}^+(\text{aq}) + \text{Ag(s)} + \text{Cl}^-(\text{aq}) \tag{3.5}$$

is given by

$$E = E^{\ominus} - \frac{RT}{F}\ln a_{\text{H}^+} \cdot a_{\text{Cl}^-}$$

Note that activities, a_i, are used in preference to concentrations. This equation may be re-expressed in terms of mean ionic activities which may, in turn be separated into concentrations, m, and activity coefficients, γ:

$$E = E^{\ominus} - \frac{RT}{F}\ln a_{\pm}^2 = E^{\ominus} - \frac{RT}{F}[\ln m_{\pm}^2 + \ln \gamma_{\pm}^2]$$

where

$$m_{\pm} = \sqrt{(m_{\text{H}^+} \cdot m_{\text{Cl}^-})} = m \text{ (see cell diagram) and}$$
$$\gamma_{\pm} = \sqrt{(\gamma_{\text{H}^+} \cdot \gamma_{\text{Cl}^-})}.$$

After rearrangement, we obtain:

$$E + \frac{2RT}{F}\ln m - E^{\ominus} = -\frac{2RT}{F}\ln \gamma_{\pm}^2 \tag{3.6}$$

One form of Debye–Huckel theory yields an expression for $\log \gamma_{\pm}$:

$$\log \gamma_{\pm} = -A\sqrt{m} + Cm$$

where A and C are constants, A being equal to 0.509 for dilute aqueous solutions at 298 K. Combining this equation with (3.6), inserting numerical values of the constants and rearranging we obtain:

$$E + 0.1183 \log m - 0.0602\sqrt{m} = E^{\ominus} - 0.1183\,Cm \tag{3.7}$$

This equation can form a graphical basis for the determination of $E^{\ominus}$ by plotting the left-hand side as a function of m. The intercept should be $E^{\ominus}$. Substitution of $E^{\ominus}$ into (3.6) would then yield values of $\gamma_{\pm}$. Experimentally, this is not straightforward since there is often considerable experimental scatter owing to the influence of small amounts of impurities. Good experimental data are, however, available in the literature.[1]

Rather than write a program for the analysis of this specific problem it is preferable to recast it in a more general form for any suitable cell.

For a general electrolyte, we can define its mean ionic concentration $m_\pm$ (and, in a similar manner $\gamma_\pm$) as follows:

$$m_\pm^\nu = m^\nu (\nu_+^{\nu_+} \cdot \nu_-^{\nu_-})$$

where

m = concentration of the electrolyte

ν_+ = number of cations

ν_- = number of anions

ν = total number of ions

Alternatively,

$$m_\pm^\nu = f m^\nu$$

where $f = (\nu_+^{\nu_+} \cdot \nu_-^{\nu_-})$. Thus, for $CaCl_2$: $\nu_+ = 1; \nu_- = 2; \nu = 3$ and $f = 4$.

In the specific case discussed at the beginning of this section the potential of one of the electrodes was zero since this was the standard hydrogen electrode. In the general case the reference potential may be non-zero and we will denote it by E_{ref}. Making use of these definitions, we can now express equation (3.7) in the general form:

$$E + E_{ref} + A \log m + B \log f - 0.0602 \sqrt{m} = E^{\ominus} - A.C.m.$$

where $A = 0.059156\nu/n$ and n = number of electrons involved in the cell reaction, per mole.

This could still be used for a graphical method but a linear least squares method is preferable in view of the greater complexity of the equation and, also, because of the tendency of the experimental data to have considerable scatter. The program POTN (Table 3.2) calculates $E^{\ominus}$ from the above equation, using the left-hand side as the dependent variable. Inspection of the program reveals that, after this calculation, $\gamma_\pm$ is calculated at each data point. Input to the program is as follows:

Card 1: The number of data points, IN; values of f and v (see above); the reference electrode potential, E_{ref}, in volts and n, the number of electrons involved in the reaction (Format 3I0,F0.0,I0).

Cards 2 to IN+1: pairs of molarity, e.m.f. readings (2F0.0).

The least-squares analysis is performed in lines 28–41 of the program. The values of $\gamma_\pm$ are calculated in the DO loop commencing at line 45.

The reader will find the data listed in Table 3.2 suitable for program testing. The data refer to the reaction (3.5) and, therefore, the mean activity coefficients are those of the $H^+(aq)$ and $Cl^-(aq)$ ions.

In view of the experimental difficulties associated with obtaining sufficiently accurate e.m.f. values for this experiment it is worthwhile to use the program as an exercise in data analysis (a so-called 'dry-lab' experiment). Other examples are to be found in Chapter 4.

Table 3.2
Program POTN — calculation of activity coefficients from potentiometric data

```
0008                    MASTER POTN
0009                    REAL MOLALM
0010                    INTEGER EXPAC
0011                    DIMENSIONMOLALM(100),ECELL(100)
0012                    READ(1,1)IN,NOEFF,EXPAC,EREF,NELECT
0013                    WRITE(2,2)IN,NOEFF,EXPAC,EREF,NELECT
0014                  2 FORMAT(1H1,'   DETERMINATION OF MEAN ACTIVITY COEFFICIENTS AND STAN
0015                   1DARD ELECTRODE POTENTIAL ',///,' SETS OF DATA ',I3,',COEFF,EXPAC ',
0016                   2I3,',REFERENCE POTENTIAL',F5.3,' V,  NO OF ELECTRONS ',I3)
0017                  1 FORMAT(3I0,F0.0,I0)
0018          C
0019                    READ(1,3)(MOLALM(I),ECELL(I),I=1,IN)
0020                    COEFF=NOEFF
0021                    WRITE(2,5)
0022                  5 FORMAT(///,'   MOLARITY       POTENTIAL')
0023                    WRITE(2,4)(MOLALM(I),ECELL(I),I=1,IN)
0024                  3 FORMAT(2F0.0)
0025                  4 FORMAT(5X,F9.4,5X,F9.4)
0026                    C=NFLECT
0027                    D=EXPAC
0028                    A=D*0.059156/C
0029                    B=A/D
0030                    SX=0
0031                    SY=0
0032                    SX2=0
0033                    SXY=0
0034                    DO 6 I=1,IN
0035                    S=MOLALM(I)
0036                    T=ECELL(I)+EREF+A*ALOG10(MOLALM(I))+B*ALOG10(COEFF)-0.0602*SQRT(
0037                   1MOLALM(I))
0038                    SX=SX+S
0039                    SY=SY+T
0040                    SX2=SX2+S*S
0041                  6 SXY=SXY+S*T
0042                    B=(SX2*SY-SX*SXY)/(IN*SX2-SX*SX)
0043                    WRITE(2,7)B
0044                  7 FORMAT(1H ,///,' STANDARD ELECTRODE POTENTIAL ',F7.4,' VOLTS')
0045                    WRITE(2,8)
0046                  8 FORMAT(1H ,///,'   MOLARITY       ACTIVITY COEFFICIENT')
0047                    DO 9 I=1,IN
0048                    G=10.**((B-EREF-ECELL(I)-A*ALOG10(MOLALM(I))-B*ALOG10(COEFF))/A)
0049                  9 WRITE(2,10)MOLALM(I),G
0050                 10 FORMAT(3X,F8.4,12X,F8.6)
0051                    STOP
0052                    END
```

DETERMINATION OF MEAN ACTIVITY COEFFICIENTS AND STANDARD ELECTRODE POTENTIAL

SETS OF DATA 10,COEFF,EXPAC   1 2,REFERENCE POTENTIAL0.000 V. NO OF ELECTRONS   1

MOLARITY	POTENTIAL
0.0045	0.5038
0.0056	0.4926
0.0073	0.4795
0.0091	0.4666
0.0112	0.4586
0.0134	0.4497
0.0171	0.4378
0.0256	0.4182
0.0559	0.3822
0.1238	0.3420

STANDARD ELECTRODE POTENTIAL 0.2222 VOLTS

MOLARITY	ACTIVITY COEFFICIENT
0.0045	0.927563
0.0056	0.922558
0.0073	0.914778
0.0091	0.904486
0.0112	0.896740
0.0134	0.889878
0.0171	0.879694
0.0256	0.859527
0.0559	0.823544
0.1238	0.784637

3.1.3 Thermoanalytical Data Analysis

As our third group of examples, we will consider thermal methods, particularly DSC (differential scanning calorimetry) and TG (thermogravimetry) which have become popular means of analysis in recent years. DSC is used frequently for the determination of the purity of solid organic compounds whilst both DSC and TGA have been used in evaluating kinetic parameters for solid state decompositions.

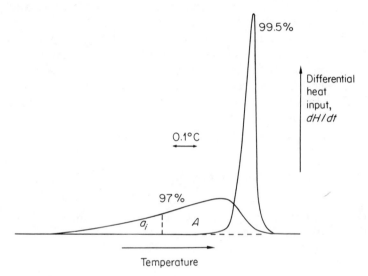

Figure 3.1   Dependence of DSC melting curves on purity

Purity from DSC Data   In this technique,[2] the signal on the pen recorder of a DSC instrument is proportional to the differential heat flow required to keep the temperature of a sample equal to that of an inert reference. If the fusion behaviour of a compound is being studied (Figure 3.1) the DSC record is a peak, the width of which depends on the purity of the compound.[3] The abscissa (temperature) readings should be corrected for the thermal lag of the instrument; these corrected temperatures are, theoretically, related to the mole fraction, X, of impurity:

$$T_i = T_0 - XRT_0^2/\Delta H_f^\ominus \alpha_i \tag{3.8}$$

where

T_i = corrected temperature
T_0 = melting temperature of the pure compound
R = the gas constant
$\Delta H_f^\ominus$ = enthalpy of fusion of the pure compound
α_i = fraction melted at T_i (see Figure 3.1 where α_i is defined as the ratio of the partial area, a_i, to the total area, A)

It is not possible to obtain X directly from a least squares analysis of (3.8) because the partial peak area measurements are usually in error. This is simply because of the difficulty in deciding precisely where a broad fusion peak begins. To allow for this, the α_i values are corrected by the formula

$$\alpha_i = (a_i + fA)/A$$

where f is varied (usually between 0.0 and 0.3) until the linear behaviour predicted by (3.8) is obtained.

This laborious manual procedure is avoided by the program BEE7 (Table 3.3). This program increments f in a stepwise manner and, after each increment, performs a least squares fit on (3.8) using T_i as the dependent variable and $1/\alpha_i$ as the independent variable to calculate T_0 and X. The 'best' values of these parameters are those which give the smallest residual, R:

$$R = \sum_i [T_i(\text{obs}) - T_i(\text{calc})]^2$$

Data is read into BEE7 in line numbers 11–16. The order of data input is as follows:

Card 1: Repetition integer ($\neq 0$ for more than one run) .I5 format.

Card 2: Title card (20A4).

Card 3: Absolute temperatures corrected for instrumental lag (10F6.2).

Card 4: Partial area measurement corresponding to each temperature (a_i in Figure 3.1) 10F6.2.

Table 3.3

Program BEE7 for the calculation of absolute melting points and purities of compounds from DSC curves

```
00U8                MASTER  BEE7
0009         C      PROGRAM TO CALCULATE MELTING POINTS AND PURITIES FROM DSC CURVES
0010         C          BY  SUCCESSIVE APPROXIMATIONS
0011                DIMENSIONDEV(50),TT(50),FFRAC(50)
0012                DIMENSION T(50),AREA(50),TITLE(50),FRAC(50),AT(50),GDT(50),RES(50)
0013                COMMON/XVAR/M,X(20,20)
0014          94  READ(1,22)NREP
0015                READ(1,90)(TITLE(I),I=1,20)
0016                READ(1,1)N
0017                READ(1,2)(T(I),I=1,N)
0018                READ(1,2)(DEV(I),I=1,N)
0019                READ(1,4)FACTOR,WEIT,FWT,RANGE,TAREA,SPEED,WIDTH,SCAN,DFLTAH,DMAX
0020                I2=0
0021                FR=0
0022         222  FORMAT(1H1,20A4)
0023          22  FORMAT(I5)
0024          90  FORMAT(20A4)
0025           1  FORMAT(I5)
0026           2  FORMAT(10F6.2)
0027                DELTAH=DELTAH*1000.
0028                WRITE(2,222)(TITLE(I),I=1,20)
0029         C      IF DELTAH IS KNOWN THE NEXT STEP IS SKIPPED
0030           4  FORMAT(10F7.3)
0031                IF(DELTAH.GT.0)GO TO 3
```

```
                    DELTAH=(TAREA*FACTOR*FWT*60.0*RANGE)/(SPEED*WIDTH*WEIT)
0052        C       WRITE ORIGINAL DATA
0053          5 WRITE(2,6)WEIT,SCAN,DELTAH
0054          6 FORMAT(1H ,7HWEIGHT=,F5.2,2HMG,3X,13HHEATING RATE=,F4.0,/HK MIN-1,
0055            13X,/HDELTAH=,E15.8,7HJ MOL-1)
0056            R=8.314
0057            WRITE(2,43)
0058            WRITE(2,9)
0059         99 DO 7 I=1,N
0060          7 FRAC(I)=DMAX/DEV(I)
0061            IF(I2.EQ.0)WRITE(2,8)(T(I),FRAC(I),I=1,N)
0062          8 FORMAT(1H ,4X,F6.2,11X,F6.2)
0063          9 FORMAT(32H TEMPERATURE    1/FRACTION MELTED)
0064            DO 95 I=1,4
0065            DO 95 J=1,4
0066         95 X(I,J)=0
0067            DO 96 I=1,N
0068            X(1,3)=X(1,3)+T(I)
0069            X(1,2)=X(1,2)+FRAC(I)
0070            X(2,3)=X(2,3)+FRAC(I)*T(I)
0071         96 X(2,2)=X(2,2)+FRAC(I)**2
0072            X(1,1)=N
0073            X(2,1)=X(1,2)
0074        C   CALL SUBROUTINE FOR SIMULTANEOUS EQUATIONS
0075            M=2
0076            CALL SIML
0077        C   CALCULATE RESIDUAL
0078            I2=I2+1
0079        C   AT IS ABSOLUTE MPT.
0080            AT(I2)=X(1,3)
0081        C   GDT IS SLOPE OF GRAPH
0082            GDT(I2)=X(2,3)
0083            RES(I2)=0
0084            DO 10 I=1,N
0085         10 RES(I2)=RES(I2)+(T(I)-GDT(I2)*FRAC(I)-AT(I2))**2
0086            IF(I2.EQ.1)GO TO 11
0087            IF(RES(I2).GT. RES(I2-1))GO TO 12
0088         11 TO=AT(I2)
0089            DO 70 I=1,N
0070            TT(I)=T(I)
0071         70 FFRAC(I)=FRAC(I)
0072            DT=-GDT(I2)
0073            PP=100.*DFLTAH*DT/(R*(TO**2))
0074            NR=FR*100
0075         12 CONTINUE
0076            FR=FR+0.01
0077            IF(FR.GT.0.30)GO TO 15
0078            DO 14 I=1,N
0079         14 DEV(I)=DEV(I)+FR*DMAX
0080            GO TO 99
0081         15 WRITE(2,92)
0082         93 FORMAT(10X,'INPUT DATA')
0083         92 FORMAT(10X,'FINAL RESULTS')
0084            WRITE(2,9)
0085            WRITE(2,8)(TT(I),FFRAC(I),I=1,N)
0086            WRITE(2,91)NR,TO,DT,PP
0087         91 FORMAT(1H0,22HPERCENT CORRECTION WAS,I4,3X,10HABS MPT IS,F7.2,1HK,
0088            13X,10HFPT DEPRESSION IS ,F5.3,1HK,3X,24HMOLE PERCENT IMPURITY IS,F
0089            27.3)
0090            IF(NREP.GT.0)GO TO 94
0091            STOP
0092            END
```

```
      PURITY OF COMPOUND Y
WEIGHT= 5.00MG    HEATING RATE=  2.K MIN-1    DELTAH= 0.24400000E 05J MOL-1
            INPUT DATA
  TEMPERATURE   1/FRACTION MELTED
     392.30         8.00
     392.80         4.20
     393.40         2.30
     394.10         1.20
            FINAL RESULTS
  TEMPERATURE   1/FRACTION MELTED
     392.30         1.48
     392.80         1.20
     393.40         1.02
     394.10         0.72

PERCENT CORRECTION WAS  10   ABS MPT IS 395.81K   FPT DEPRESSION IS 2.368K
MOLE PERCENT IMPURITY IS  4.435
```

Before describing the final data card, we should explain that in our laboratory we choose to measure the areas with a planimeter. Since our pen recorder operates in inches per minute and the chart width is an exact number of inches, it is convenient for us to measure areas in square inches — a distinctly non-S.I. convention! This may not be true of your equipment but it is, of course, important to ensure that the areas, chart speed and chart width are in *compatible* units. The final data card is:

Card 5:

FACTOR — planimeter conversion factor (i.e. in our case, to square inches)
WEIT — mass of sample /mg.
FWT — molar mass of sample /g mol^{-1}.
RANGE — sensitivity for a full scale deflection /mJ s^{-1}.
TAREA — total peak area / planimeter units.
SPEED — chart speed per minute — see above.
WIDTH — chart width
DELTAH — enthalpy of fusion (if known)/kJ mol^{-1}.
DMAX — total area / planimeter units (see below).

Rather than use partial areas, we have found it to be quicker, and almost as accurate, to approximate α_i by

$$\alpha_i = d_i/d_m$$

where d_i is the pen deviation from the base line at the temperature T_i and d_m is the maximum deviation. If the reader wishes to use this approximation he should replace partial areas (card 4) by pen deflections and DMAX on card 5 by the maximum deflection.

The enthalpy of fusion, DELTAH, is calculated in line 32 of the program (the factor of 60 should be omitted if the chart speed is in units per second). FRAC(I) is $1/\alpha_i$ and is calculated in the DO loop in lines 40 and 41. After printing the initial data the normal equations are calculated (lines 48–54) and solved (line 57). The goodness-of-fit is calculated in lines 65 and 66 and stored. The correction procedure, described above, is carried out in lines 79 and 80 and the least squares analysis is repeated. During each cycle the values of T_0 and X are stored and, finally, the 'best' values are output as depicted in Table 3.3.

Suitable test data for BEE7 are as follows:

Card 1: NREP = 0.
Card 2: TITLE = TEST.
Card 3: N = 7.
Card 4: T(I) values:

393.38   393.43   393.5   393.55   393.59   393.64   393.71

Card 5: DEV(I) values:

10.0   17.0   25.0   34.0   46.0   55.0   66.0

Card 6: With FACTOR, WEIT, FWT, RANGE, TAREA, SPEED, WIDTH all = 0, SCAN = 1.0, DELTAH = 25.1, DMAX = 76.0.

These data gave the following results:

Input data

Temperature	1/Fraction melted
393.38	7.60
393.43	4.47
393.50	3.04
393.55	2.24
393.59	1.65
393.64	1.38
393.71	1.15

Final Results

Temperature	1/Fraction melted
393.38	0.45
393.43	0.43
393.50	0.41
393.55	0.39
393.59	0.37
393.64	0.35
393.71	0.34

Percent correction was 20.ABS, MPT was 394.65 K, FPT depression was 2.831 K Mole percent, impurity was 5.488.

Kinetic parameters from DSC and TG   These convenient techniques can yield valuable information on solid decompositions such as (3.9):

$$A(s) \longrightarrow B(s) + C(g)$$
(3.9)

Under suitable conditions, the rate of decomposition of A, $d\alpha/dt$, is given by:

$$d\alpha/dt = k(1-\alpha)^n$$
(3.10)

where k is a rate constant, α is the extent of decomposition and n must, in theory,[4] take particular values between 0 and 1. For a DSC instrument, the rate of reaction is related to the pen deflection d, from the baseline by[5]

$$d\alpha/dt = M\,d\,r/(\Delta H\,w\,W) \tag{3.11}$$

where M is the molar mass, r is the sensitivity of the DSC for a full scale deflection, ΔH is the enthalpy of reaction, w is the sample mass before decomposition and W is the chart width of the pen recorder. The rate is also given by (from (3.10))

$$d\alpha/dt = A\,[\exp(-E^*/RT)]\,(1-\alpha)^n \tag{3.12}$$

where E^* is the activation energy and A is the pre-exponential factor for the reaction. Combining (3.11) and (3.12) and taking the logarithms of both sides we obtain:

$$\log\{Mdr/(\Delta H\,w\,W)\} = \log A - E^*/(2.303RT) + n\log(1-\alpha)$$

The program BEE5 (Table 3.4) can be used to evaluate $\log A$, E^* and n by a three-parameter least squares fit. In this program, it is assumed

Table 3.4

Program BEE5 — calculation of kinetic parameters of solid-state decompositions from DSC data

```
0008            MASTER BEE5
0009            DIMENSION T(50),AREA(50),DEV(50),ALFA(50),AT(50)
0010            DIMENSION TITLE(20)
0011            COMMON/XVAR/N,X(20,20)
0012        999 READ(1,53)NREP
0013          1 READ(1,90)(TITLE(I),I=1,20)
0014            WRITE(2,91)(TITLE(I),I=1,20)
0015         90 FORMAT(20A4)
0016         91 FORMAT(1H1,20A4)
0017            READ(1,53)NP
0018         53 FORMAT(I5)
0019            READ(1,155)FACTOR,WEIT,FWT,RANGE,TAREA,SPEED,WIDTH
0020        155 FORMAT(7F7.3)
0021            TAREA=TAREA*FACTOR
0022            DELH=(TAREA*FWT*60.0*RANGE)/(SPEED*WIDTH*WEIT)
0023            WRITE(2,80)DELH
0024         80 FORMAT(1H ,7HDELTAH=,E15.8,7HJ MOL-1)
0025            READ(1,54)(T(I),I=1,NP)
0026            READ(1,55)(AREA(I),I=1,NP)
0027            READ(1,54)(DEV(I),I=1,NP)
0028            WRITE(2,52)
0029         52 FORMAT(1X,2H T,4X,5H AREA,4X,5H ALFA,3X,4H DEV)
0030            DO 50 I=1,NP
0031            AREA(I)=AREA(I)*FACTOR
0032            ALFA(I)=AREA(I)/TAREA
0033            WRITE(2,100)T(I),AREA(I),ALFA(I),DEV(I)
0034        100 FORMAT(1H ,F4.0,3X,F5.1,3X,F5.2,3X,F5.1)
0035         50 CONTINUE
0036         54 FORMAT(10F5.1)
0037         55 FORMAT(10F6.2)
0038            EN=NP
0039            R=0.0001
0040            X(1,1)=0.0
0041            X(1,2)=0.0
0042            X(1,3)=0.0
0043            X(1,4)=0.0
0044            X(2,2)=0.0
0045            X(2,3)=0.0
0046            X(2,4)=0.0
0047            X(3,3)=0.0
0048            X(3,4)=0.0
0049            DO 111 I=1,NP
0050            TR=-1./(R*T(I))
```

```
0051                 AL=ALOG(ALFA(I))
0052                 FL=ALOG(1.-ALFA(I))
0053                 ALDOT=ALOG((DEV(I)*RANGE*FWT)/(WIDTH*25.4*WEIT*DELH))
0054                 SQ=((ALDOT)**2)
0055                 X(1,1)=X(1,1)+(1./SQ)
0056                 X(1,2)=X(1,2)+(TR/SQ)
0057                 X(1,3)=X(1,3)+(FL/SQ)
0058                 X(1,4)=X(1,4)+(ALDOT/SQ)
0059                 X(2,2)=X(2,2)+((TR**2)/SQ)
0060                 X(2,3)=X(2,3)+((TR*FL)/SQ)
0061                 X(2,4)=X(2,4)+((TR*ALDOT)/SQ)
0062                 X(3,3)=X(3,3)+((FL**2)/SQ)
0063                 X(3,4)=X(3,4)+((FL*ALDOT)/SQ)
0064             111 CONTINUE
0065                 DO 112 J2=1,3
0066                 DO 112 I2=1,3
0067             112 X(I2,J2)=X(J2,I2)
0068       C
0069       C         END OF SUMMATION
0070       C         START OF RECURSION
0071       C
0072                 DO 900 J=1,4
0073                 DO 900 I=1,3
0074             900 X(I,J)=(10.**7)*X(I,J)
0075                 N=3
0076                 CALL SIML
0077                 AN=EXP(X(1,4))
0078                 WRITE(2,185)AN
0079             185 FORMAT(1H ,23HPRE-EXPONENTIAL FACTOR=,E15.8)
0080                 ACT=8.314*(10.**4)*X(2,4)
0081                 WRITE(2,179)ACT
0082             179 FORMAT(1H ,18HACTIVATION ENERGY=,E15.8,7HJ MOL-1)
0083                 EN=X(3,4)
0084                 WRITE(2,178)EN
0085             178 FORMAT(1H ,3HEN=,E15.8)
0086                 WRITE(2,907)
0087                 FL=0.0
0088                 ALDOT=0.0
0089             907 FORMAT(1X,'TEMP    OBSERVED RATE   CALCD RATE     1/T    BEST LIN
0090            1EAR FIT    RATE CONSTANT')
0091                 DO 909 I=1,NP
0092                 AT(I)=1./T(I)
0093            3500 ADOBS=((DEV(I)*RANGE*FWT)/(WIDTH*25.4*WEIT*DELH)
0094                 ADCAL=AN*(EXP(-X(2,4)/(R*T(I))))*((1.-ALFA(I))**X(3,4))
0095                 RCONS=ADCAL/((1.-ALFA(I))**X(3,4))
0096                 ALDOT=ALOG(ADOBS)-EN*ALOG(1.-ALFA(I))
0097             909 WRITE(2,899)T(I),ADOBS,ADCAL,AT(I),ALDOT,RCONS
0098             899 FORMAT(1H ,F7.3,E12.5,3X,E12.5,3X,F7.5,3X,E12.5,3X,E12.5)
0099              29 IF(NREP.GT.0)GO TO 999
0100                 STOP
0101                 END
```

```
      TEST
DELTAH= 0.11495987E 06J MOL-1
 T     AREA    ALFA    DEV
480.    1.0    0.10    15.0
485.    1.4    0.14    16.0
490.    1.9    0.19    23.0
495.    2.4    0.24    29.0
500.    3.2    0.32    38.0
505.    4.3    0.45    50.0
510.    5.6    0.55    56.0
515.    7.0    0.69    55.0
520.    8.2    0.81    52.0
525.    9.4    0.93    41.0
PRE-EXPONENTIAL FACTOR= 0.18430926E 12
ACTIVATION ENERGY= 0.12887326E 06J MOL-1
EN= 0.63522645E 00
TEMP    OBSERVED RATE   CALCD RATE      1/T    BEST LIN EAR FIT    RATE CONSTANT
480.000 0.16879E-02    0.16294E-02    0.00208   -0.63181E 01   0.17409E-02
485.000 0.20775E-02    0.22092E-02    0.00206   -0.60819E 01   0.24286E-02
490.000 0.29864E-02    0.29481E-02    0.00204   -0.56814E 01   0.33651E-02
495.000 0.37654E-02    0.38992E-02    0.00202   -0.54097E 01   0.46320E-02
500.000 0.49340E-02    0.49744E-02    0.00200   -0.50698E 01   0.63553E-02
505.000 0.64921E-02    0.60561E-02    0.00198   -0.46852E 01   0.86113E-02
510.000 0.72711E-02    0.69658E-02    0.00196   -0.44109E 01   0.11635E-01
515.000 0.71413E-02    0.73877E-02    0.00194   -0.41926E 01   0.15628E-01
520.000 0.67518E-02    0.72568E-02    0.00192   -0.39386E 01   0.20874E-01
525.000 0.53235E-02    0.51186E-02    0.00190   -0.35461E 01   0.27727E-01
```

that α_i is given by

$$\alpha_i = a_i/A$$

where a_i is the area swept out by the pen from the start of the peak to point i and A is the total peak area.

Input to BEE5 is as follows:

Card 1: Repetition integer ($\neq 0$ for more than one run), I5 format.
Card 2: Title card (40A4).
Card 3: Number of data points (I3).
Card 4: FACTOR, WEIT, FWT, RANGE, TAREA, SPEED, WIDTH — as for card 5 of the input to program BEE7 (7F7.3).
Card 5: Absolute temperatures (10F6.2).
Card 6: Areas/planimeter units (10F6.2) — see notes to program BEE7.
Card 7: Pen deflections/mm (10F5.1).

The first part of the program (to line 35) is similar to BEE7. We then set up the normal equations. Note that the factor of 25.4 in line 53 should be deleted if both the pen deviations and the chart width are in the same units. An interesting feature of the normal equations is that the relative deviations method (Section 2.4.4) is used. Also, experience has shown that the equations can sometimes become ill-conditioned because of the smallness of some of the computed $X(I,J)$ terms and, for this reason, a scaling factor is used (line 74). From here, the program is straightforward — calculating the required parameters and, finally, listing the output data given in Table 3.4. Note that the 'best linear fit' is simply the difference between the logarithms of the observed rate and $(1 - \alpha)^n$. This difference is the natural logarithm of the rate constant which should vary linearly with $1/T$. Inspection of Table 3.4 reveals that this is true and it is a useful means of checking whether the computed values of A, E^* and n are reasonable. The input data required by BEE5 to produce the output in Table 3.4 are:

Card 1: Title (see table).
Card 2: 010.
Card 3: 000.016006.500353.000035.000630.000001.000010.000.
Card 4: 480.0485.0490.0495.0500.0505.0510.0515.0520.0525.0.
00585.00.
Card 5: 060.00085.00120.00150.00200.00270.00348.00440.00515.
Card 6: 013.0016.0023.0029.0038.0050.0056.0055.0052.0041.00.

It is also possible to use TG in studies of reaction kinetics. For example, Schlempf et al.[6] have described a program to determine A, the pre-exponential factor, and E^*, the activation energy. Their program performed a least-squares fit of time, t, and sample weight, w, to a polynomial:

$$w = \sum_{i=0}^{n} c_i t^i$$

For a zero-order reaction, the rate constant at any temperature, T, on the curve is given by

$$k = \left[\sum_{i=1}^{n} i c_i t^{i-1} \right] \bigg/ \left[\sum_{i=0}^{n} c_i t^i \right] = d\alpha/dt$$

E^* was obtained from a least squares fit to the equation:

$$\log k = \log A - (E^* 2.303 R)(1000/T)$$

where E^* is in kJ mol^{-1}.

Many other approaches have been suggested to compute kinetic parameters from TG traces, each method having its own proponents.[7,8,9] The reader is referred to the literature if he wishes to adapt one of these approaches. The recent literature contains lively discussions of the relative merits of the various methods.[10,11]

3.1.4 Other Applications of Linear Least Squares Methods

Numerous opportunities for the applications of least squares methods arise in physical chemistry. In some cases the data may be analysed from equations which are already in a linear form; in other cases it is necessary to transform the equation into a linear form by, for example, a logarithmic transformation. A few further examples of experiments amenable to this type of treatment are listed in Table 3.5.

3.2 Miscellaneous Instrumental Applications

The following sections make use of programming techniques additional to that of linear least squares fitting of data. The techniques include: solution of sets of simultaneous equations; calculation of first derivatives; the use of multiple-nested DO loops and the application of non-linear-squares analysis.

3.2.1 Multicomponent Analysis

This method of analysis can be applied to ultraviolet, visible or infrared spectroscopy. Mass spectrometric data can be analysed by an almost identical method, as will be shown in Problem 3.4.5. We shall restrict our attention in this section to the analysis of spectroscopic data.

If two absorbing solutes are present in a solution, their concentrations may be determined by the solution of two simultaneous equations. Similarly, the concentrations of n absorbing solutes can be determined by solving n simultaneous equations, subject to the

Table 3.5

Examples of relationships amenable to Linear Least Squares Analysis

Experiment	Relationship	Standard form	Normal equations
Beer's Law	$A_i = ac_il$	$y_i = b_0 + b_1x_i$ where $y_i = A_i$ (absorbance) $b_1 = c_i$(concentration) . l(path length) $x_i = a$ (absorptivity)	$\Sigma y_i = b_0 N + b_1 \Sigma x_i$ $\Sigma xy_i = b_0 \Sigma x_i + b_1 \Sigma x_i$
Thermodynamic parameters	$-RT \ln K = \Delta G^{\ominus}$ $= \Delta H^{\ominus} - T\Delta S^{\ominus}$ or $\ln K = -\dfrac{\Delta H^{\ominus}}{RT} + \dfrac{\Delta S^{\ominus}}{R}$	$y_i = b_0 + b_1x_i$ where $b_0 = \Delta S^{\ominus}/R$ $b_1 = \Delta H^{\ominus}/R$ $y_i = \ln K; x_i = 1/T$	As above
Radioactive decay	$N_t = N_0 e^{-\lambda t}$ or $\log N_t = \log N_0 - 0.693t/t_{1/2}$	$y_i = b_0 + b_1x_i$ where $b_0 = \log N_0$ $b_1 = 0.693/t_{1/2}$ $y_i = \log N$ (t = time; $t_{1/2}$ = half-life)	As above
Surface chemistry	$\theta = aP/(1 + aP)$ or $1/\theta = 1 + (1/aP)$	$y_i = b_0 + b_1x_i$ where $b_0 = 1$ (in theory) $b_1 = 1/a; x_i = 1/P$ (P = gas pressure); $y_i = 1/\theta$ (θ is fraction of surface covered)	As above

Table 3.5 (continued)

Experiment	Relationship	Standard form	Normal equations
Heat capacities	$C_p = a + bT + cT^2$ (other equations are used)	$y_i = b_0 + b_1 x_i + b_2 x_i^2$ b_0, b_1, b_2 are constants $x_i = T_i$ (absolute temperature) $y_i = C_p$ = heat capacity at T_i	$\Sigma y_i = b_0 N + b_1 \Sigma x_i + b_2 \Sigma x_i^2$ $\Sigma x_i y_i = b_0 \Sigma x_i + b_1 \Sigma x_i^2 + b_2 \Sigma x_i^3$ $\Sigma x_i^2 y_i = b_0 \Sigma x_i + b_1 \Sigma x_i^3 + b_2 \Sigma x_i^4$
Saturation vapour pressure	$\ln p_0 = b_0 + b_1 T^{-1} + b_2 T$	$y_i = b_0 + b_1/x_i + b_2 x_i$; b_0, b_1, b_2 are constants; $x_i = T_i$ (absolute temperature); $y_i = p_0$ = saturation vapour pressure at T_i	$\Sigma y_i = b_0 N + b_1 \Sigma \dfrac{1}{x_i} + b_2 \Sigma x_i$ $\Sigma \dfrac{y_i}{x_i} = b_0 \Sigma \dfrac{1}{x_i} + b_1 \Sigma \dfrac{1}{x_i^2} + b_2 N$ $\Sigma y_i x_i = b_0 \Sigma x_i + b_1 N + b_2 \Sigma x_i^2$

conditions:

(1) Each component obeys the Beer–Lambert law.
(2) The wavelengths of maximum absorbance are sufficiently different for each substance.

The method is most easily illustrated for a system of two solutes: the absorbances at the wavelengths i and j for the substances x and y are given by:

$$A_i = a_{i,x} \, bc_x + a_{i,y} \, bc_y$$
$$A_j = a_{j,x} \, bc_x + a_{j,y} \, bc_y \tag{3.13}$$

where $a_{p,q}$ is the absorptivity of substance q at wavelength p, b is the path length and c_q is the concentration of substance q. If we can measure the absorbances, A_p, and if we know or can calculate the absorptivities, then the concentrations can be found.

For a more complex system of n solutes, n absorbance measurements at n different wavelengths (subject to the above conditions) must be made. The simultaneous equations (3.14) then require solution:

$$A_1 = a_{11}bc_1 + a_{12}bc_2 + \ldots + a_{1n}bc_n$$
$$A_2 = a_{21}bc_1 + a_{22}bc_2 + \ldots + a_{2n}bc_n$$

$$\vdots \qquad \vdots \qquad \vdots \tag{3.14}$$

$$A_n = a_{n1}bc_1 + a_{n2}bc_2 + \ldots + a_{nn}bc_n$$

As an example of this method, it was found that two compounds, tyrosine and tryptophan, have wavelengths of maximum absorbance at 240 nm and 280 nm, respectively. The absorbances of standard solutions of each compound are tabulated in Table 3.6, together with the corresponding absorbances of a mixture.

Program MULT (Table 3.6) calculates the concentration of each compound by solving a set of two equations, as described above. The method of operation of MULT is as follows: firstly, the data cards are input:

Card 1: Title card (10A8).
Card 2: The number of components, N(I0) and the path length (F0.0).
Card 3: The concentration of each standard substance (20F0.0).
Card 4: The wavelengths used (20F0.0).
Cards 5 to (4+N): Absorbances of each of the N standard substances at the N wavelengths — one card per substance (20F0.0).
Card (5+N): The N absorbances of the mixture in the same wavelength order as for the standards (20F0.0).

Table 3.6
Multicomponent spectroscopic analysis (with an example for two components) using program MULT

```
0008                MASTER MULT
0009                REAL L
0010                DIMENSIONC(20),ABS(20,20),E(20,20),ABSX(20),H(10),WL(20)
0011                COMMON/XVAR/M,X(20,20)
0012                READ(1,14)(H(I),I=1,10)
0013                WRITE(2,16)(H(I),I=1,10)
0014             14 FORMAT(10A8)
0015             16 FORMAT(1H1,10A8//)
0016                READ(1,1)N,L
0017              1 FORMAT(IU,F0.0)
0018                READ(1,5)(C(I),I=1,N)
0019                READ(1,5)(WL(I),I=1,N)
0020                READ(1,5)((ABS(I,J),J=1,N),I=1,N)
0021          C     I=SUBSTANCE,J=WAVELENGTH
0022              5 FORMAT(20F0.0)
0023                DO 4 I=1,N
0024                DO 4 J=1,N
0025              4 E(I,J)=ABS(I,J)/(C(I)*L)
0026                WRITE(2,2)N,L
0027              2 FORMAT(1H ,1X,I3,' SUBSTANCES      , PATH LENGTH =',F9.7)
0028                DO 6 I=1,N
0029                WRITE(2,7)I,C(I)
0030                WRITE(2,8)(WL(J),J=1,N)
0031                WRITE(2,9)(ABS(I,J),J=1,N)
0032              6 WRITE(2,5)(E(I,J),J=1,N)
0033              7 FORMAT(1H ,//,' COMPOUND ',I3,' CONC./   DM-3 =',E10.4)
0034              8 FORMAT(1H ,//,' WAVELENGTH /NM =',10F10.0)
0035              9 FORMAT(1H ,//,' ABSORBANCE =    ',10F10.3)
0036              5 FORMAT(1H ,//,' ABSORBTIVITY  =',10(1X,E9.3))
0037             17 READ(1,10,END=15)(ABSX(J),J=1,N)
0038             10 FORMAT(20F0.0)
0039                WRITE(2,11)(ABSX(J),J=1,N)
0040             11 FORMAT(1H ,//////,' ABS. OF UNKNOWN = ',10F10.3)
0041                DO 12 I=1,N
0042                X(I,(N+1))=ABSX(I)
0043                DO 12 J=1,N
0044             12 X(I,J)=F(J,I)*L
0045                M=N
0046                CALL SIML
0047                WRITE(2,20)
0048             20 FORMAT(1H ,'  ***** CONCENTRATIONS ******')
0049                WRITE(2,15)(X(I,(N+1)),I=1,N)
0050             15 FORMAT(1H ,//,10(2X,E10.4))
0051                GO TO 17
0052             15 STOP
0053                END
```

```
TYROSINE-TRYPTOPHAN

 2 SUBSTANCES      PATH LENGTH =1.0000000

COMPOUND  1 CONC./    DM-3 =0.1300E-01

WAVELENGTH /NM =     240.     280.

ABSORBANCE    =      0.768    0.097

 ABSORBTIVITY = 0.591E 02 0.746E 01


COMPOUND  2 CONC./    DM-3 =0.1800E-01

WAVELENGTH /NM =     240.     280.

ABSORBANCE    =      0.165    0.465

 ABSORBTIVITY = 0.917E 01 0.258E 02


ABS. OF UNKNOWN =    1.020    0.408
  ***** CONCENTRATIONS ******

0.1551E-01  0.1151E-01
```

(The reader should note the use of free formats for input to this program — very useful because of the differing magnitudes of absorbances and concentrations in different branches of spectroscopy.)

From the data, the absorptivities are calculated and, together with the input data, these are listed on the line printer. A set of simultaneous equations of the form of (3.14) is then set up and solved for the concentrations using subroutine SIML. Finally, the concentrations are listed. Specimen data are listed with the program in Table 3.6 for the tyrosine/tryptophan system. This is almost a trivial example but the principles extend to more complex systems.

3.2.2 Detection of End Points in Potentiometric Titrations

In a potentiometric titration, a voltage is obtained from an electrode that is sensitive to an ionic species such as $H^+(aq)$. In this particular case, a signal proportional to the pH of the solution is obtained. In the following discussion, we shall represent this by E whether it is a true voltage or pH. We begin by considering the pH titration of an acid with a base.

As base is added to an acidic solution, E changes and when the acid is just neutralized (an equivalence point) E changes very rapidly with a small addition of base, as seen in Figure 3.2 for the titration of phosphoric acid with sodium hydroxide. It can be seen in the figure that the differential, dE/dV, is at a maximum at each end point. This permits a graphical determination of the end points by the tedious process of calculating the value of dE/dV at each data point.

A similar technique can be used in a computer program. In Table 3.7 we list the values of pH for the titration of phosphoric acid with sodium hydroxide solution. Program CRVY (Table 3.7) fits the pH, volume (V) data in sets of three to a quadratic equation:

$$pH_i = a_0 + a_1 V_i + a_2 V_i^2$$

The derivative at point i is simply:

$$d(pH)/dV = D_i = a_1 + 2a_2 V_i$$

The derivatives are stored and listed. End points are firstly detected in an approximate fashion when:

$$D_{i-1} < D_i > D_{i+1}$$

and

$$D_i > THOLD$$

The latter condition ensures that small 'jiggles' in pH do not become end points. A suitable value of THOLD (card 2) is about 0.5 for most pH titrations. The accurate end points are found by fitting the three derivatives defining an approximate end point to a quadratic:

$$D_i = b_0 + b_1 V_i + b_2 V_i^2$$

The maximum value of D_i is interpolated between the three points; the corresponding volume, V_i, is an end point.

A listing of CRVY and specimen data are given in Table 3.7. Input is as follows:

Card 1: Title card(10A8).
Card 2: THOLD value (F5.2) — usually 0.5.
Card 3: The number of points, M(I3).
Cards 4 to (3+M): volume, pH (or voltage). 2F5.2.

Note that subroutine POLY is required for this program (Section 2.2.2.) This is used for the quadratic data fitting. Derivatives are calculated in line 30 of the program. Approximate end points, as described above, are located by the logical IF (line 56) and accurate end

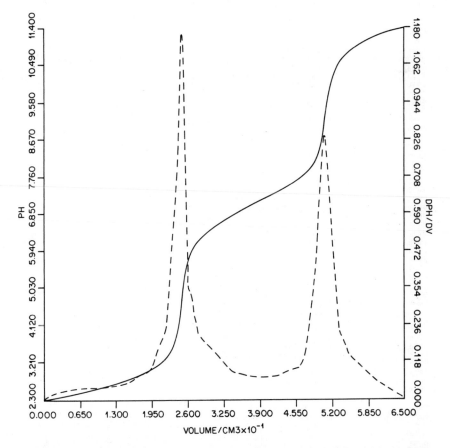

Figure 3.2   pH titration curve for the titration of H_3PO_4 with NaOH
$(--------= d(\text{pH})/dV)$

82

Table 3.7
Program CRVY for end-point detection in pH titrations

```
0008                    MASTER CRVY
0009                    DIMENSIONVOL(100),PH(100),DPH(100),D2PH(100),END(20)
0010                    DIMENSIONV2(100),DPHP(100),EP(100),H(10)
0011                    COMMON/XVAR/N,X(20,20)
0012                    COMMON/A/A(100),Y(100),NP,YCALC(100)
0013               5    READ(1,200)(H(I),I=1,10)
0014                    WRITE(2,300)(H(I),I=1,10)
0015             200    FORMAT(10A8)
0016             300    FORMAT(1H1,10A8//)
0017                    READ(1,100)THOLD
0018             100    FORMAT(F5.2)
0019                    READ(1,1)M
0020                    IO=2
0021                    READ(1,2)(VOL(I),PH(I),I=1,M)
0022              50    DO 11 I=1,3
0023                    J=IO+I-1
0024                    A  (I)=VOL(J)
0025              11    Y  (I)=PH(J)
0026                    NP=3
0027                    N=5
0028                    CALL POLY
0029                    K=IO+1
0030                    DPH(K)=X(2,4)+2.*X(3,4)*VOL(K)
0031              80    FORMAT(1H ,F7.3)
0032                    IF((IO+2).EQ.M)GO TO 12
0033                    IO=IO+1
0034                    GO TO 50
0035              12    CONTINUE
0036                    WRITE(2,101)
0037             101    FORMAT(1H ,'  VOL/CM 3          PH          DPH/DV ',//)
0038                    DO 4 I=1,M
0039                    WRITE(2,3)VOL(I),PH(I), DPH(I)
0040               3    FORMAT(1H ,4X,F5.2,8X,F5.2,11X,F5.2)
0041               4    CONTINUE
0042               1    FORMAT(I5)
0043               2    FORMAT(2F5.2)
0044               6    FORMAT(1H ,10X,F5.2,6X,F5.2)
0045                    IP=1
0046            C       FIND THE END POINT
0047                    DO 67 INDP=1,100
0048              67    EP(INDP)=0
0049                    I2=M-5
0050                    DO 60 IPT=2,I2
0051                    ITEST=0
0052                    ONE=ABS(DPH(IPT))
0053                    TWO=ABS(DPH(IPT+1))
0054                    THREE=ABS(DPH(IPT-1))
0055                    IF(ONE.GT.THOLD.AND.TWO.LT.ONE.AND.THREE.LT.ONE)ITEST=1
0056                    IF(ITEST.EQ.0)GO TO 60
0057                    DO 62 IT=1,3
0058                    JT=IPT+IT-2
0059                    A (IT)=VOL(JT)
0060              62    Y  (IT)=ABS(DPH(JT))
0061                    CALL POLY
0062                    PSTORE=0
0063                    BNK=(A (5)-A (1))/100
0064                    DO 63 ISTEP=1,101
0065                    VEND=A (1)+(ISTEP*BNK)
0066                    PEND=X(1,4)+X(2,4)*VEND+X(3,4)*VEND**2
0067                    IF(PEND.LT.PSTORE)GO TO 63
0068                    VSTORE=VEND
0069                    PSTORE=PEND
0070              63    CONTINUE
0071                    IF((VSTORE-EP(IP)).GT.2.0)IP=IP+1
0072                    EP(IP)=VSTORE
0073              60    CONTINUE
0074                    IPN=IP-1
0075                    WRITE(2,66)IPN
0076              66    FORMAT(1H ,' NO. OF END POINTS WAS ',I5)
0077                    WRITE(2,64)(EP(IK),IK=2,IP)
0078              64    FORMAT(1H ,' END POINTS/CM 3 AT   ',6F8.2)
0079                    STOP
0080                    END
0081                    SUBROUTINE POLY
0082                    COMMON/A/A(100),Y(100),NP,YCALC(100)
0083                    COMMON/XVAR/N,X(20,20)
0084            C       FITS A POLYNOMIAL UP TO 20 TH  DEGREE
```

```
0085                    NVAR=N
0086                    DO 8 I=1,NVAR+1
0087                    DO 8 J=1,NVAR+1
0088                  8 X(I,J)=0
0089                    DO 2 I=1,NVAR
0090                    DO 2 J=1,NVAR
0091                    IF((I+J).EQ.2)GO TO 2
0092                    DO 6 K=1,NP
0093                  6 X(I,J)=X(I,J)+(A(K))**(I+J-2)
0094                  2 CONTINUE
0095                    X(1,1)=NP
0096                    DO 3 I=2,NVAR
0097                    DO 3 K=1,NP
0098                  3 X(I,NVAR+1)=X(I,NVAR+1)+(Y(K))*(A(K)**(I-1))
0099                    DO 4 K=1,NP
0100                  4 X(1,NVAR+1)=X(1,NVAR+1)+Y(K)
0101                    CALL SIML
0102                    DO 5 J=1,NP
0103                    YCALC(J)=0
0104                    DO 5 I=1,NVAR
0105                  5 YCALC(J)=YCALC(J)+X(I,(NVAR+1))*((A(J))**(I-1))
0106                    RETURN
0107                    END
```

ORTHOPHOSPHORIC ACID

VOL/CM 3	PH	DPH/DV
0.00	2,50	0,00
3.00	2,35	0.00
7.00	2,45	0.03
11.00	2,60	0.04
15.00	2,75	0.05
19.00	3,00	0.08
22.00	3,30	0.22
24.00	3,90	0.70
24.50	4,30	1,10
25.00	5,00	1,20
25.50	5,50	0,70
26.00	5,70	0.37
28.00	6,20	0,21
34.00	6,80	0,09
38.00	7,10	0,07
42.00	7,35	0,07
44.00	7,50	0,09
46.00	7,70	0,10
47.00	7,80	0,13
49.00	8,20	0,40
50.00	8,70	0,77
50.50	9,15	0,85
51.00	9,55	0,77
52.00	10,25	0,48
53.00	10,50	0,23
55.00	10,85	0,14
60.00	11,20	0,05
65.00	11,40	0,03
70.00	11,55	0,03
72.00	11,60	0,01
74.00	11,60	0,00
75.00	11,60	0,00
NO. OF END POINTS WAS	2	
END POINTS/CM 3 AT	24,83	50,50

points are located using the statements in lines 57–60 (quadratic fitting of the derivatives) and lines 64–73.

3.2.3 Molecular Formulae from Mass Spectrometry
One of the major applications of high-resolution mass spectrometry is to calculate possible molecular formulae from a given value of m/e for a parent or fragment ion. With a C, H, N, O containing compound the possible formulae are represented by $C_i H_j N_k O_l$. Given the accurate

atomic masses, it is possible to compute the anticipated mass and to compare this with the observed value. If we also know the resolution* of the instrument (usually, in parts per million) we can reject formulae which are in insufficiently good agreement. Some *impossible* formulae can also be rejected by recalling that if the compound contains an odd number of N-atoms then its nearest integer m/e value must also be odd.

The simple program MASS (Table 3.8) takes account of the above. The heart of the program is a set of nested DO loops, lines 26–43, that generate all possible combinations of C, H, N and O up to either specified or calculated maxima. Maxima are *calculated* for the number of atoms of C, H, N or O when the control integers MC, MH, MN or MO

Table 3.8

Calculation of elemental composition using program MASS with mass spectrometric data

```
0008                    MASTER MASS
0009                    READ(1,1)MC,MH,MN,MO,MX
0010                    READ(1,2)P
0011                    C=12,0
0012                    H=1,007825
0013                    AN=14,003073
0014                    AO=15,994914
0015                    IF(MC,LT,0)MC=IFIX(P/C)
0016                    IF(MH,LT,0)MH=2*MC+2
0017                    IF(MO,LT,0)MO=IFIX(P/AO)
0018                    IF(MN,LT,0)MN=2*MC
0019                  1 FORMAT(5I0)
0020                  2 FORMAT(FU.0)
0021                    WRITE(2,7)
0022                  7 FORMAT(1H1,'  MASS SPECTROMETRY CALCULATION',/,'   M/E       C
0023                  1 H       N       U')
0024                    WRITE(2,3)MC,MH,MN,MO,MX
0025                  3 FORMAT(1H ,'MAXIMUM',4I6,6X,I5,' PPM')
0026                    DO 4 I=1,MC
0027                    DO 4 J=1,(MH+1)
0028                    DO 4 K=1,(MO+1)
0029                    DO 4 L=1,(MN+1)
0030                    J1=J-1
0031                    K1=K-1
0032                    L1=L-1
0033                    W=C*FLOAT(I)+H*FLOAT(J1)+AN*FLOAT(L1)+AO*FLOAT(K1)
0034                    IF(L1,EQ,0)L1=2
0035                    IS=(-1)**L1
0036                    IT=(-1)**(IFIX(W))
0037                    IF(IS,NE,IT)GO TO 4
0038                  8 IP=IFIX(ABS(W-P)/P)
0039                    IF(IP,GT,1)GO TO 4
0040                    IP=IFIX(ABS((W-P)*(10**6)/P))
0041                    IF(IP-MX)5,5,4
0042                  5 WRITE(2,6)W,I,J1,L1,K1,IP
0043                  4 CONTINUE
0044                  6 FORMAT(1H ,F7,3,4I6,6X,I5)
0045                    STOP
0046                    END
```

MASS SPECTROMETRY CALCULATION					
M/E	C	H	N	U	
MAXIMUM	12	26	3	3	90 PPM
147,043	7	5	3	1	76
147,032	8	5	1	2	0

*i.e. the ability of the instrument to resolve two peaks separated by ΔM. This is the equivalent to specifying the accuracy of m/e detection to be $\Delta M/M$ for an m/e value of M.

are negative numbers. For example, the maximum number of carbon atoms is the molar mass (from the parent peak) divided by the atomic mass of carbon. Otherwise they are the specified maximum numbers of each atom. The status of these integers is tested in lines 15–18 and appropriate calculations made. Input to MASS is as follows:

Card 1: Control integers MC, MH, MN, MO (see above) and the resolution in p.p.m. − FORMAT 5 I0.

Card 2: Parent peak m/e value (F0.0).

As an example, the following data can be used:

Card 1: −1 −1 3 3 90 [The chemist knew that this compound had no more than 3 N or O atoms].

Card 2: 147.032.

From the output in Table 3.8 it was deduced that the compound was isatin ($C_8 H_5 NO_2$):

Isatin

3.2.4 Use of the Standard Addition Method With Ion-selective Electrodes

This application has a number of interesting features, including a method of fitting data to a non-linear equation by the use of the first derivatives (see Section 2.3). The experiment is based on the use of ion-selective electrodes[12] which have an electrode response, E, given by

$$E = E^\ominus + S \log a \qquad (3.15)$$

where $E^\ominus$ is the standard electrode potential, S is the slope ($= RT/nF$ ideally) and a is the ionic activity. This latter cannot be determined directly from (3.15) because for many ion-selective electrodes $E^\ominus$ and, possibly, S are unknown. Either 'Gran's plots' or a standard addition method can, however, give good results. In this latter method v_i cm^3 of a solution containing the ion to be determined are added to a solution containing an unknown ionic concentration and the change in E noted. For the nth addition to V cm^3 of a solution of concentration C, we have,

$$E_n^\ominus = E^\ominus + S \log \frac{CV + c \sum\limits_{i=1}^{n} v_i}{V + \sum\limits_{c=1}^{n} v_i} \qquad (3.16)$$

where the values of v_i are the volumes of each standard addition of concentration c. Note that, in the absence of suitable data, we have substituted concentrations for activities — an approximation which is valid in dilute solutions. Our problem is to find $E^{\ominus}$, S, and, in particular, C — the initial concentration of the ion before any standard additions. A fairly efficient method[13] of solution is to use a Taylor's expansion formula. To apply this method to our problem we can write, for the nth data point

$$E_n^k \approx E_k^{\ominus} + S_k \log \frac{C_k V + c \sum\limits_{i=1}^{n} v_i}{V + \sum\limits_{i=1}^{n} v_i}$$

where k refers to the kth approximation to $E^{i=1}$, S and C. For brevity, this may be written as $E_n^k \approx E_k^{\ominus} + S_k X_k$.

The difference between the calculated and measured potentials can be defined as ΔE_n:

$$E_n^k = E_k^{\ominus} + S_k X_k + \Delta E_n$$

ΔE_n is then expanded as a function of $E^{\ominus}$, S and C:

$$\Delta E_n = (\partial E_n / \partial E^{\ominus}) \delta E_k^{\ominus} + (\partial E_n / \partial S) \delta S_k + (\partial E_n / \partial C) \delta C_k$$

where δE_k, δS_k and δC_k denote the corrections to be made to the kth approximations of E, S and C. Making use of equation (3.16) we obtain

$$\partial E_n / \partial E^{\ominus} = 1$$

$$\partial E_n / \partial S = \log \left[\frac{CV + c \sum\limits_{i=1}^{n} v_i}{V + \sum\limits_{i=1}^{n} v_i} \right] = x_n$$

$$\partial E_n / \partial C = SV \log \left\{ e / \left[CV + c \sum\limits_{i=1}^{n} v_i \right] \right\} = y_n$$

from which

$$E_n^k = E_k^{\ominus} + S_k x_k + \delta E_k^{\ominus} + x_n \delta' S_k + y_n \delta C_k$$

By defining

$$Z_n = E_n^k - E_k^{\ominus} - S_k x_k$$

we obtain

$$Z_n = \delta E_k^{\ominus} + x_n \delta S_k + y_n \delta C_k$$

We can then obtain the corrections δE_k, δS_k, δC_k by using z as the dependent variable and x and y as independent variables in a three parameter least squares fit. After making the corrections to E, S and C the process can be repeated until convergence occurs. This is a tedious process manually but can be made much easier with the aid of the program ADDFIT (Table 3.9). Input to this program is as follows:

Card 1: J, K, N, VOL (I3, 5X, I3, 5X, I2, 5X, I2, 5X, F.6.2).

    J = number of measurements (J−1 standard additions).
    K = experiment number (identifier).
    N = charge on ion to be determined (negative for anions).
  VOL = initial volume of solution.

Cards 2 to (2+J): electrode potential, volume of standard addition, concentration of the standard (F6.1, 5X, F6.2, 5X, E9.3).

Initial estimates of S, C and E are made in lines 38–41. There follows a least squares analysis based on the above description. The result of each iteration is printed in the section of output headed 'RESULTS'. Suitable test data may be extracted from the output in Table 3.9. Convergence is rapid in most cases, here requiring only five iterations.

3.3 Conclusion − Further Areas of Study in Practical Chemistry

As we explained at the beginning of this chapter, no attempt has been made to provide an exhaustive coverage. There are, for example, many problems in physical chemistry which require least squares analysis (Section 3.1).

Other examples suitable for introductory undergraduate study include:

  (1) Calculation of Unit Cell dimensions from X-ray data.[14]
  (2) Analysis of magnetic susceptibility data.[15]
  (3) Calculation of polymer molar masses from Zimm plots.[16]
  (4) Solution of ionic equilibria problems.[17]
  (5) Computation of equilibrium constants.[18,19]

The theoretical background to these and other areas is well covered by standard physical chemistry texts. Other examples are to be found in Chapters 4 and 7.

A useful source of information on chemically orientated programs is the *Journal of Chemical Education*. Although it is rare that complete programs are listed, authors are usually willing to supply listings free of charge. Also, the programs discussed in this chapter can be augmented by use of program exchanges such as operated by the NCC [National Computing Centre (UK)] and the QCPE [Quantum Chemistry Program Exchange]. This is discussed at greater length in Chapter 7.

88

Table 3.9
Program ADDFIT for analysis of ion-selective electrod data. (Reprinted with
permission from *Anal. Chem*, 42, 1176 (1970) Copyright by the American
Chemical Society)

```
0008                    MASTER ADDFIT
0009                    DIMENSION E(20),V(20),C(20),X(20),Y(20),Z(20),
0010                   1    SUMV(20),SUMCV(20)
0011          C
0012                  1 READ(1,1001) J,K,N,VOL
0013               1001 FORMAT(I3,5X,I3,5X,I2,5X,F6.2)
0014                    READ(1,1002) (E(I), V(I), C(I), I = 1,J)
0015               1002 FORMAT(F6.1,5X,F6.2,5X,E9.3)
0016                    KABS = IABS(K)
0017                    WRITE(2,2004) KABS
0018               2004 FORMAT(///19X,19H*****  EXPERIMENT   ,I3,7H *****//
0019                   1    5X,8HDATA....)
0020                    WRITE(2,2001) J, N, VOL
0021               2001 FORMAT(15X,4HJ = ,I3,5X,4HN = ,I2,5X,6HVOL = ,F6.2//)
0022                    WRITE(2,2002) (E(I),V(I),C(I),I=1,J)
0023               2002 FORMAT(15X,1HE,11X,1HV,10X,1HC//(12X,F6.1,5X,F6.2,5X,E9.3))
0024                    WRITE(2,2005)
0025               2005 FORMAT(//5X,11HRESULTS...//23X,3HCIN,12X,1HS,8X,5HESTAN//)
0026          C
0027          C         CALCULATE SUMV(I), SUMCV(I)
0028          C
0029                    SUMV(1) = V(1)
0030                    SUMCV(1) = V(1)*C(1)
0031                    DO 2 I = 2,J
0032                    SUMV(I) = SUMV(I-1) + V(I)
0033                    SUMCV(I) = SUMCV(I-1) + C(I)*V(I)
0034                  2 CONTINUE
0035          C
0036          C         INITALISE S, CIN, ESTAN
0037          C
0038                    S = 59/N
0039                    CIN = (C(2)*V(2))/(((VOL + V(2))*(EXP(ALOG(10.0)*
0040                   1    ((E(2) - E(1))/S))))- VOL)
0041                    ESTAN = E(1) - (S*ALOG10(CIN))
0042          C
0043          C         INITIALISE VARIABLES
0044          C
0045                    L = 1
0046                    DELC = 0.0
0047                    DELS = 0.0
0048                    DELE = 0.0
0049          C
0050          C         CALCULATE NEW VALUES OF S, CIN, ESTAN
0051          C
0052                  3 CIN = CIN + DELC
0053                    S = S + DELS
0054                    ESTAN = ESTAN + DELE
0055                    WRITE(2,2003) CIN, S, ESTAN
0056               2003 FORMAT(20X,E9.3,5X,F7.3,5X,F8.3)
0057                    IF(L - 1)5,5,4
0058                  4 IF(((100.0*ABS(DELC))/CIN) - 0.01)6,6,5
0059                  6 IF(K)9,9,1
0060          C
0061                  5 DO 7 I = 1,J
0062                    X(I) = ALOG10(((CIN*VOL) + SUMCV(I))/(VOL + SUMV(I)))
0063                    Y(I) = (S*VOL*ALOG10(EXP(1.0)))/(CIN*VOL + SUMCV(I))
0064                    Z(I) = E(I) - ESTAN - (S*X(I))
0065                  7 CONTINUE
0066          C
0067          C         THIS ROUTINE CALCULATES THE LEAST SQUARES COEFFICIENTS OF
0068          C         Z = A + B*X + C*Y
0069          C
0070          C         INITIALISE VARIABLES
0071          C
0072                    SUMX = 0.0
0073                    SUMY = 0.0
0074                    SUMZ = 0.0
0075                    SUMXY = 0.0
0076                    SUMXZ = 0.0
0077                    SUMYZ = 0.0
0078                    SUMX2 = 0.0
0079                    SUMY2 = 0.0
0080          C
0081                    DO 8 I = 1,J
```

```
0082            SUMX  =  SUMX  +  X(I)
0083            SUMY  =  SUMY  +  Y(I)
0084            SUMZ  =  SUMZ  +  Z(I)
0085            SUMXY =  SUMXY +  (X(I)*Y(I))
0086            SUMXZ =  SUMXZ +  (X(I)*Z(I))
0087            SUMYZ =  SUMYZ +  (Y(I)*Z(I))
0088            SUMX2 =  SUMX2 +  (X(I)*X(I))
0089            SUMY2 =  SUMY2 +  (Y(I)*Y(I))
0090          8 CONTINUE
0091       C
0092            DNUM = J*((SUMX2*SUMYZ) - (SUMXZ*SUMXY))
0093          1       -SUMX*((SUMX*SUMYZ) - (SUMY*SUMXZ))
0094          2       +SUMZ*((SUMX*SUMXY) - (SUMY*SUMX2))
0095       C
0096            DNOM = (J*SUMX2*SUMY2) - (J*SUMXY*SUMXY)
0097          1       -(SUMX*SUMX*SUMY2) + (SUMX*SUMXY*SUMY)
0098          2       +(SUMY*SUMX*SUMXY) - (SUMY*SUMX2*SUMY)
0099       C
0100            DELC = DNUM/DNOM
0101            ERRM = SUMZ - (DELC*SUMY)
0102            ERRN = SUMXZ - (DELC*SUMXY)
0103            DELS = ((J*ERRN)-(ERRM*SUMX))/((J*SUMX2)
0104          1       -(SUMX*SUMX))
0105            DELE = (ERRM - DELS*SUMX) / J
0106            L = L + 1
0107            GO TO 3
0108          9 STOP
0109            END
```

```
                ***** EXPERIMENT   0  *****

DATA,...
            J =   6     N = -1      VOL = 50,00


            E            V           C

          -4,0         0,00      0,000E 00
          -9,5         0,50      0,200E-01
         -14,0         0,50      0,200E-01
         -17,0         0,50      0,200E-01
         -20,5         0,50      0,200E-01
         -23,5         0,50      0,200E-01


RESULTS,...

                CIN            S          ESTAN

          0,794E-03       -59,000      -186,904
          0,846E-03       -60,685      -190,520
          0,853E-03       -60,973      -191,215
          0,853E-03       -60,993      -191,262
          0,853E-03       -60,994      -191,265
```

3.4 Problems

3.4.1 Write a program to calculate the activation enthalpy, $\Delta H\ddagger$, and entropy, $\Delta S\ddagger$, of a chemical reaction. Sample data for a first-order reaction are:

Rate constant/s^{-1}:	4.07×10^{-4}	1.13×10^{-3}	3.09×10^{-3}	7.92×10^{-3}
Temperature/K:	288	298	308	318

3.4.2 Write a program to calculate the Gibbs free energy change for a chemical reaction at any temperature, T. Use standard heats of formation, $\Delta_f H°$, standard free energies, $\Delta_f G°$, and the coefficients a, b

and c for the molar heat capacity, C_p:

$$C_p = a + bT + cT^2$$

Sample data are:

Reaction	N_2	$+$	$3H_2$	$2NH_3$
$\Delta_f H$ /kJ mol^{-1}	0		0	-16.20
$\Delta_f G$ /kJ mol^{-1}	0		0	-16.6
a/J mol^{-1} K^{-1}	28.3		27.7	25.9
b/J mol^{-1} K^{-1}	2.54		3.39	33.0
c/J mol^{-1} K^{-1}	5.44		0.00	-30.5

For test purposes, we used (answers section) a temperature of 500 K.

3.4.3   For the distribution of a solute between two immiscible solvents, we have:

$$nS_{(\text{solvent 1})} \overset{K}{\rightleftharpoons} S_n \text{ (solvent 2)}$$

Devise a program to calculate K and n. For test data we use the system

$$2CH_3 CO_2 H_{(\text{aqueous})} \rightleftharpoons (CH_3 CO_2 H)_{2 \text{ (organic)}}$$

Concentrations of acid were determined by titration with 0.4987 mol dm^{-3} sodium hydroxide solution. The results were:

	Titration Values/cm^3	
Experiment no.	Organic layer	Aqueous layer
1	3.595	6.570
2	7.000	12.845
3	14.995	25.010

3.4.4   Modify program KNTT (Section 3.1.1) so that it will analyse data for any first-order reaction.

3.4.5   When the mass spectra of four pure hydrocarbons were recorded (under identical conditions) the following peak heights were obtained at the indicated mass (m/e) values:

Hydrocarbon $m/e =$	31	34	45	58
A	0.0	24.1	22.4	10.8
B	3.2	0.0	0.0	38.9
C	6.4	0.0	100.0	0.0
D	100.0	0.0	4.2	0.0

From each peak height, we can define the sensitivity of the spectrometer, S_{ij}, to component j at m/e equal to i. At the same m/e values as quoted for the pure compounds, a mixture of A, B, C and D gave relative peak heights of 100.0, 9.3, 51.4 and 12.2. Use a multicomponent analysis method to obtain the partial pressure of each component in the mixture. (See Reference 20 for further details.)

3.4.6   Use either the line printer or a digital plotter to record pH and first derivative values for program CRVY (Figure 3.2).

3.4.7   For the titration of a dibasic acid, $H_2 A$, with a base, B, the following relationships can be obtained:

Total concentration of acid, $A_T = [H_2 A] + [HA^-] + [A^{2-}]$

Total concentration of base, $B_T = [HA^-] + 2[A^{2-}] - [H^+]$

Defining X, Y, Z as:

$$X = B_T [H^+] + [H^+]^2 - [H^+] A_T$$
$$Y = 2C_T - B_T - [H^+]$$
$$Z = [H^+]^2 (B_T + [H^+])$$

We then obtain expressions for K_1 and K_2, the successive ionization constants:

$$K_1 = (Y_1 Z_2 - Y_2 Z_1)/(X_1 Y_2 - X_2 Y_1)$$
$$K_2 = (X_1 Z_2 - X_2 Z_1)/(Y_1 Z_2 - Y_2 Z_1)$$

The subscripts 1 and 2 denote the addition of less than and more than one equivalent of base respectively. Select an appropriate acid (e.g. oxalic acid), record appropriate experimental data and calculate K_1 and K_2. Test data are to be found in Reference 21.

3.4.8   This problem requires that you have access to thermogravimetric equipment. Use the polynomial method of Schlempf (Section 3.1.3) to obtain the pre-exponential factor, A, and activation energy, E, for a thermal decomposition. A suitable test substance is calcium oxalate $(A = 10^6 \, s^{-1}; E = 85 \, kJ \, mol^{-1})$.

References

1. Glastone, S., *Thermodynamics for Chemists*, Van Nostrand, Princeton, N.J., 1958, p. 392.
2. O'Neill, M. J., *Anal. Chem.*, 36, 1238 (1964).
3. Marti, E., *Thermochim. Acta*, 5, 173 (1972).
4. Galwey, A. K., *Chemistry of Solids*, Chapman and Hall, London, 1967, p. 163 *et seq.*
5. Beech, G., *J. Chem. Soc. (A)*, 925 (1969).

6. Schlempf, J. M., Freeburg, F. E., Rodgers, D. J., and Angeloni, F. M., *Anal. Chem.,* **38**, 520 (1960).
7. Freeman, E. S., and Carroll, B., *J. Phys. Chem.,* **62**, 394 (1958).
8. Coats, A. W., and Redfern, J. P., *Nature*, **201**, 68 (1964).
9. Sharp, J. H., and Wentworth, S. A., *Anal. Chem.*, **41**, 2060 (1969).
10. Carroll, B. and Manche, E. P., *Anal. Chem.*, **42**, 1296 (1970).
11. Judd, M. D., and Pope, M. I., *J. Thermal. Anal.,* **5**, 501 (1973).
12. Durst, R. A., (Ed.), 'Ion selective Electrodes', National Bureau of Standards, Special Publication 314, Washington, D.C., 1969, p. 375.
13. Branch, M. J. D., and Rechnitz, G. A., *Anal. Chem.*, **42**, 1172 (1970).
14. Norris, A. C., Broadbent, S. E., and Davies, N. R., *J. Chem. Ed.*, **50**, 7 (1973).
15. Crawford, T. H., and Swanson, J., *J. Chem. Ed.,* **48**, 382, (1971).
16. Kerker, M., *The Scattering of Light,* Academic Press, New York, 1969, p. 434.
17. Haglund, E., Moss, D., and Flynn, J., *J. Chem. Ed.,* **43**, 582 (1966).
18. Erickson, L. E., *J. Chem. Ed.,* **46**, 383 (1969).
19. Ramette, R. W., *J. Chem. Ed.,* **44**, 647 (1967).
20. Kiser, R. W., *Introduction to Mass Spectrometry and its Applications*, Prentice-Hall, Englewood Cliffs, N.J., 1965, p. 221.
21. Jensen, R. E., Garvey, R. G., and Poulson, B. A., *J. Chem. Ed.*, **47**, 147 (1970).

Chapter 4

Tutorial and 'Dry-lab' Applications

Until recently, computers have been mainly used in chemical education for the treatment of experimental data. An increasingly important area, however, is concerned with

(i) Analysis of data collected by another experimenter (e.g. reported in a research paper) and which would be difficult to collect in the student laboratory for reasons of time or accuracy. This is sometimes called a 'Dry-Lab' application.

(ii) Simulation or modelling of chemical systems. This can be *ab initio*, in which a student actually develops a description of a system or, as in our case, uses a developed program to illustrate various aspects of the behaviour of the system.

In the following sections, we draw upon examples in kinetics, thermodynamics and spectroscopy as illustrations of (i) and (ii) above.

4.1 Kinetics

Although computers have, for many years, been associated with kinetics from the point of view of data treatment, it is interesting to observe the recent interest in the computer simulation of kinetic systems; References 1 to 7 are typical. Reference 6 is of interest since it describes the applications of analog computers to kinetic studies; in some respects, analog computers are more suitable than digital in this area and the reader should, therefore, decide which is more appropriate to his work. The first of the two applications in this section is equally amenable to analog and digital solution whereas the second is a purely digital exercise.

94

4.1.1 Successive Radioactive Decay

This concerns the decay sequence

$A \rightarrow B \rightarrow C \rightarrow D$, etc.

Such series occur naturally and we will describe a program that will calculate the amount of each radionuclide in the truncated series

$A \rightarrow B \rightarrow C$ (4.1)

A more complex series can be analysed but no new principles are involved. Also, although the program is specifically oriented to radioactive decay, it is a simple matter to modify it to deal with, for example, gas phase kinetics.

For a simple decay of the type

$A \rightarrow B$

we have

$$\frac{dN_A}{dt} = -\lambda_A N_A \tag{4.2}$$

where N_A is the number of atoms of A at time t and λ_A is the 'decay constant'. Integration of (4.2) yields the equation

$N_A^t = N_A^o e^{-\lambda_A t}$

where N_A^t is the number of A atoms at time t and N_A^o was the number at zero time. The constant λ_A is related to the half-life, $t_{1/2}$, of A (the time at which $N_A^t = N_A^o/2$) by

$\lambda_A = 0.693/t_{1/2}$

For successive decay, as in (4.1), we find

$N_A^t = N_A^o e^{-\lambda_A t}$

$$N_B^t = \frac{\lambda_A}{\lambda_B - \lambda_A} N_A^o (e^{-\lambda_A t} - e^{-\lambda_B t}) + N_B^o e^{-\lambda_B t} \tag{4.3}$$

$N_C^t = N_A^o + N_B^o + N_C^o - N_A^t - N_B^t$

where the superscripts have their previous meaning. A program which uses these equations is RADK, listed in Table 4.1. This program accepts the numbers of atoms at zero times, together with the appropriate half-lives (for which the λs are calculated) and plots the amounts of A, B and C on the line printer at the desired times. The absolute numbers of A, B and C atoms are calculated in lines 38–40. The relative numbers are scaled to lie between 1 and 101 and the resultant integer is used as the index of the array IK(101). The characters A, B or C are stored in this array (line 46) which is then printed at each time interval from zero to the desired maximum. Input to the program is by a single card:

Input card: Number of A, B and C atoms at zero time; half-lives of A and B; maximum time for the simulation; (Format 3I0, 3F0.0).

Note that all times must be in the same units in this version of the program (see Problem 4.5.1). In Table 4.1 the reader will find an example of output from RADK for the system

$$^{218}\text{Po} \xrightarrow{t_{1/2} = 3\,\text{m}\,(\alpha)} {}^{214}\text{Pb} \xrightarrow{t_{1/2} = 26.8\,\text{m}\,(\beta)} {}^{214}\text{Bi}$$

The program can be used to investigate the various common situations which include:

$t_{1/2}$ (parent) $> t_{1/2}$ (daughter)

$t_{1/2}$ (parent) $\gg t_{1/2}$ (daughter)

Table 4.1

Program RADK — program to plot the numbers of atoms in the sequence $(A \to B \to C)$ as a function of time

```
0008            MASTER RADK
0009            DIMENSION IK(101),KAR(4),N(4),NX(4)
0010            DATA KAR/1H ,1HA,1HB,1HC/
0011            READ(1,1)NAJ,NBO,NCO,TA,TB,TT
0012          1 FORMAT(3I0,4F0.0)
0013      C     ALL HALF LIVES AND TOTAL TIME MUST HAVE THE SAME DIMENSIONS
0014            WRITE(2,2)NAO,NBO,NCO,TA,TB,TT
0015          2 FORMAT(1H1,'         RADIOACTIVE   DECAY   SIMULATION ',//,1X,' NAO ='
0016           1,I10,5X,'NBO =',I10,5X,'NCO =',I10,'        ATOMS AT TIME ZERO',//,1
0017           2X,'    RESPECTIVE  HALF LIVES (ALL SAME UNITS) ARE',2(F10.3,4X),//
0018           3,1X,'    OBSERVATION TIME IS ',F10.3,//,120(1H-),//////)
0019      C     CLEAR LINE BUFFER
0020            DO 3 I=1,101
0021          3 IK(I)=KAR(1)
0022            WRITE(2,4)
0023          4 FORMAT(1X,'   TIME                      RELATIVE NUMBERS OF ATOMS',/,9
0024           2X,101(1H-))
0025            T=0
0026            C1=0.693/TA
0027            C2=0.693/TB
0028            DT=TT/100
0029            N(1)=NAO
0030            N(2)=0
0031            N(3)=0
0032            NSUM=NAO+NBO+NCO
0033            DO 5 I=1,101
0034            IF(I.EQ.1)GO TO 66
0035            T=(I-1)*DT
0036            IF(T.GT.TT)GO TO 9
0037            N(1)=NAO*EXP(-C1*T)
0038            N(2)=NAO*(C1/(C2-C1))*(EXP(-C1*T)-EXP(-C2*T))+NBO*EXP(-C2*T)
0039            N(3)=NAO+NBO+NCO-N(1)-N(2)
0040         66 DO 6 J=1,3
0041      C     THIS IS FOR GRAPHICAL OUTPUT
0042            NP=100*N(J)/NSUM +1
0043            NX(J)=NP
0044            K=J+1
0045          6 IK(NP)=KAR(K)
0046            WRITE(2,7)T,(IK(K),K=1,101)
0047          7 FORMAT(1X,F7.1,' I',101A1)
0048      C     CLEAR LINE BUFFER
0049            DO 8 L=1,3
0050            NL=NX(L)
0051          8 IK(NL)=KAR(1)
0052          5 CONTINUE
0053          9 STOP
0054            END
```

RADIOACTIVE DECAY SIMULATION

NAU = 2000   NBO = 0   NCO = 0   ATOMS AT TIME ZERO

RESPECTIVE HALF LIVES (ALL SAME UNITS) ARE   5,000   26,800

OBSERVATION TIME IS   84,000

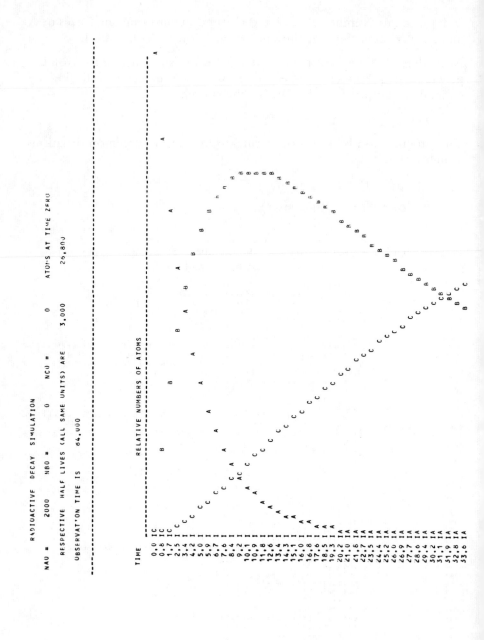

TIME   RELATIVE NUMBERS OF ATOMS

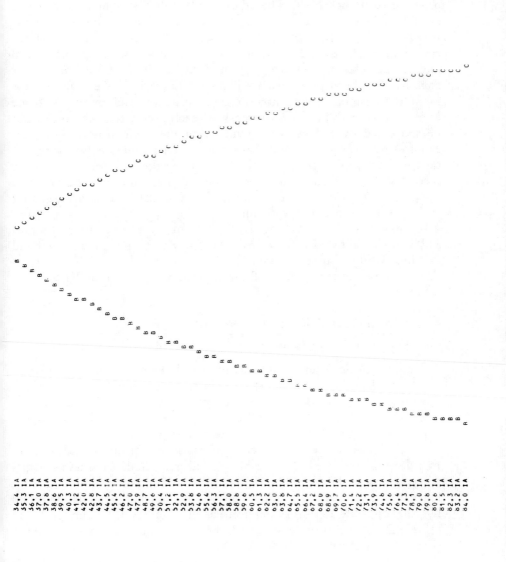

```
34.4 IA
35.3 IA
36.0 IA
37.6 IA
37.8 IA
38.6 IA
39.5 IA
40.3 IA
41.2 IA
42.0 IA
42.8 IA
43.7 IA
44.5 IA
45.4 IA
46.2 IA
47.0 IA
47.9 IA
48.7 IA
49.6 IA
50.4 IA
51.2 IA
52.1 IA
52.9 IA
53.8 IA
54.6 IA
55.4 IA
56.3 IA
57.1 IA
58.0 IA
58.8 IA
59.6 IA
60.5 IA
61.3 IA
62.2 IA
63.0 IA
63.8 IA
64.7 IA
65.5 IA
66.4 IA
67.2 IA
68.0 IA
68.9 IA
69.7 IA
70.6 IA
71.4 IA
72.2 IA
73.1 IA
73.9 IA
74.6 IA
75.6 IA
76.4 IA
77.3 IA
78.1 IA
79.0 IA
79.8 IA
80.6 IA
81.5 IA
82.3 IA
83.2 IA
84.0 IA
```

Systems in which these relationships are true are said to display transient and secular equilibrium, respectively.

4.1.2 Monte Carlo Simulation of Kinetic Schemes

The preceding discussion was limited to a specific decay scheme; the program to be described is based on the 'Monte Carlo' method which is considerably more attractive in that it is easily adaptable to systems which are so complex that it is not possible to write down closed-form equations such as (4.3). The mathematical basis of the Monte Carlo method is described elsewhere[5,8] and an excellent general review for chemists has been published by Para and Lazzarini.[9] We shall use this method to simulate the outcome of a series of random events. If we know the probability of occurrence of each event, then the series can be simulated by the use of a random number generator: to quote an example given by Para and Lazzarini,[9] let us generate two series of random numbers in the interval $(0, 1)$ calling the first series x, and the second y. From x and y we calculate d, a third random number, given by $(x^2 + y^2)^{1/2}$. We then define a limit of d, equal to 1, and recall that the equation of a circle with unit radius is $x^2 + y^2 = 1$ as seen in Figure 4.1. When d is $\leqslant 1$, it belongs to the first quadrant of a circle of radius 1. When d is allowed to take any value for x and y lying between 0 and 1 it lies within a square of side 1. If we generate a total of N_0 values of d, we should find that, if N of the values are $\leqslant 1$, then, as N_0 tends to infinity

$$\frac{N}{N_0} \rightarrow \frac{\text{area of quadrant of circle}}{\text{area of square}} = \frac{(\pi/4)}{1} = \frac{\pi}{4}$$

A program to perform this calculation would be a good introduction to the Monte Carlo method and would illustrate that the 'best' answer (i.e. the value of π) is obtained with a larger number of random events. At this level, the Monte Carlo method can be seen to be intuitively simple and we shall now examine its application to a problem in kinetics.

A chemical reaction is clearly a statistical process and should, therefore, be well suited to the Monte Carlo method. We shall use digits to represent molecules. The digits are placed in some form of grid or matrix[10] so that the position of any digit (molecule) can be identified by the appropriate grid location, as in Figure 4.2. The grid can represent the reaction vessel and the percentage occupation with a particular digit represents the percentage of any one molecule in the vessel. The grid in Figure 4.2 therefore represents a reaction vessel with 20% of molecule '1' and 10% of molecule '2'. Any degree of filling of the grid is allowed and the fraction of positions occupied by any particular digit is equivalent to the mole fraction of the molecule in the simulated system, if we regard the blank positions as being unreactive

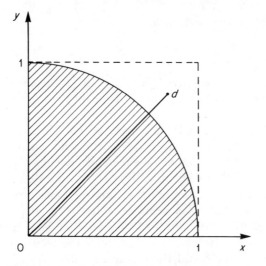

Figure 4.1 A quadrant of a circle of radius 1 and a square of side 1. Random numbers x and y are combined to give d; $d = (x^2 + y^2)^{1/2}$

	1	2	3	4	5	6	7	8	9	10
0		1				1	2	1		2
10	2				1					
20				1					1	
30		1				2			2	1
40				1		1	1			
50		2							1	
60			1			2	1			
70			2			1		2		1
80		1							1	
90				1	2			1		

Figure 4.2 A numbered grid containing 100 locations for the molecules '1' and '2'

'solvent' molecules. Therefore, in Figure 4.2 we have a 0.2 mole fraction of A, 0.1 mole fraction of B, and 0.7 mole fraction of solvent. Note that we have filled the grid randomly but this is by no means necessary.

We can use this diagram to assist in the simulation of a one-step reaction, having a rate constant k:

$$A \xrightarrow{k} B$$

with A molecules represented by '1' and B molecules by '2'. We generate random numbers from 1 to 100 and when the number coincides with the number of a grid position occupied by '1' then a 'fruitful event' occurs and '1' is converted to '2'. (This would be true for grid positions 2, 6, 8 etc.) When the grid position is empty or occupied by a 2, the event is non-fruitful. Eventually, all 1s will be converted to 2s and the reaction will be 100% complete.

The total number of attempts (i.e. random numbers generated) will be an integer, N_T. This also represents the total elapsed time for the real chemical reaction. As a unit of time we could use a smaller integer, N, such that N_T is a multiple of N. After each series of N attempts to convert 1s into 2s, we could then note the numbers of each digit and, perhaps, display the result graphically. In the case of a bimolecular reaction we would take pairs of randomly chosen grid positions and only when the occupants of these grid positions are the chosen molecules would we allow them to 'react'. We can also allow our product to react further, for example:

$$A \xrightarrow{k_1} B \xrightarrow{k_2} C, \text{ etc.}$$

To simulate this system, we would introduce the digit 3 into our grid to represent C molecules; we then generate, for each reaction step, a set of grid positions, N_1 and N_2 in order that N_1/N_2 is equal to k_1/k_2. This is permissible if N_1 and N_2 are each small compared to N_T, the total number of grid positions generated. Similarly, we could simulate branching reactions or consecutive and branching reactions of any degree of complexity.

By now it should be clear that the Monte Carlo method is remarkably similar to our normal statistical description of chemical reactions. We can even construct our model so that the computer units of time (i.e. N, the number of tries) coincide with a real time scale. For example, a single-step first-order reaction

$$A \xrightarrow{k_1} B$$

is described by the chemical rate equation

$$\frac{-d[A]}{dt} = k_1[A]$$

or, in integrated form

$$\ln \frac{[A]_0}{[A]} = k_1 t \qquad (4.4)$$

In Monte Carlo terms, we can consider the rate of disappearance of the digit 1 from our grid in Figure 4.2. The rate of disappearance will be proportional to the number of 1s and also, to an increase in the number of tries, N_1, at converting 1s into 2s:

$$\frac{-d[1]}{dN_1} = \beta[1] \qquad (4.5)$$

where dN_1 is a small increase in the number of tries and β is a proportionality constant. Integration of (4.5) yields

$$\ln \frac{[1]_0}{[1]} = \beta N_1$$

Comparison with (4.4) shows that

$$\beta N_1 \equiv k_1 t$$

Therefore, real time (t) and computer time (N_1) must be in the simple ratio

$$\frac{N_1}{t} = \frac{k_1}{\beta}$$

thus permitting us to superimpose simulated and experimental graphs by simple scale-adjustment of the axes. Similar arguments can be applied[10,11] to reactions of higher order and the results again allow us to superimpose experimental and computed data.

We shall now describe the program MONT (Table 4.2) which will plot the course of a chemical reaction in units of 'computer time' (v.s.). The system that we will simulate is:

$$A \underset{k_2}{\overset{k_1}{\rightleftarrows}} B \xrightarrow{k_3} C$$

We choose this because the closed-form solutions for the system are of moderate complexity for the average student; the Monte Carlo simulation is, however, very simple and can be easily extended to other systems. The program begins by reading the data, which we define at this point.

Card input: The number of atoms of A, B and C; as integers NA, NB and NC, the sum of which must be $\leqslant 2000$; IMAX, the total number of simulations to be made of the system, from zero time; NLOOP, the number of sets of tries to be made in each simulated computer-time period; N1, N2 and N3 are integers which are proportional to k_1, k_2

Table 4.2
Monte Carlo simulation program for the system A ⇌ B → C

```
0008              MASTER MONT
0009              DIMENSIONM(2000)
0010              DIMENSION KAR(12),IPAGE(60,120)
0011              DATA KAR/1H ,1HA,1HB,1HC,1HI,1H-,1HY,1HM,1HA,1HX,1HO,1H+/
0012          C   READ INITIAL CONCNS FOR FIRST-ORDER CONSECUTIVE REACTION
0013        200   READ(1,1,END=99)NA,NB,NC,NT,IMAX,NLOOP,N1,N2,N3
0014          1   FORMAT(9I0)
0015              WRITE(2,2)NA,NB,NC,IMAX,NLOOP,NT
0016          2   FORMAT(1H1,' NA,NB,NC = ',3(2X,I5),'    IMAX = ',I4,' NLOOP',I4,'
0017          1   NT= ',I6)
0018              WRITE(2,50)N1,N2,N3
0019         50   FORMAT(//,1H ,' VALUES OF N1 N2 AND N3 ARE ',3I7)
0020              WRITE(2,55)NT
0021         55   FORMAT(1H ,' TOTAL NUMBER OF MOLECULAR LOCATIONS IS ',I6)
0022              WRITE(2,10)
0023         10   FORMAT(///,'    IM       NA       NB       NC',/)
0024              DO 100 I=1,60
0025              DO 100 J=1,120
0026        100   IPAGE(I,J)=KAR(1)
0027              IK=0
0028          C   FILL MATRIX
0029              S=0.2
0030              CALL FPMCRV(S)
0031              DO 3 I=1,NA
0032          3   M(I)=1
0033              DO 4 I=1,NB
0034          4   M(I)=2
0035              DO 5 I=1,NC
0036          5   M(I)=3
0037              IM=0
0038              XMAX=IMAX
0039              YMAX=NA+NB+NC
0040              GO TO 80
0041          6   IM=IM+1
0042              IF(IM.GT.IMAX)GO TO 103
0043              DO 7 I=1,NLOOP
0044              DO 8 J=1,N1
0045              CALL RAND(NT,INEW,S)
0046              IF(INEW.GT.2000)GO TO 8
0047              IF(M(INEW).NE.1)GO TO 8
0048              NA=NA-1
0049              NB=NB+1
0050              M(INEW)=2
0051          8   CONTINUE
0052              DO 9 K=1,N2
0053              CALL RAND(NT,INEW,S)
0054              IF(INEW.GT.2000)GO TO 9
0055              IF(M(INEW).NE.2)GO TO 9
0056              NB=NB-1
0057              NC=NC+1
0058              M(INEW)=3
0059          9   CONTINUE
0060              DO 15 L=1,N3
0061              CALL RAND(NT,INEW,S)
0062              IF(INEW.GT.2000)GO TO 15
0063              IF(M(INEW).NE.2)GO TO 15
0064              NB=NB-1
0065              NC=NC+1
0066              M(INEW)=3
0067         15   CONTINUE
0068          7   CONTINUE
0069         80   WRITE(2,11)IM,NA,NB,NC
0070         11   FORMAT(1H ,I6,3(4X,I6))
0071              IA=(NA/YMAX)*50
0072              IB=(NB/YMAX)*50
0073              IC=(NC/YMAX)*50
0074              IT=(IM/XMAX)*100+10
0075              IK=IK+1
0076              IF(IK.EQ.51)GO TO 103
0077              IF(IA.EQ.0)IA=1
0078              IF(IB.EQ.0)IB=1
0079              IF(IC.EQ.0)IC=1
0080              IF(IT.EQ.0)IT=1
```

```
0081                    IPAGE(IA,IT)=KAR(2)
0082                    IPAGE(IB,IT)=KAR(3)
0083                    IPAGE(IC,IT)=KAR(4)
0084                    GO TO 6
0085              103 WRITE(2,105)
0086              105 FORMAT(1H1)
0087                    DO 107 J=4,7
0088              107 IPAGE(50,J)=KAR(J+3)
0089                    MARK=50
0090                    IPAGE(1,7)=KAR(11)
0091                    DO 101 I=1,50
0092                    J=51-I
0093                    IPAGE(J,9)=KAR(5)
0094                    IPAGE(MARK,9)=KAR(12)
0095                    IF(J.EQ.MARK)MARK=MARK-10
0096              101 WRITE(2,102)(IPAGE(J,K),K=1,120)
0097              102 FORMAT(120A1)
0098                    WRITE(2,106)
0099              106 FORMAT(1H ,8X,1H+,10(9(1H-),1H+))
0100                    WRITE(2,108)
0101              108 FORMAT(1H ,106X,'XMAX')
0102                    WRITE(2,104)
0103              104 FORMAT(/,1H ,'                    TIME/COMPUTER UNITS')
0104                    GO TO 200
0105               99 STOP
0106                    END
0107                    SUBROUTINE RAND(J,INEW,S)
0108                1 CALL FPMCRV(S)
0109                    INEW=J*S
0110                    IF(INEW.EQ.0)GO TO 1
0111                    RETURN
0112                    END
```

NA,NB,NC =   2000   0   0   IMAX =   30 NLOOP 150 NT=   9000

VALUES OF N1 N2 AND N3 ARE   10   10   5
TOTAL NUMBER OF MOLECULAR LOCATIONS IS   9000

IM	NA	NB	NC
0	2000	0	0
1	1675	294	31
2	1406	466	128
3	1157	557	286
4	986	577	437
5	841	565	594
6	725	544	731
7	597	528	875
8	505	495	1000
9	422	462	1116
10	353	429	1218
11	285	405	1310
12	239	352	1409
13	208	306	1486
14	171	263	1566
15	144	234	1622
16	122	192	1686
17	101	159	1740
18	87	156	1777
19	70	120	1810
20	62	98	1840
21	49	87	1864
22	44	74	1882
23	37	66	1897
24	33	55	1912
25	30	47	1923
26	28	38	1934
27	26	25	1949
28	25	20	1955
29	23	12	1965
30	21	11	1968

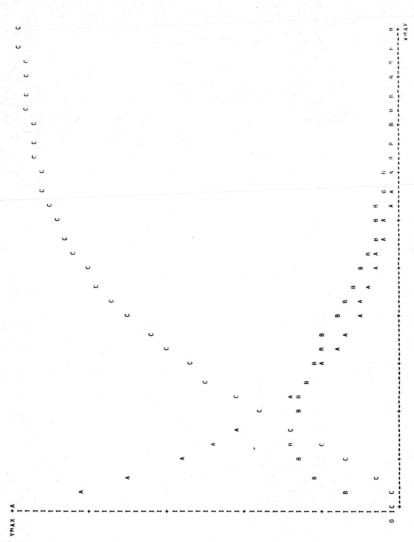

TIME/CUMPUTER UNITS.

and k_3. (They should be small by comparison with (NA + NB + NC) for an accurate simulation.[10]) NT, the largest random number to be generated (corresponding to the total number of molecular locations). FORMAT is 8I0.

A few words of explanation are in order here about the use of the two control integers IMAX and NLOOP both of which affect the simulated time-scale (as also, of course do N1, N2 and N3). IMAX is simply used to terminate the simulation after a set number of complete program cycles; NLOOP, however, can be regarded as an axis-scaling variable so that we only print a result after the 'reaction loops' using N1, N2 and N3 have been entered NLOOP times. Therefore, if N1, N2 and N3 are kept constant, an increase in NLOOP will have the effect of speeding up all the steps in the reaction in that there will be NLOOP times N1 attempts to convert A into B and so on, before the output WRITE statement. NLOOP can be regarded as a crude temperature control!

We next fill a page buffer (lines 24—26) and then fill the array M(2000) with the correct digits (lines 31—36). Random numbers are generated by subroutine RAND(J,INEW,S) and stored in the argument INEW. Note that this subroutine also calls FPMCRV(S). This is a random number generator program which generates numbers from 0 to 1 and is specific to ICL 1900 series machines. We used it for convenience — there was no point in developing a special routine because almost every moderate-sized computer will have a random number generator in its function library. In the unlikely event of this not being so, consult Reference 9. ICL users should note that FPMCRV(S) has to be initialized (in line 30) before calling RAND. The program then proceeds in the expected manner; for example, in the loop from line 45—51, M(INEW) is compared with the integer 1, and if a 'hit' is scored, '1' is converted into a '2' and so on. Output is displayed numerically (line 69) and symbolically (lines 71—96). The latter uses a page buffer, described in Section 1.3.2.

The power of this method is more apparent when applied to systems which are either prohibitively difficult or impossible to solve in a closed form, such as

$$A \underset{k_3}{\overset{k_1}{\longrightarrow}} B + C \overset{k_2}{\longrightarrow} D \quad \text{(difficult)}$$

$$A + B \overset{k_1}{\longrightarrow} C + E$$
$$A + C \overset{k_2}{\longrightarrow} D + E \quad \text{(impossible except in special circumstances)}$$

The conceptual simplicity and ease of application of the Monte Carlo method permits the investigation of both simple and complex schemes.

4.2 Thermodynamics and Equilibria

The subject of thermodynamics is not generally loved by most chemistry students. In this section we present three examples of computer applications which can be used to enhance student interest and appreciation. There are many other examples of applications in this general field, as can be seen from References 12–21 at the end of this chapter.

For a chemical reaction;

$$cC + dD \ldots = pP + qQ \ldots \tag{4.6}$$

we may wish to calculate data relating to molar changes in thermodynamic functions such as enthalpy, ΔH, entropy, ΔS, and free energy, ΔG; the extent to which a reaction has progressed *at equilibrium* is given by the equilibrium constant, K where

$$K_g = \pi_i (f_i)^{\nu_i}$$

i.e.

$$K_g = \frac{(f_P)^p (f_Q)^q \ldots}{(f_C)^c (f_D)^d \ldots}$$

K_g refers to a gas-phase reaction using f_i for the fugacity of species; ν_i is the coefficient for each substance in (4.6) and is positive for reactants, negative for products.

$$K_c = \pi_i (a_i)^{\nu_i}$$

i.e.

$$K_c = \frac{(a_P)^p (a_Q)^q \ldots}{(a_C)^c (a_D)^d \ldots}$$

K_c refers to a solid and/or liquid phase reaction in which the a_i activities (approximately equal to concentrations in dilute solution).

Equilibrium constants can be obtained from standard data or experimental measurements,[22] and are related to ΔG by:

$$\Delta G = -RT \ln K$$

Bearing the above in mind, we now proceed to the first of our examples.

4.2.1 Solution Equilibria

An understanding of solution equilibria is particularly important in the study of acids and bases and of complex formation. Although our examples will be concerned solely with acids and bases, the ideas are easily extended to complex formation which is, in any case, dealt with in the literature.[20,21]

Let us presume that we have a solution of a weak acid, HA, and a strong monoacid base, B, such as sodium hydroxide. We wish to write a computer program to calculate the pH ($-\log_{10}[H^+]$) of any given solution. At equilibrium we have:

(i) the total amount of HA added is $[HA]_T$, so that:

$$[HA]_T = [HA] + [A^-] \qquad (4.7)$$

(ii) The positive and negative charges are equal:

$$[B] + [H^+] = [A^-] + [OH^-] \qquad (4.8)$$

(presuming B to contain a cation with one positive charge).

(iii) The dissociation constant, K_a, of the acid is:

$$K_a = [H^+] [A^-]/[HA] \qquad (4.9)$$

and of water is:

$$K_w = [H^+] [OH^-] \quad (10^{-14} \text{ at } 298 \text{ K}) \qquad (4.10)$$

Note that we are using concentrations, rather than activities which is permissible in dilute solutions.

From (4.7) and (4.9) we obtain:

$$[HA]_T = [HA] (1 + K_a/[H^+]) \qquad (4.11)$$

Whilst from (4.8), (4.9) and (4.10):

$$[B][H^+] + [H^+]^2 = [HA] K_w \qquad (4.12)$$

On combining (4.11) and (4.12), we finally obtain the cubic equation, (4.13)

$$[H^+]^3 + [H^+]^2 (K_a + [B]) + [H^+] ([B] K_a - [HA]_T K_a - K_w)$$
$$- K_a K_w = 0 \qquad (4.13)$$

Although it is possible to solve this equation analytically, we shall use an approximation procedure similar to that of Jurs and Isenhour.[23] Later in this section the same procedure will be applied to a more complex set of equilibria, thereby ensuring continuity.

The program written for this purpose is PHCALC in Table 4.3. It begins (lines 25 and 26) by calculating the minimum possible $[H^+]$ value, XMIN, which is zero, and the maximum, XMAX, which is the sum of $[HA]_T$ and $(K_w)^{\frac{1}{2}}$. Increments are then added to XMIN, to obtain estimates of X (i.e. $[H^+]$) and the value of the function on the left-hand side of (4.13) is computed at each step; so long as this value is negative, the increment is doubled for the next step; when the function changes sign, the two X-values, between which the change in sign occurred are used to bracket the new range of X, with the lower value as XMIN. The step size is then halved and the process is repeated until

Table 4.3
PHCAL — program to calculate the pH of a solution of a weak acid and a strong base

```
0008            MASTER PHCAL
0009       C    THIS IS A PROGRAM TO CALCULATE THE PH OF A SOLUTION CONTAINING A
0010       C    WEAK ACID AND A STRONG BASE
0011       C    THE PH IS CALCULATED BY AN APPROXIMATION PROCEDURE BASED ON
0012       C    LOCATING THE POINT AT WHICH A DEFINED FUNCTION CHANGES SIGN
0013            READ(1,1)AHT,BT,PC
0014          1 FORMAT(3F0.0)
0015            READ(1,10)AKA,AKW
0016         10 FORMAT(2E10.4)
0017            WRITE(2,2)AHT,BT,AKA,AKW,PC
0018          2 FORMAT(1H1,20X,'CALCULATION OF THE PH OF A SOLUTION OF A WEAK ACID
0019          1 AND STRONG BASE',//,120(1H*),//,'  TOTAL ACID CONC = ',F7.3,' MOL
0020          2DM-3,    TOTAL  BASE  CONC, =',F7.3,//,'   ACID  DISSOCIATION CONST
0021          3ANT IS ',F10.4,'   IONIC PRODUCT OF WATER IS ',E10.4,//,'
0022          4          H+  CONCENTRATION  WILL BE DETERMINED  TO AN ACCURACY O
0023          5F ',F5.2,' PER CENT',//,120(1H*))
0024            N=0
0025            XMIN=0
0026            XMAX=AHT+SQRT(AKW)
0027            DELTA=0.1*(XMAX-XMIN)
0028          3 X=XMIN+DELTA
0029            FUNC=X**3+(AKA+BT)*X**2-(AKW+AHT*AKA-BT*AKA)*X-AKA*AKW
0030            N=N+1
0031            IF(FUNC)4,5,6
0032          4 XMIN=X
0033            DELTA=2*DELTA
0034            GO TO 3
0035          6 XMAX=X
0036            IF(((XMAX-XMIN)/XMAX).LE.(PC/100).OR.N.GT.100)GO TO 5
0037            DELTA=DELTA/2.0
0038            GO TO 3
0039          5 PH=-ALOG10(X)
0040            HA=AHT/(1+(AKA/X))
0041            AM=AHT-HA
0042            PHA=-ALOG10(HA)
0043            PAM=-ALOG10(AM)
0044            WRITE(2,7)PH,HA,PHA,AM,PAM
0045          7 FORMAT(1X,'   PH        HA        PHA        AM        PAM',//,1X,5(
0046          1F8.3,2X),//,120(1H*))
0047            STOP
0048            END
```

```
         CALCULATION OF THE PH OF A SOLUTION OF A WEAK ACID AND STRONG BASE
****************************************************************************

TOTAL ACID CONC =   0.355 MOL DM-3,   TOTAL  BASE  CONC, = 0.133

  ACID  DISSOCIATION CONSTANT IS 0.1850E-04    IONIC PRODUCT OF WATER IS 0.1000E-13
            H+  CONCENTRATION  WILL BE DETERMINED  TO AN ACCURACY OF  0.01 PER CENT

****************************************************************************
    PH        HA        PHA        AM        PAM
  4.557     0.200     0.699     0.133     0.875

****************************************************************************
```

the change in X at each iteration is acceptably small. This procedure is illustrated in Figure 4.3. (There are, of course, three solutions to our cubic equation but only the real, positive one is acceptable chemically).

From the calculated solution, X, the pH is calculated, together with the concentrations of HA and A^- (lines 39—44). Input to the program consists of 2 cards:

Card 1: Total concentrations of HA and B; final convergent percentage change in $[H^+]$ (Format 3F0.0).

Card 2: Dissociation constants of HA and of water (Format 2E10.4).

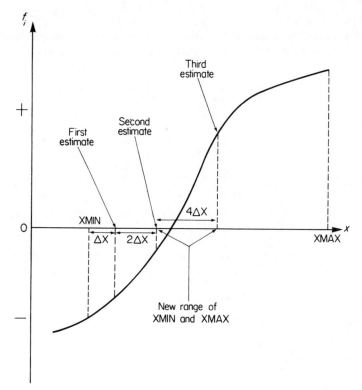

Figure 4.3   Iterative solution of a polynomial, f_i by doubling or halving the step size until f_i is acceptably close to zero

In Table 4.3, we give an example of the use of this program in calculating the pH of a solution of acetic acid ($K_a = 1.85 \times 10^{-5}$) and a strong base. Similar programs can be written for any combination of weak or strong acids or bases.

We will now use the same method in a subroutine to service a more complex program. This program simulates the progress of a titration of a solution of any acid (up to tribasic, H_3A) with a solution of a strong or weak monoacid base. So long as the change in pH with volume, $d(\mathrm{pH})/dV$, is small (<0.5), a further one cubic centimetre of the solution of base is added. If $d(\mathrm{pH})/dV$ is greater than 0.5, the volume of base added is 0.1 cm^3. This nicely follows the procedure adapted in the laboratory and an example of a real-life titration of this type can be seen in Figure 3.2.

A listing of this simulation program, PJBT, is given in Table 4.4. It will deal with either of the following titrations:

(i) Weak acid/strong base.
(ii) Weak acid/weak base.

Table 4.4

PJBT — a program to simulate the titration of an acid (up to triprotic). The example given is for oxalic acid being titrated with sodium hydroxide

```
0008              MASTER PJBT
0009        C     PROGRAM TO FOLLOW THE PROGRESS OF THE TITRATION OF AN ACID(UP TO
0010        C     TRIBASIC) WITH A MONOACID BASE,
0011        C     VARIABLES-K1=1ST DISSOCIATION CONSTANT OF ACID
0012        C               K2=2ND DISSOCIATION CONSTANT OF ACID50.0 FOR MONOBASIC)
0013        C               K3=3RD DISSOCIATION CONSTANT(0.0 FOR MONOOR DI-)
0014        C               KB=DISS.CONST. OF BASE(1000,0 FOR A STRONG BASE)
0015        C               KW=IONIC PRODUCT OF WATER
0016        C               C =ORIGINAL MOLAR CONCENTRZTION OF ACID
0017        C               CN=ORIGINAL MOLAR CONCENTRATION OF BASE
0018        C               V =ORIGINAL VOLUME OF ACID
0019        C               TITLEA = FORMULA OF ACID
0020        C               TITLEB = FORMULA OF BASE
0021        C               QUEST = FINAL VOL. OF BASE = QUEST TIMES VOL. OF ACID
0022              REAL K1,K2,K3,KB,KW
0023              DIMENSION CF(7),X(200),Y(200,9),XNAM(10),YNAM(10)
0024              COMMON CF,K1,K2,K3,CC,CX,CA,CH3A,CH2A,CHA,AM,NUM,KW
0025              READ(1,5)ND,CODE
0026            5 FORMAT(I2,A8)
0027              WRITE(2,6)CODE,ND
0028            6 FORMAT(1H ,5HCODE ,A8,23HNUMBER OF SETS OF DATA ,I2///)
0029              DO 7 JJ=1,ND
0030              READ(1,1)K1,K2,K3,KB,KW,V,C,CN,TITLEA,TITLEB,QUEST
0031            1 FORMAT(8E10.3/2A8,F3.1)
0032              WRITE(2,2)V,TITLEA,C,TITLEB,CN
0033            2 FORMAT(1X,' TITRATION OF ',F6.2,' CM3 OF',A8,' CONC. =',E11.4,' MO
0034           1L DM-3 , WITH ',A8,' CONC. =',E11.4,' MOL DM-3')
0035              DO 10 I=1,200
0036              X(I)=0.0
0037              DO 11 J=1,9
0038              Y(I,J)=0.0
0039           11 CONTINUE
0040           10 CONTINUE
0041              AM=0.0
0042              CREM=1.0
0043              PHP=0.0
0044              N=0
0045              CHECK=QUEST*V
0046              IF(KB-100.0)4,4,3
0047            3 IF(K2,EQ,0)GO TO 301
0048              IF(K3,EQ,0)GO TO 201
0049              WRITE(2,102)K1,K2,K3
0050          102 FORMAT(1H ,5H K1= ,E11.4,5H K2= ,E11.4,5H K3= ,E11.4,15H BASE IS S
0051           1TRONG//)
0052              WRITE(2,103)
0053          103 FORMAT(3X,3HCM3,7X,4HPH3A,10X,5HPH2A-,9X,4HPHA-,10X,4HPA2-,25X,7HP
0054           1H,9X,6HDPH/DV,6X,10HITERATIONS/)
0055          104 N=N+1
0056              CC=C*V/(V+AM)
0057              CCN=CN*AM/(V+AM)
0058              CF(7)=0.0
0059              CF(6)=1.0
0060              CF(5)=K1+CCN
0061              CF(4)=K1*(CCN+K2-CC)-KW
0062              CF(3)=K1*(K2*(CCN+K3-2.0*CC)-KW)
0063              CF(2)=K1*K2*(K3*(CCN-3.0*CC)-KW)
0064              CF(1)=-(K1*K2*K3*KW)
0065              CALL PPHT
0066              Y(N,1)=-(ALOG10(CX))
0067              DPH=(Y(N,1)-PHP)/CREM
0068              X(N)=AM
0069              Y(N,2)=-(ALOG10(CH3A))
0070              Y(N,3)=-(ALOG10(CH2A))
0071              Y(N,4)=-(ALOG10(CHA))
0072              Y(N,5)=-(ALOG10(CA))
0073              IF(AM)105,105,107
0074          105 WRITE(2,106)AM,Y(N,2),Y(N,3),Y(N,4),Y(N,5),Y(N,1),NUM
0075          106 FORMAT(1H ,F6.2,4F14.4,14X,F7.4,25X,I5)
0076              GO TO 109
0077          107 WRITE(2,108)AM,Y(N,2),Y(N,3),Y(N,4),Y(N,5),Y(N,1),DPH,NUM
0078          108 FORMAT(1H ,F6.2,4F14.4,14X,F7.4,9X,F7.4,9X,I5)
0079          109 PHP=Y(N,1)
0080              IF(DPH-0.5)110,110,111
0081          110 CREM=1.0
0082              GO TO 112
```

```
0083          111 CREM=0,1
0084          112 AM=AM+CREM
0085              IF(N,GT,200)GO TO 7
0086              IF(AM,GE,CHECK)GO TO 7
0087              GO TO 104
0088          201 WRITE(2,202)K1,K2
0089          202 FORMAT(1H ,5H K1= ,E11,4,5H K2= ,E11,4,15H BASE IS STRONG//)
0090              WRITE(2,203)
0091          203 FORMAT(3X,3HCM3,7X,4HPH2A,10X,4HPHA-,10X,4HPA2-,25X,2HPH,9X,6HDPH/
0092             1DV,6X,10HITERATIONS/)
0093          204 N=N+1
0094              CC=C*V/(V+AM)
0095              CCN=CN*AM/(V+AM)
0096              CF(7)=0,0
0097              CF(6)=0,0
0098              CF(5)=1,0
0099              CF(4)=CCN+K1
0100              CF(3)=K1*(K2+CCN-CC)-KW
0101              CF(2)=K1*(K2*(CCN-2,0*CC)-KW)
0102              CF(1)=-(K1*K2*KW)
0103              CALL PPHT
0104              Y(N,1)=-(ALOG10(CX))
0105              DPH=(Y(N,1)-PHP)/CREM
0106              X(N)=AM
0107              Y(N,2)=-(ALOG10(CH2A))
0108              Y(N,3)=-(ALOG10(CHA))
0109              Y(N,4)=-(ALOG10(CA))
0110              IF(AM)205,205,207
0111          205 WRITE(2,206)AM,Y(N,2),Y(N,3),Y(N,4),Y(N,1),NUM
0112          206 FORMAT(1H ,F6,2,3F14,4,14X,F7,4,25X,I5)
0113              GO TO 209
0114          207 WRITE(2,208)AM,Y(N,2),Y(N,3),Y(N,4),Y(N,1),DPH,NUM
0115          208 FORMAT(1H ,F6,2,3F14.4,14X,F7,4,9X,F7,4,9X,I3)
0116          209 PHP=Y(N,1)
0117              IF(DPH-0,5)210,210,211
0118          210 CREM=1,0
0119              GO TO 212
0120          211 CREM=0,1
0121          212 AM=AM+CREM
0122              IF(N,GT,200)GO TO 7
0123              IF(AM,GE,CHECK)GO TO 7
0124              GO TO 204
0125          301 WRITE(2,302)K1
0126          302 FORMAT(1H ,5H K1= ,E11,4,15H BASE IS STRONG//)
0127              WRITE(2,303)
0128          303 FORMAT(3X,3HCM3,7X,3HPHA,11X,3HPA-,25X,2HPH,9X,6HDPH/DV,6X,10HIT
0129             1ATIONS/)
0130          304 N=N+1
0131              CC=C*V/(V+AM)
0132              CCN=CN*AM/(V+AM)
0133              CF(7)=0,0
0134              CF(6)=0,0
0135              CF(5)=0,0
0136              CF(4)=1,0
0137              CF(3)=K1+CCN
0138              CF(2)=KW+K1*(CCN-CC)
0139              CF(1)=-(K1*KW)
0140              CALL PPHT
0141              Y(N,1)=-(ALOG10(CX))
0142              DPH=(Y(N,1)-PHP)/CREM
0143              X(N)=AM
0144              Y(N,2)=-(ALOG10(CHA))
0145              Y(N,3)=-(ALOG10(CA))
0146              IF(AM)305,305,307
0147          305 WRITE(2,306)AM,Y(N,2),Y(N,3),Y(N,1),NUM
0148          306 FORMAT(1H ,F6,2,2F14.4,14X,F7,4,25X,I5)
0149              GO TO 309
0150          307 WRITE(2,308)AM,Y(N,2),Y(N,3),Y(N,1),DPH,NUM
0151          308 FORMAT(1H ,F6,2,2F14.4,14X,F7,4,9X,F7,4,9X,I3)
0152          309 PHP=Y(N,1)
0153              IF(DPH-0,5)310,310,311
0154          310 CREM=1,0
0155              GO TO 312
0156          311 CREM=0,1
0157          312 AM=AM+CREM
0158              IF(N,GT,200)GO TO 7
0159              IF(AM,GE,CHECK)GO TO 7
0160              GO TO 304
0161            4 IF(K2,EQ,0,0)GO TO 601
0162              IF(K3,EQ,0,0)GO TO501
0163              WRITE(2,402)K1,K2,K3,KB
0164          402 FORMAT(1H ,5H K1= ,E11,4,5H K2= ,E11,4,5H K3= ,E11,4,5H KB= ,E11,4
```

```
0165                    1//)
0166                    WRITE(2,103)
0167            404 N=N+1
0168                    CC=C*V/(V+AM)
0169                    CCN=CN*AM/(V+AM)
0170                    CF(7)=1.0
0171                    CF(6)=CCN+KW/KB+K1
0172                    CF(5)=K1*(CCN+KW/KB+K2-CC)-KW
0173                    CF(4)=K1*K2*(CCN+KW/KB+K3-2.0*CC)-K1*KW-KW*KW/KB-CC*K1*KW/KB
0174                    CF(3)=K1*(K2*(K3*(CCN+KW/KB-3.0*CC)+(KW-2.0 *CC*KW/KB))-KW*KW/KB)
0175                    CF(2)==(K1*K2*(KW*KW/KB+ K3*KW+3.0*CC*K3*KW/KB))
0176                    CF(1)==(K1*K2*K3*KW*KW/KB)
0177                    CALL PPHT
0178                    Y(N,1)=-(ALOG10(CX))
0179                    DPH=(Y(N,1)-PHP)/CREM
0180                    X(N)=AM
0181                    Y(N,2)=-(ALOG10(CH3A))
0182                    Y(N,3)=-(ALOG10(CH2A))
0183                    Y(N,4)=-(ALOG10(CHA))
0184                    Y(N,5)=-(ALOG10(CA))
0185                    IF(AM)405,405,407
0186            405 WRITE(2,106)AM,Y(N,2),Y(N,3),Y(N,4),Y(N,5),Y(N,1),NUM
0187                    GO TO 409
0188            407 WRITE(2,108)AM,Y(N,2),Y(N,3),Y(N,4),Y(N,5),Y(N,1),DPH,NUM
0189            409 PHP=Y(N,1)
0190                    IF(DPH-0.5)410,410,411
0191            410 CREM=1.0
0192                    GO TO 412
0193            411 CREM=0.1
0194            412 AM=AM+CREM
0195                    IF(N.GT.200)GO TO 7
0196                    IF(AM.GE.CHECK)GO TO 7
0197                    GO TO 404
0198            501 WRITE(2,502)K1,K2,KB
0199            502 FORMAT(1H ,5H K1= ,E11.4,5H K2= ,E11.4,5H KB= ,E11.4//)
0200                    WRITE(2,203)
0201            504 N=N+1
0202                    CC=C*V/(V+AM)
0203                    CCN=CN*AM/(V+AM)
0204                    CF(7)=0.0
0205                    CF(6)=1.0
0206                    CF(5)=KW/KB+CCN+K1
0207                    CF(4)=K1*(KW/KB+CCN+K2-CC)
0208                    CF(3)=K1*(K2*(KW/KB+CCN-2.0*CC)-KW*(1.0+CC/KB))-KW*KW/KB
0209                    CF(2)==(K1*(KW*(KW/KB+K2*(1.0+2.0*CC/KB))))
0210                    CF(1)==(K1*K2*KW*KW/KB)
0211                    CALL PPHT
0212                    Y(N,1)=-(ALOG10(CX))
0213                    DPH=(Y(N,1)-PHP)/CREM
0214                    X(N)=AM
0215                    Y(N,2)=-(ALOG10(CH2A))
0216                    Y(N,3)=-(ALOG10(CHA))
0217                    Y(N,4)=-(ALOG10(CA))
0218                    IF(AM)505,505,507
0219            505 WRITE(2,206)AM,Y(N,2),Y(N,3),Y(N,4),Y(N,1),NUM
0220                    GO TO 509
0221            507 WRITE(3,208)AM,Y(N,2),Y(N,3),Y(N,4),Y(N,1),DPH,NUM
0222            509 PHP=Y(N,1)
0223                    IF(DPH-0.5)510,510,511
0224            510 CREM=1.0
0225                    GO TO 512
0226            511 CREM=0.1
0227            512 AM=AM+CREM
0228                    IF(N.GT.200)GO TO 7
0229                    IF(AM.GE.CHECK)GO TO 7
0230                    GO TO 504
0231            601 WRITE(2,602)K1,KB
0232            602 FORMAT(1H ,5H K1= ,E11.4,5H KB= ,E11.4//)
0233                    WRITE(2,503)
0234            604 N=N+1
0235                    CC=C*V/(V+AM)
0236                    CCN=CN*AM/(V+AM)
0237                    CF(7)=0.0
0238                    CF(6)=0.0
0239                    CF(5)=1.0
0240                    CF(4)=K1+CCN+KW/KB
0241                    CF(3)=K1*(CCN-CC+KW/KB)-KW
0242                    CF(2)==(KW*(K1+KW/KB+K1*CC/KB))
0243                    CF(1)==(KW*KW*K1/KB)
0244                    CALL PPHT
0245                    Y(N,1)=-(ALOG10(CX))
0246                    DPH=(Y(N,1)-PHP)/CREM
```

```
0247                     X(N)=AM
0248                     Y(N,2)=-(ALOG10(CHA))
0249                     Y(N,3)=-(ALOG10(CA))
0250                     IF(AM)605,605,607
0251              605    WRITE(2,506)AM,Y(N,2),Y(N,3),Y(N,1),NUM
0252                     GO TO 609
0253              607    WRITE(2,508)AM,Y(N,2),Y(N,3),Y(N,1),DPH,NUM
0254              609    PHP=Y(N,1)
0255                  !  IF(DPH-0,5)610,610,611
0256              610    CREM=1,0
0257                     GO TO 612
0258              611    CREM=0,1
0259              612    AM=AM+CREM
0260                     IF(N,GT,200)GO TO 7
0261                     IF(AM,GE,CHECK)GO TO 7
0262                     GO TO 604
0263                7    CONTINUE
0264                     STOP
0265                     END



0266                     SUBROUTINE  PPHT
0267             C       REQUIRED BY PROGRAM #PJBT
0268             C       TO CALCULATE IONIC CONCENTRATIONS
0269                     REAL K1,K2,K3
0270                     DIMENSION CF(7)
0271                     COMMON CF,K1,K2,K3,CC,CX,CA,CH3A,CH2A,CHA,AM,NUM,KW
0272                     NUM=0
0273                     PC=1,0
0274                     XMIN=0
0275                     XMAX=CC+SQRT(KW)
0276                     DELTA=0,1*(XMAX-XMIN)
0277                3    X=XMIN+DELTA
0278                     Y=X*(X*(X*(X*(X*(CF(7)*X+CF(6))+CF(5))+CF(4))+CF(3))+CF(2))+CF(1)
0279                     NUM=NUM+1
0280                     IF(Y)4,809,6
0281                4    XMIN=X
0282                     DELTA=2*DELTA
0283                     GO TO 3
0284                6    XMAX=X
0285                     IF((((XMAX-XMIN)/XMAX),LE,(PC/100),OR,NUM,GT,100)GO TO 809
0286                     DELTA=DELTA/2,0
0287                     GO TO 3
0288              809    IF(K2,EQ,0,0) GO TO 811
0289                     IF(K3,EQ,0,0)GO TO 810
0290                     CH3A=CC*X**3/(K1*K2*K3+K1*K2*X+K1*X*X+X**3)
0291                     CH2A=K1*CH3A/X
0292                     CHA=K2*CH2A/X
0293                     CA=K3*CHA/X
0294                     GO TO 812
0295              810    CH2A=CC*X*X/(K1*K2+K1*X+X*X)
0296                     CHA=K1*CH2A/X
0297                     CA=K2*CHA/X
0298                     GO TO 812
0299              811    CHA=CC*X/(K1+X)
0300                     CA=K1*CHA/X
0301              812    CX=X
0302                     RETURN
0303                     END
```

TITRATION OF   5.00 CM3 OF C2O4H2 CONC. = 0.1000E 00 MOL DM-3 , WITH NAOH   CONC. = 0.1000E 00 MOL DM-3
K1= 0.5890E-01 K2= 0.5250E-04 BASE IS STRONG

CM3	PH2A	PHA-	PA2-	PH	DPH/DV	ITERATIONS
0.00	1.3256	1.2782	4.2808	1.2773	0.1638	14
0.10	1.3429	1.2792	4.2654	1.2936	0.1740	16
1.10	1.5228	1.2850	4.0972	1.4677	0.1898	14
2.10	1.7185	1.2909	3.9133	1.6574	0.2377	14
3.10	1.9612	1.2960	3.6807	1.8951	0.3549	15
4.10	2.3235	1.3034	3.3332	2.2500	0.7846	12
5.10	3.1405	1.3358	2.5810	3.0546	1.0062	23
5.20	3.2501	1.3447	2.4893	3.1352	0.9202	17
5.30	3.3521	1.3547	2.4073	3.2273	0.8327	21
5.40	3.4517	1.3660	2.3303	3.3155	0.7747	19
5.50	3.5410	1.3778	2.2647	3.3930	0.7148	22
5.60	3.6249	1.3903	2.2057	3.4645	0.6332	18
5.70	3.7010	1.4030	2.1551	3.5278	0.6108	16
5.80	3.7755	1.4166	2.1075	3.5889	0.5659	20
5.90	3.8460	1.4304	2.0648	3.6455	0.5164	18
6.00	3.9117	1.4444	2.0272	3.6971	0.4942	21
6.10	3.9756	1.4589	1.9922	3.7465	0.3953	19
7.10	4.5336	1.6216	1.7596	4.1419	0.3510	22
8.10	5.1018	1.8389	1.6259	4.4928	0.4435	21
9.10	5.8997	2.1933	1.5368	4.9363	5.8838	24
10.10	17.6106	8.0203	1.4800	10.8202	2.9840	44
10.20	18.2103	8.5216	1.4829	11.1186	1.7082	47
10.30	18.5547	8.4953	1.4857	11.2894	1.2510	47
10.40	18.8078	8.6232	1.4886	11.4145	0.9115	44
10.50	18.9929	8.7171	1.4914	11.5056	0.7743	42
10.60	19.1505	8.7974	1.4942	11.5830	0.5522	50
10.70	19.2837	8.8654	1.4969	11.6483	0.5503	45
10.80	19.3966	8.9231	1.4997	11.7053	0.4794	49
10.90	19.4952	8.9738	1.5024	11.7512	0.2977	47
11.90	20.1171	9.2981	1.5289	12.0490	0.5587	48
12.90	20.4595	9.4817	1.5539	12.2077	0.1056	50
13.90	20.6944	9.6110	1.5775	12.3133	0.0773	46
14.90	20.8715	9.7107	1.5999	12.3907	0.0595	49
15.90	21.0118	9.7915	1.6212	12.4502	0.0478	47
16.90	21.1277	9.8596	1.6415	12.4980	0.0397	47
17.90	21.2266	9.9187	1.6609	12.5377	0.0306	49
18.90	21.3064	9.9679	1.6794	12.5683	0.0302	49
19.90	21.3846	10.0159	1.6972	12.5985		47

In either case, the weak acid which is being titrated may be mono-, di-, or tri-basic. For those who wish to follow the operation of PJBT,[24] we can examine the case of a titration of a weak tribasic acid, H_3A, with a strong base, B. The equilibria involved are:

$$H_3A \rightleftharpoons H^+ + H_2A^- \quad : \quad K_1 = [H^+][H_2A^-]/[H_3A]$$
$$H_2A^- \rightleftharpoons H^+ + HA^{2-} \quad : \quad K_2 = [H^+][HA^{2-}]/[H_2A^-]$$
$$HA^{2-} \rightleftharpoons H^+ + A^{3-} \quad : \quad K_3 = [H^+][A^{3-}]/[HA^{2-}]$$

By writing down the equations of charge balance and of mass balance (and after some manipulation), the reader should be able to verify that the H^+ concentration is given by:

$$
\begin{aligned}
[H^+]^5 &+ [H^+]^4(K_1 + [B]) + [H^+]^3(K_1K_2 + K_1[B] - K_w - [H_3A]_T K_1) \\
&+ [H^+]^2(K_1K_2K_3 + K_1K_2[B] - K_1K_w - 2K_1K_2[H_3A]_T) \\
&+ [H^+](K_1K_2K_3 - K_1K_2K_w - 3K_1K_2K_3[H_3A]) \\
&- K_1K_2K_3K_w = 0
\end{aligned}
\tag{4.14}
$$

This fifth-order polynomial is incapable of simple analytical solution and this is why we developed the program PHCALC since it forms the basis of the subroutine PPHT to service PJBT. Inspection of lines 58–64 shows that the coefficients of (4.14) are stored in the array CF before calling PPHT. Most of the notation for the variables is fairly obvious, and much is explained in comment cards. CREM is the volume increment of base; C is the total concentration of acid, and CCN is the total concentration of base, both corrected for dilution. Note also that the subroutine PPHT calculates the concentrations of H_3A, H_2A^- and A^{3-} by use of the equilibria relationships used in the development of (4.14). Lines 274–287 are identical to lines in program PHCALC, with the exception of line 278 in which the function to be minimized is defined.

Input to PJBT is as follows:

Card 1: Number of sets of data (ND) and short title (CODE) in Format (I2, A8).

Card 2: First, second and third dissociation constants K_1, K_2 and K_3 (set equal to zero where appropriate); dissociation constant of water; dissociation constant, K of the base (set equal to 1000.0 if base is strong); original volume, V, of acid; concentration of acid; concentration of base; Format 8E10.3.

Card 3: Formula of acid; formula of base; factor which, when multiplied by V, will give the total volume of base to be added. (Format 2A8, F3.1).

PJBT can be a very effective teaching aid. An example of the output for a typical simulation is given in Table 4.4. A useful extension to

PJBT would be to provide graphical output, using a line printer or plotter, to display the concentration of each species.

4.2.2 Ionic Solids

Almost every undergraduate course in inorganic chemistry includes a study of ionic bonding in reasonable depth. Textbooks[25,26] also discuss experimental determinations and theoretical calculations of thermodynamic parameters involved in ionic bonding, with particular emphasis being placed on the usefulness and accuracy of suitable equations. The most important thermodynamic property is the 'Lattice Energy' which is defined as the energy change, ΔU, for the conversion of one mole of a crystalline ionic compound into the gasious ions at infinite separation. For a 1:1 compound, this process is:

$$MX(cryst) \longrightarrow M^{n^+}_{(g)} + X^{n^-}_{(g)}$$

Ionic compounds crystallize in a variety of forms, differing in their arrangements of cations and anions (see Figure 4.4). For example, in the rock-salt structure, as adopted by many 1:1 compounds such as NaCl, each cation is surrounded by six anions at a distance, r, the sum of the cation and anion radii. This interaction is attractive and gives an electrostatic energy, U, per cation of

$$U = \frac{-6z^+z^-e^2}{r \cdot 4\pi\epsilon_o}$$

where z^+ and z^- are numerical charges on the cation and anion in terms of e, the electronic charge; ϵ_0 is the permittivity of free space. By considering the destabilizing effect of the next set of nearest neighbours (12 cations at $\sqrt{2}r$), the stabilizing effect of the next set of anions and so on, we eventually obtain a series expansion for the total electrostatic energy per ion of the form:

$$U_T = \frac{-z^+z^-e^2}{r \cdot 4\pi\epsilon_o} \left(6 - \frac{12}{\sqrt{2}} + \frac{8}{\sqrt{3}} - \frac{6}{2} \ldots \text{etc.} \right) \tag{4.15}$$

The slowly converging series in (4.15) has a fixed value for this, the rock salt, and other crystal structures. The value of the series is called the Madelung Constant and it has been calculated with high precision by computer for all of the common crystal structures[27] and even for some of the less common ones. The value of the Madelung constant will be denoted by A, typical values being given in Table 4.5.

We also have to consider the repulsive effects arising from the interactions between the electron clouds and between the nuclei of adjacent ions. These forces dominate at short internuclear distances, and one simple expression which is fairly satisfactory is of the form:

$$U_R = + B/r^n \tag{4.16}$$

In this equation, B is a constant and n is typically between 7 and 11. The lattice energy U_1, is the sum of U_T and U_R. For one mole we have:

$$U_1 = U_T + U_R = \frac{N_A A z^+ z^- e^2}{r \cdot 4\pi\epsilon_o} + \frac{B}{r^n} \qquad (4.17)$$

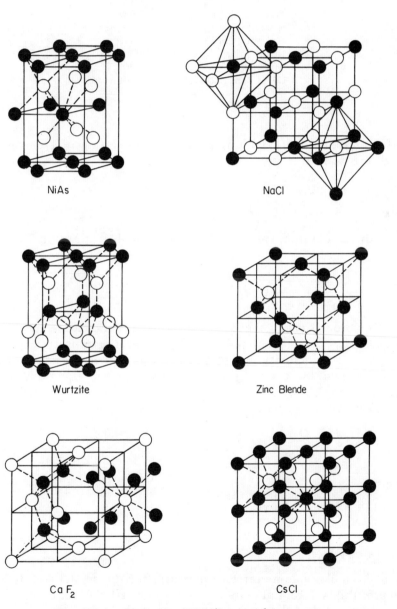

NiAs

NaCl

Wurtzite

Zinc Blende

Ca F$_2$

CsCl

Figure 4.4   Some representative crystal structures

Table 4.5
Values of the madelung constant
for some common crystal structures

Structure	A
Rock salt	1.74756
Caesium chloride	1.76267
Zinc blende	1.63805
Wurtzite	1.64132
Fluorite (CaF_2)	2.51939
Rutile (TiO_2)	2.3850
β-Quartz	2.201
Corundum (Al_2O_3)	4.040

Differentiation of U_1 with respect to r and setting $\partial U_1/\partial r$ to zero, yields the value of B

$$B = -\frac{N_A A z^+ z^- e^2}{4\pi\epsilon_0 n \cdot r_0^{(1-n)}}$$

(r_0 is the equilibrium value of r.) On inserting the expression for B into (4.17), we find

$$U_1 = \frac{N_A A z^+ z^- e^2}{r_0 \cdot 4\pi\epsilon_0}(1^{-1}/n)$$

This equation is called the Born expression for the lattice energy. It is undoubtedly the most popular equation for theoretical calculations because of its simplicity. In order to proceed further we shall adopt the notation used by Lister[28] (the derivation of the equations in this reference are clarified by reading some background material[29]); in Lister's notation, we obtain

$$U_1 = \frac{M}{r_0}\left(\frac{Q}{Q+M}\right) \tag{4.18}$$

where

$$M = N_A A z^+ z^- e^2/4\pi\epsilon_0$$

and $\tag{4.19}$

$$Q = 9V_0 r_0/\beta_0$$

V_0 is the molar volume (at zero pressure) for the crystal and β_0 is the compressibility (again at zero pressure). Some tabulations of V_0 and β_0 are available[30] but these are rather sparse so that it is convenient for us to note that:

$$n = 1 + Q/M \tag{4.20}$$

where n is the exponent in (4.16). Equation (4.18) is relatively insensitive to the value of n which is typically 7 to 11, so from an assumed value of n, Q can be estimated from equation (4.20) (although (4.19) should be used whenever data is available). The usefulness of the Lister notation is that we can take more complex equations and express them in a simple standardized form. For example, the Born–Mayer equation:

$$U_2 = \frac{M}{r_0}(1 - \rho_r)$$

becomes

$$U_2 = \frac{M}{r_0}\left(\frac{Q + M}{Q + 2M}\right) \tag{4.21}$$

Clearly this equation is related to (4.18) and uses the same parameters, Q and M. Similarly, the Born equation with Van der Waals term has the inconvenient (non-equilibrium) form:

$$E = \frac{M}{r} - \frac{C'}{r^6} + \frac{B''}{r^2} \tag{4.22}$$

but this becomes:

$$U_3 = \frac{M}{r_0}\left[\frac{Q + DQ/M + 25D}{Q + M + 36D}\right] \tag{4.23}$$

and the Born–Mayer equation with Van der Waals term is

$$U_4 = \frac{M}{r_0}\left[\frac{Q + M + D(Q/M + 32) + 6D^2/M}{Q + 2M + 42D}\right] \tag{4.24}$$

In (4.23) and 4.24), D is C'/r_0^5 in which C' is the same parameter as in equation (4.22). Values of C' are available in the literature[31] and some values of D, quoted by Lister [28] are given in Table 4.6.

The equations are now starting to become fairly complex but are still able to be solved on a calculator rather than a computer. To make it a more feasible computer problem, we will write a simple simulation program that will:

(i) Deal with any of the common crystal types.
(ii) Calculate the lattice energy of that crystal for fixed values of z^+, z^- and n over a defined range of internuclear distances.
(iii) Calculate the equilibrium values of U_1, U_2, U_3 and U_4 for

Table 4.6
Values of Q and D for some alkali halides

Compound	$10^3 \times Q/Jm$	$10^6 \times D/Jm$
NaCl	1.667	6.144
NaBr	1.848	6.933
NaI	1.931	8.244
CsCl	2.574	16.11
CsBr	2.767	17.67
CsI	2.833	18.60

comparison purposes (U_3 and U_4 will only be calculated if D in equations (4.23) and (4.24) is known).

(iv) Calculate the lattice energy from the approximate Kapustinskii equation:[25]

$$U_5 = \frac{108z^+z^-\nu}{r_0} \text{ kJ mol}^{-1}$$

where ν is the total number of ions in the simplest formula of the compound (e.g. $\nu = 3$ for CaF_2).

A program that fulfills these objectives is LATT in Table 4.7. Input to this program is as follows:

Card 1: Descriptive title (Format 10A8).

Card 2: MAD = lattice index (MAD = 1 (NaCl), 2 (CsCl), 3 (zinc blende), 4 (wurtzite), 5 (CaF_2), 6 (rutile), 7 (β-quartz), 8 (corundum)) N = exponent in the Born Repulsive term (equation (4.16)); NP = number of positive ions; NN = number of negative ions; NU = total number of ions; RO = equilibrium internuclear separation in nanometers; RL, RU = lower and upper values of internuclear separation in nanometres; D = value of C'/r_0^5 in equations (4.23) and (4.24); ID = index which causes U_3 and U_4 to be calculated if ID is greater than zero. (Format 5I0, 4F0.0, I0).

We can test this program with data for potassium chloride. This has a sodium chloride (rock salt) type lattice with MAD equal to 1. We also know that NP and NN are 1 and r_0 is 0.314 nm. An acceptable value for N is 9 and D can be estimated to be approximately 8×10^{-4} J m. The output from this program shows:

(a) The steeper rise of E, the total energy for $r < r_0$ compared with the shallower rise for $r > r_0$. (Note that E is the total electrostatic energy and is, therefore, mainly negative.)

(b) The precise agreement in finding the minimum value of E at the equilibrium internuclear distance.

Table 4.7
LATT — for the calculation of lattice energies and the effects of ionic separation

```
0008              MASTER LATT
0009              REAL N0,M(8),MD
0010              DIMENSION H(10)
0011              READ(1,1)(H(I),I=1,10)
0012            1 FORMAT(10A8)
0013              READ(1,3)MAD,N,NP,NN,NU,R0,RL,RU,D,ID
0014            3 FORMAT(5I0,4F0,0,I0)
0015              WRITE(2,2)(H(I),I=1,10),MAD,N,NP,NN,D
0016            2 FORMAT(1H1,10A8,//,' LATTICE TYPE;',I2,'; BORN EXPONENT=',I2,';NP,N
0017           1N=',2I2,'; D COEFFICIENT=',E10,4)
0018              WRITE(2,4)RL,RU,RU
0019            4 FORMAT(1X,100(1H*), /,' DYNAMIC BEHAVIOUR OF THE BORN EQUATION FOR
0020           1 R= ',F7,3,' TO',F7,3,' NM',/,' EQUILIBRIUM VALUE =',F7,3,' NM')
0021         C    KEY
0022         C
0023         C    MAD=1(NACL),2(CSCL),3(ZN BLENDE),4(WURTZITE),5(CAF2),6(RUTILE),
0024         C    7(B-QUARTZ),8(CORUNDUM
0025         C
0026         C    N=EXPONENT IN BORN REPULSIVE TERM; NP AND NN = CATION AND ANION
0027         C    CHARGES; NU = NUMBER OF IONS; R0 =EQBM DISTANCE
0028         C                                 RL AND RU LOWER AND UPPER LIMITS
0029         C                                  ( ALL  IN NM)
0030         C    D IS  C/R**5 IN THE BORN EQTN WITH VAN DER WAAL TERM BUT NEED ONLY
0031         C    BE INPUT WHEN ID>0
0032         C
0033              M(1)=1,74756
0034              M(2)=1,76267
0035              M(3)=1,65805
0036              M(4)=1,64132
0037              M(5)=2,51939
0038              M(6)=2,385
0039              M(7)=2,201
0040              M(8)=4,04
0041              MD=M(MAD)
0042              E2=2,54*(10,**(-38))
0043              N0=6,023*(10,**23)
0044              PI=3,1416
0045              E0=8,854*(10,**(-12))
0046              CT=(NU*MD*NP*NN*E2)/(4*E0*PI*1000)
0047              WRITE(2,50)
0048           50 FORMAT(1H ,20X,'    R/NM              ENERGY/KJ MOL-1',//)
0049              RS=RU*(10,**(-9))
0050              DO 51 I=1,41
0051              R=RL+(I-1)*(RU-RL)/40
0052              U=-CT*(10,**9)*((1/R)-((RU**(N-1))*(R**(-N)))/N)
0053           51 WRITE(2,52)R,U
0054           52 FORMAT(21X,F6,3,16X,F9,1)
0055              WRITE(2,53)
0056           53 FORMAT(////,100(1H*),///,1X,'   HAVING  LOOKED AT THE DYNAMIC BEHAV
0057           1IOUR OF THE BORN EQUATION WE NOW COMPARE',/,1X,'   THE EQUILIBRIUM
0058           2 VALUES PREDICTED BY SOME OTHER FUNCTIONS',////,'
0059           3ALL ANSWERS IN KJ MOL-1',///,'          BORN          BORN MAYER
0060           4          KAPUSTINSKII')
0061         C
0062         C    DEFINE FM AND Q
0063         C
0064              EM=NU*MD*E2*NP*NN/(4*PI*E0)'
0065              Q=EM*(N-1)
0066              U1=(EM/RS)*(Q/(Q+EM))/1000
0067              U2=(EM/RS)*(Q+EM)/(Q+2*EM)/1000
0068              IF(ID)54,54,55
0069           55 U3=(EM/RS)*(Q+D*Q/EM+25*D)/(Q+EM+36*D)/1000
0070              U4=(EM/RS)*(Q+EM+D*(Q/EM+32)+6*D*D/EM)/(Q+2*EM+42*D)/1000
0071           54 U5=108*NP*NN*NU/R0
0072              WRITE(2,56)U1,U2,U5
0073           56 FORMAT(1X,3(5X,F12,1))
0074              IF(ID,GT,0)WRITE(2,59)
0075           59 FORMAT(1X,'BORN WITH VAN DER WAALS      BORN MAYER WITH VAN DER WAA
0076           1LS')
0077              IF(ID,GT,0)WRITE(2,58)U3,U4
0078           58 FORMAT(1X,2(5X,F12,1))
0079              STOP
0080              END
```

122

```
KCL AS AN EXAMPLE

LATTICE TYPE 1; BORN EXPONENT= 9;NP,NN= 1 1; D COEFFICIENT=0.8000E-05
*********************************************************************************
DYNAMIC BEHAVIOUR OF THE BORN EQUATION FOR R=    0.200 TO   0.400 NM
EQUILIBRIUM VALUE =   0.314 NM
                 R/NM                    ENERGY/KJ MOL-1


                 0.200                    3726.4
                 0.205                    2773.7
                 0.210                    2032.3
                 0.215                    1452.6
                 0.220                     997.7
                 0.225                     639.2
                 0.230                     356.1
                 0.235                     131.8
                 0.240                     -46.2
                 0.245                    -187.5
                 0.250                    -299.7
                 0.255                    -388.9
                 0.260                    -459.5
                 0.265                    -515.3
                 0.270                    -559.1
                 0.275                    -593.3
                 0.280                    -619.7
                 0.285                    -639.7
                 0.290                    -654.7
                 0.295                    -665.4
                 0.300                    -672.8
                 0.305                    -677.4
                 0.310                    -679.7
                 0.315                    -680.2
                 0.320                    -679.2
                 0.325                    -677.0
                 0.330                    -673.8
                 0.335                    -669.8
                 0.340                    -665.2
                 0.345                    -660.0
                 0.350                    -654.5
                 0.355                    -648.7
                 0.360                    -642.6
                 0.365                    -636.4
                 0.370                    -630.0
                 0.375                    -623.6
                 0.380                    -617.1
                 0.385                    -610.5
                 0.390                    -604.0
                 0.395                    -597.5
                 0.400                    -591.1


*********************************************************************************

        HAVING  LOOKED AT THE DYNAMIC BEHAVIOUR OF THE BORN EQUATION WE NOW COMPARE
        THE EQUILIBRIUM VALUES PREDICTED BY SOME OTHER FUNCTIONS


                ALL ANSWERS IN KJ MOL-1

            BORN            BORN MAYER          KAPUSTINSKII
            680.2             688.7               687.9
    BORN WITH VAN DER WAALS     BORN MAYER WITH VAN DER WAALS
            682.7                 694.1
```

(c) The fact that, on going from U_1 to U_4 (adding more terms) we find that the computed value of U *increases*. The value of U_4 (694.1 kJ mol^{-1}) is closest to the experimental value (Reference 22, p. 42) of 699 kJ mol.

Several changes could be made to this program to suit the needs of the user — for example, you could:

123

Input the value of Q rather than calculate it from 4.20 as we do.
Allow z^+ and z^- to vary in the simulation to show their effect.
Provide graphical output (printer or plotter) to show the dependence
of U on r.

With the program as it stands, however, you should be able to
improve the teaching or learning of energetics of solids, so that a feeling
for the relative importance of the variables can be obtained. The
program is so simple that no explanation is necessary. The main
characteristic is the branching point which transfers control to the
various parts of the program.

4.2.3 Statistical Mechanics
One particular aspect of thermodynamics which is particularly suited to
computer-assistance is statistics. By this we mean the statistical
calculation of thermodynamic functions. The unique power of a
computer in this area is its ability to simulate large numbers of
statistical occurrences in a short time. This can effectively reinforce
one's understanding of a statistical model.

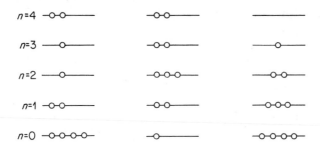

Figure 4.5   Distribution of 10 particles amongst 5
energy levels (n is the quantum number)

A suitable problem is concerned with a statistical modelling of the
Boltzmann distribution. We usually consider the distribution of
particles (atoms or molecules) among discrete energy levels which, for
simplicity we shall presume are equally spaced. Some examples of
possible distributions for a small system are shown in Figure 4.5. The
classical method maximizes W, the number of ways in which a
particular distribution can be obtained, subject to the constraints of
constant energy and constant number of particles. The derivation
normally requires a knowledge of Stirling's approximation and the
method of undetermined multipliers to obtain the result:

$$n_i = n_0 e^{-\epsilon i/kT}$$

where n_i is the number of particles in energy level i, with energy ϵ_i, and

n_0 is the number in the lowest ($\epsilon_0 = 0$) level, T is the absolute temperature and k is Boltzmann's constant.

We can write a program to simulate the statistical behaviour of such a system. For simplicity the program will use five equally spaced energy levels (this number can easily be increased) and therefore, corresponds to a harmonic oscillator with five energy levels. Any number of particles having a defined total energy are to be used for the simulation. The program STAT1 (Table 4.8) explores, in a systematic way, the possible distributions which can arise when the particles are placed in the energy levels. The number of ways in which any particular distribution W_i can be obtained is given by:

$$W_i = \frac{n_T!}{n_4!n_3!n_2!n_1!n_0!}$$

Each distribution is printed out with the number of particles in each level and the value of $\log_{10} W_i$ (which is more convenient and manageable than W_i itself). Examination of STAT1 shows it to be straightforward. The input consists of one card:

Input card: NTOT (total number of particles) and ETOT (total energy) in 2I0 format. No specific energy units are used.

In line 16, NR4 is calculated which is one more than the maximum possible number of particles in level 4, which is arbitrarily assigned an 'energy' of 4 units. Similarly in line 25 the maximum number (+1) in level 3, for a given number in level 4 is calculated. These numbers (with those for levels 2, 1 and 0) are used in a set of nested DO loops to finally calculate $\log_{10} W_i$ (ANW in line 36). This makes use of the function subroutine FAC(N), which calculates the value of any factorial $n!$ As can be seen, the output consists of the number of particles in each level and, in the final column, $\log_{10} W$. Visual inspection of this column reveals the maximum value(s) of $\log_{10} W_i$ for the particular system being simulated. For example, in Table 4.8, with NTOT equal to 20 and ETOT equal to 24, the maximum value of $\log_{10} W$ is 10.145, given by either of the distributions:

E0	E1	E2	E3	E4
8	5	3	3	1
7	6	4	2	1

With such a small number of levels and particles, it is not possible to say which is the more correct. It can be seen, however, that either corresponds closely to the expected Boltzmann distribution and that we have demonstrated effectively that this is the distribution of maximum likelihood. If we wished to, we could list the distributions in increasing (or decreasing) values of $\log_{10} W$.

Table 4.8

Program STAT1 — statistical distribution of particles in quantized energy levels

```
0008                    MASTER STAT1
0009                    REAL NU
0010                    INTEGER ETOT,ETOT1,ETOT2,ETOT3,ETOT4
0011              10 CONTINUE
0012                    READ(1,1001)NTOT,ETOT
0013            1001 FORMAT(2I0)
0014                    EBAR=ETOT/NTOT
0015                    IF(NTOT.EQ.0)GO TO 25
0016                    NR4=ETOT/4+1
0017                    WRITE(2,1991)NTOT,ETOT,EBAR
0018            1991 FORMAT(1X,'NTOT=',I3,1X,'ETOT=',I14   ,1X,'EBAR=',F10.5,/)
0019                    WRITE(2,1002)
0020            1002 FORMAT(6X,3H E0,6X,3H E1,6X,3H E2,6X,3H E3,6X,3H E4,6X,8HLOG10 NW,
0021              1//)
0022                    DO 4 I4=1,NR4
0023                    II4=NR4-I4
0024                    ETOT4=4*II4
0025                    NR3=(ETOT-ETOT4)/3+1
0026                    DO 3 I3=1,NR3
0027                    II3=NR3-I3
0028                    ETOT3=ETOT4+3*II3
0029                    NR2=(ETOT-ETOT3)/2+1
0030                    DO 2 I2=1,NR2
0031                    II2=NR2-I2
0032                    ETOT2=ETOT3+2*II2
0033                    II1=ETOT-ETOT2
0034                    II0=NTOT-II4-II3-II2-II1
0035                    IF(II0.LT.0)GO TO 2
0036                    ANW=FAC(NTOT)-FAC(II0)-FAC(II1)-FAC(II2)-FAC(II3)-FAC(II4)
0037                    WRITE(2,1003)II0,II1,II2,II3,II4,ANW
0038            1003 FORMAT(5(6X,I3),6X,F10.3)
0039               2 CONTINUE
0040               3 CONTINUE
0041               4 CONTINUE
0042                    WRITE(2,1990)
0043            1990 FORMAT(11H END OF RUN,///)
0044                    GO TO 10
0045              25 STOP
0046                    END


0047                    FUNCTION FAC(N)
0048                    FAC=0.0
0049                    IF(N.EQ.0)GO TO 2
0050                    DO 3 J=1,N
0051               3 FAC=FAC+ALOG10(FLOAT(J))
0052               2 RETURN
0053                    END
```

NTOT= 20 ETOT= 24 EBAR= 1.00000

E0	E1	E2	E3	E4	LOG10 NW
14	0	0	0	6	4.588
13	1	0	1	5	6.513
13	0	2	0	5	6.212
12	2	1	0	5	7.326
11	4	0	0	5	7.326
13	0	1	2	4	6.911
12	2	0	2	4	7.724
12	1	2	1	4	8.025
11	3	1	1	4	8.627
10	5	0	1	4	8.367
12	0	4	0	4	6.945
11	2	3	0	4	8.326
10	4	2	0	4	8.765
9	6	1	0	4	8.589
8	8	0	0	4	7.795
13	0	0	4	3	6.433
12	1	1	3	3	8.149
11	3	0	3	3	8.451
12	0	3	2	3	7.848
11	2	2	2	3	9.104
10	4	1	2	3	9.367
9	6	0	2	3	8.890

11	1	4	1	3	8.627
10	3	5	1	3	9.492
9	5	2	1	3	9.668
8	7	1	1	3	9.300
7	9	0	1	3	8.346
11	0	6	0	3	7.149
10	2	5	0	3	8.668
9	4	4	0	3	9.288
8	6	3	0	3	9.367
7	8	2	0	3	8.999
6	10	1	0	3	8.191
5	12	0	0	3	6.848
12	1	0	5	2	7.326
12	0	2	4	2	7.724
11	2	1	4	2	8.805
10	4	0	4	2	8.765
11	1	3	3	2	8.928
10	3	2	3	2	9.668
9	5	1	3	2	9.668
8	7	0	3	2	8.999
11	0	5	2	2	8.104
10	2	4	2	2	9.543
9	4	3	2	2	10.066
8	6	2	2	2	10.020
7	8	1	2	2	9.476
6	10	0	2	2	8.367
10	1	6	1	2	8.668
9	3	5	1	2	9.668
8	5	4	1	2	10.020
7	7	3	1	2	9.902
6	9	2	1	2	9.367
5	11	1	1	2	8.405
4	13	0	1	2	6.911
10	0	8	0	2	6.920
9	2	7	0	2	8.522
8	4	6	0	2	9.242
7	6	5	0	2	9.446
6	8	4	0	2	9.242
5	10	3	0	2	8.668
4	12	2	0	2	7.724
3	14	1	0	2	6.367
2	16	0	0	2	4.463
12	0	1	6	1	6.848
11	2	0	6	1	7.627
11	1	2	5	1	8.405
10	3	1	5	1	8.969
9	5	0	5	1	8.668
11	0	4	4	1	8.025
10	2	3	4	1	9.367
9	4	2	4	1	9.765
8	6	1	4	1	9.543
7	8	0	4	1	8.698
10	1	5	3	1	8.969
9	3	4	3	1	9.890
8	5	3	3	1	10.145
7	7	2	3	1	9.902
6	9	1	3	1	9.191
5	11	0	3	1	7.928
10	0	7	2	1	7.823
9	2	6	2	1	9.367
8	4	5	2	1	10.020
7	6	4	2	1	10.145
6	8	3	2	1	9.844
5	10	2	2	1	9.145
4	12	1	2	1	8.025
3	14	0	2	1	6.367
9	1	8	1	1	8.221
8	3	7	1	1	9.300
7	5	6	1	1	9.747
6	7	5	1	1	9.747
5	9	4	1	1	9.367
4	11	3	1	1	8.627
3	13	2	1	1	7.515
2	15	1	1	1	5.969
1	17	0	1	1	3.835
9	0	10	0	1	6.267
8	2	9	0	1	7.920
7	4	8	0	1	8.698
6	6	7	0	1	8.969
5	8	6	0	1	8.844
4	10	5	0	1	8.367
3	12	4	0	1	7.547

2	14	3	0	1	6,367
1	16	2	0	1	4,764
0	18	1	0	1	2,580
12	0	0	8	0	5,100
11	1	1	7	0	7,083
10	3	0	7	0	7,346
11	0	3	6	0	7,149
10	2	2	6	0	8,367
9	4	1	6	0	8,589
8	6	0	6	0	8,066
10	1	4	5	0	8,367
9	3	3	5	0	9,191
8	5	2	5	0	9,321
7	7	1	5	0	8,902
6	9	0	5	0	7,890
10	0	6	4	0	7,589
9	2	5	4	0	9,066
8	4	4	4	0	9,640
7	6	3	4	0	9,668
6	8	2	4	0	9,242
5	10	1	4	0	8,367
4	12	0	4	0	6,945
9	1	7	3	0	8,346
8	3	6	3	0	9,367
7	5	5	3	0	9,747
6	7	4	3	0	9,668
5	9	3	3	0	9,191
4	11	2	3	0	8,326
3	13	1	3	0	7,036
2	15	0	3	0	5,190
9	0	9	2	0	6,966
8	2	8	2	0	8,573
7	4	7	2	0	9,300
6	6	6	2	0	9,513
5	8	5	2	0	9,321
4	10	4	2	0	8,765
3	12	3	2	0	7,848
2	14	2	2	0	6,543
1	16	1	2	0	4,764
0	18	0	2	0	2,279
8	1	10	1	0	7,221
7	3	9	1	0	8,346
6	5	8	1	0	8,844
5	7	7	1	0	8,902
4	9	6	1	0	8,589
3	11	5	1	0	7,928
2	13	4	1	0	6,911
1	15	3	1	0	5,491
0	17	2	1	0	3,534
8	0	12	0	0	5,100
7	2	11	0	0	6,782
6	4	10	0	0	7,589
5	6	9	0	0	7,890
4	8	8	0	0	7,795
3	10	7	0	0	7,346
2	12	6	0	0	6,547
1	14	5	0	0	5,367
0	16	4	0	0	3,685

END OF RUN

An alternative approach to the same problem is mentioned in the Problems section.

4.3 Spectroscopy I — Curve Addition and Curve Resolution

Quantities such as the position, ν_{max}, absorptivity, ϵ_{max}, and half-width, θ, of a particular band are easy to obtain so long as no other bands are present in the near vicinity. These quantities are presented in Figure 4.6 for a single band. Such a band often approximates to the behaviour of a Gaussian function, although other functions are possible.

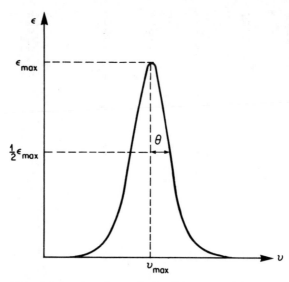

Figure 4.6 A theoretical Gaussian absorption band of maximum absorptivity ϵ_{max}, half-width at half-height, θ, at a frequency ν_{max}

In ultraviolet spectroscopy, one suitable Gaussian-type function is:

$$\epsilon = \epsilon_{max} e^{-(\bar{\nu}-\bar{\nu}_{max})^2 \theta^{-2}} \qquad (4.25)$$

where the parameters are as defined in Figure 4.6:

> $\epsilon =$ absorptivity as a function of the wavenumber, ν, in cm^{-1}
> $\epsilon_{max} =$ molar absorptivity at the wavenumber of the maximum, ν_{max}
> $\theta =$ half-width of the band at half-height in wavenumbers, cm^{-1}

Note that the units of ϵ and ϵ_{max} depend on the units of concentration and of path length. For example, if the units of concentration are mol dm^{-3} and the path length is measured in cm, then the units of ϵ and ϵ_{max} are mol^{-1} dm^3 cm^{-1}. Other functions can also be used to describe the shape of spectroscopic bands and some are discussed by Jorgensen.[32] The half-width, θ, is identical on the low- and high-energy sides of the maximum of a Gaussian peak but, experimentally, this is not always so. To allow for this, two parameters θ_+ and θ_- could be used or, as suggested by Klabuhn,[33] one could use a function of the form

$$\epsilon = \epsilon_{max} e^{-[(\bar{\nu}_{max}-\bar{\nu})\bar{\nu}_{max}/(\theta\bar{\nu})]^2} \qquad (4.26)$$

With this function, ϵ for $\bar{\nu} > \bar{\nu}_{max}$ decreases more slowly than for $\bar{\nu} < \bar{\nu}_{max}$. This asymmetric behaviour is typical of the appearance of ultraviolet bands of organic compounds and it may therefore, be preferable in these cases.

4.3.1 Peak Envelope Addition

Returning to function (4.25), it can be shown that [32] the peak area defined by the function is given by

$$\int_0^\infty \epsilon d\bar{\nu} = \epsilon_{max} \theta \sqrt{\pi}$$

This is proportional to the probability of the transition responsible for the band. The oscillator strength,[32] f, can then be calculated from (4.27).

$$f = 4.60 \times 10^{-9} \epsilon_{max} \delta \qquad (4.27)$$

where δ is the total bandwidth ($\delta = 2\theta$) at ϵ equal to $\frac{1}{2} \epsilon_{max}$.

Equations (4.25) and (4.27) can form the basis of a very useful computer program for simulating electronic absorption spectra with one or more component bands. If we are in a region of a spectrum where two or more bands overlap, then the observed ϵ_{obs} will be the sum of the individual values of ϵ for each component that absorbs in that region. Putting this mathematically, we have

$$\epsilon_{obs} = \sum_j \epsilon_j$$

and from equation (4.25), we have

$$\epsilon_{calc} = \sum_j A_j \exp - \left\{ \frac{(\bar{\nu} - C_j)^2}{B_j} \right\} \qquad (4.28)$$

where A_j, B_j and C_j replace ϵ_{max}, $\bar{\nu}_{max}$ and θ respectively. If we use known, or estimated, values of the parameters, then the composite spectrum over a given range of $\bar{\nu}$ can be computed and then compared with the actual spectrum; also oscillator strengths can be calculated from (4.27). We may then make adjustments to the values of each A, B or C until satisfactory agreement between the various ϵ_{obs} and ϵ_{calc} is obtained.

The program SPECTRUM[35] (Table 4.9) performs the computations described above, and uses the line printer to give a graphical representation of the individual component bands and of the composite spectrum, as can be seen in the example. The parameters R and W in the READ and WRITE statements are the peripheral device numbers for the card reader and line printer, respectively. In our case R is 1 and W is 2, and these can normally be defined in the program description. The program deals with up to five overlapping bands, for each of which the parameters A, B and C must be known. The program, firstly,

Table 4.9

Program SPECTRUM — graphical simulation of additivity of absorption curves to give a composite spectrum

```
0008          MASTER SPECTRUM
0009          DIMENSION MARK(7),IPLOT(101),NB(6),F(5)
0010          DIMENSION NAME(18),E(6,252),XAMDA(252),VZERO(5),EZERO(5),DELTA(5)
0011          DATA MARK/1H1,1H2,1H3,1H4,1H5,1H*,1H /
0012          INTEGER BAND(5,5)    ,R,W,TOTB
0013          R = 1
0014          W = 2
0015       15 READ(R,101) ( NAME(I),I=1,18), TOTB
0016      101 FORMAT(18A4,I8)
0017          IF(TOTB-10) 25,20,20
0018       20 WRITE(W,102)
0019      102 FORMAT(///,4X,' NUMBER OF BANDS IS TYPED INCORRECT OR EXCEEDS PROG
0020         1RAM SPECIFICATIONS')
0021          STOP
0022       25 IF(TOTB) 145,145,30
0023       50 READ(R,103)STINC,RANGE,VINC
0024      103 FORMAT(3F10,2)
0025          DO 35 IK=1,250
0026       35 J=1,6
0027          E(J,IK) = 0.0
0028       35 CONTINUE
0029          WRITE(W,104)NAME,TOTB
0030      104 FORMAT(1H1,18A4,//,10x,' NUMBER OF PRESCRIBED BANDS =',IX)
0031          WRITE(W,105)
0032      105 FORMAT(//,' IDENTIFICATION',10X,'NUZERO(CM-1)',4X,'EZERO      ',4X
0033         1,'DFLTA(CM-1)',4X,'RANGE(CM-1)',4X,'INCREMENT(CM-1)',4X,'START INC
0034         2,(CM-1)',//,41X,'CM-1 M-1')
0035          DO 40 I=1,TOTB
0036          READ(R,106)(BAND(I,K),K=1,5),VZERO(I),EZERO(I),DELTA(I)
0037      106 FORMAT(5A4,3F10,2)
0038          WRITE(W,107)(BAND(I,K),K=1,5),VZERO(I),EZERO(I),DELTA(I),RANGE,VIN
0039         1C,STINC
0040      107 FORMAT(//,1X,5A4,2X,F10,2,4X,F10,2,5X,F10,2,5X,F,0,2,8X,F10,2,10x,F
0041         110,2)
0042       40 CONTINUE
0043          DO 45 J=1,TOTB
0044          F(J)=((4,60E-9)*(EZERO(J)))*(DELTA(J))
0045          WRITE(W,108)(BAND(J,K),K=1,5),F(J)
0046      108 FORMAT(/,1X,5A4,' OSCILLATOR STRENGTH(F)= ',1X,F10,4)
0047       45 CONTINUE
0048          CALST =(STINC)+(RANGE)
0049          DO 70 K=1,TOTB
0050          KK=0
0051          STINX=STINC
0052          DO 65 L=1,250
0053          KK=KK+1
0054 C        BY DEFINITION OF THE FORMULA, THE WIDTH AT HALF-HEIGHT (DFLTA) IS
0055 C        DIVIDED BY TWO,
0056          X=-(((STINX-VZERO(K))/(DELTA(K)/2,0))**2)
0057          E(K,L) =  EZERO(K)*(EXP(X))
```

```
0058        XNMS=(1.0E07)/STINX
0059        IF(K-1) 55,50,55
0060   50   XAMDA(L) = STINX
0061   55   CONTINUE
0062        IF(STINX - CALST) 60,70,70
0063   60   CONTINUE
0064        STINX=STINX+VINC
0065   65   CONTINUE
0066   70   CONTINUE
0067   C    THIS SECTION IS USED TO HAVE A PRINTER PLOT.
0068        XMAX = 0.0
0069        DO 85 IK=1,KK
0070        E(6,IK) = 0.0
0071        DO 75 JJ=1,TOTB
0072   C    SUMMATION OF COMPONENTS.
0073        E(6,IK) = E(6,IK) + E(JJ,IK)
0074   75   CONTINUE
0075   C    E(6,IK) IS THE TOTAL SPECTRUM.
0076   C    THIS "IF" FINDS THE MAXIMUM VALUE IN E(6,IK).
0077        IF(E(6,IK) - XMAX) 85,85,80
0078   80   XMAX = E(6,IK)
0079   85   CONTINUE
0080        J1 = 1
0081        J2 = 10
0082   90   IF( J2 - KK) 110,95,95
0083   95   J2 = KK
0084   110  WRITE(W,109)(K,K=J1,J2)
0085   109  FORMAT(//,10X,10(4X,I3,5X))
0086        WRITE(W,11) ( XAMDA(I),I=J1,J2)
0087   111  FORMAT(1X,NUCCM-1)',1X,10F10,1)
0088        DO 115 IJ=1,TOTB
0089        WRITE(W,112)IJ,(E(IJ,I),I=J1,J2)
0090   115  CONTINUE
0091   112  FORMAT(1X,'BAND(',I2,')',1X,10F10,4)
0092        WRITE(W,113)(E(6,I),I=J1,J2)
0093   113  FORMAT(1X,'TOTAL',4X,10F10,3)
0094        IF(J2 - KK) 120,125,125
0095   120  J1 = J1 + 10
0096        J2 = J2 + 10
0097        GO TO 90
0098   125  CONTINUE
0099        WAVEN = STINC
0100        WRITE(W,114)
0101   114  FORMAT(1H1)
0102        WRITE(W,116) KK
0103   116  FORMAT(/,1X,'THE NUMBER OF POINTS =',I5,/)
0104        WRITE(W,150)
0105   150  FORMAT(1X,' TOTAL POSITION      ',101(1H-))
0106   C    CLEAR THE LINE BUFFER
0107        DO 200 I=1,101
0108   200  IPLOT(I)=MARK(7)
0109        DO 140 IK = 1,KK
0110        DO 130 JJ=1,6
```

```
0111        IF(JJ.GT.TOTB)JJ=6
0112   C    THIS NORMALIZES THE OUTPUT(PRINTED) SPECTRUM WHILE STILL GIVING
0113   C    THE CALCULATED VALUE.
0114        NPTS = (E(JJ,IK)/ XMAX) *100. + 1
0115        NB(JJ) = NPTS
0116        IPLOT(NPTS) = MARK(JJ)
0117   130  CONTINUE
0118   C    PRINTOUT OF PLOT
0118        WRITE(W,117) E(6,IK),WAVEN,(IPLOT(I),I=1,101)
0120   117  FORMAT(1X,F6.1,2X,F8.2,3X,101A1)
0121        WAVEN = WAVEN + VINC
0122        DO 135 JJ=1,6
0123        IF(JJ.GT.TOTB)JJ=6
0124   C    CLEAR LINE BUFFER IN FILLED AREAS ONLY
0125        NPTS = NB(JJ)
0126        IPLOT(NPTS) = MARK(7)
0127   135  CONTINUE
0128   140  CONTINUE
0129        GO TO 15
0130   145  CONTINUE
0131        STOP
0132        END
```

CR(DI-BENZYL-THIOCARBAMATE)5 IN DMF
NUMBER OF PPESCRIBED BANDS =   2

IDENTIFICATION	NUZERO(CM-1)	EZERO	DELTA(CM-1)	RANGE(CM-1)	INCREMENT(CM-1)	START INC. (CM-1)
		CM-1 M-1				
BAND 1	15650.00	51.10	5120.00	15000.00	100.00	10100.00
BAND 2	20540.00	35.35	3280.00	15000.00	100.00	10100.00

BAND 1   OSCILLATOR STRENGTH(F)= 0.4465E-03

BAND 2   OSCILLATOR STRENGTH(F)= 0.5032E-03

	1	2	3	4	5	6	7	8	9	10
NU(CM-1)	10100.0	10200.0	10300.0	10400.0	10500.0	10600.0	10700.0	10800.0	10900.0	11000.0
BAND(1)	0.0001	0.0002	0.0002	0.0004	0.0006	0.0009	0.0013	0.0020	0.0029	0.0043
BAND(2)	0.0000	0.0000	0.0000	0.0000	0.0000	0.0000	0.0000	0.0000	0.0000	0.0000
TOTAL	0.000	0.000	0.000	0.000	0.001	0.001	0.001	0.002	0.003	0.004

Index	NUCCM-1	BAND(1)	BAND(2)	TOTAL
11	11100.0	0.0063	0.0000	0.006
12	11200.0	0.0091	0.0000	0.009
13	11300.0	0.0131	0.0000	0.013
14	11400.0	0.0186	0.0000	0.019
15	11500.0	0.0263	0.0000	0.026
16	11600.0	0.0368	0.0000	0.037
17	11700.0	0.0511	0.0000	0.051
18	11800.0	0.0704	0.0000	0.070
19	11900.0	0.0962	0.0000	0.096
20	12000.0	0.1304	0.0000	0.150
21	12100.0	0.1753	0.0000	0.175
22	12200.0	0.2357	0.0000	0.254
23	12300.0	0.3090	0.0000	0.309
24	12400.0	0.4053	0.0000	0.405
25	12500.0	0.5272	0.0000	0.527
26	12600.0	0.6802	0.0000	0.680
27	12700.0	0.8704	0.0000	0.870
28	12800.0	1.1047	0.0000	1.105
29	12900.0	1.3905	0.0000	1.391
30	13000.0	1.7360	0.0000	1.736
31	13100.0	2.1495	0.0000	2.150
32	13200.0	2.6598	0.0000	2.640
33	13300.0	3.2153	0.0000	3.215
34	13400.0	3.8844	0.0000	3.884
35	13500.0	4.6541	0.0000	4.654
36	13600.0	5.5309	0.0000	5.531
37	13700.0	6.5389	0.0000	6.519
38	13800.0	7.6206	0.0000	7.621
39	13900.0	8.8356	0.0000	8.836
40	14000.0	10.1604	0.0000	10.160
41	14100.0	11.5882	0.0000	11.588
42	14200.0	13.1085	0.0000	13.109
43	14300.0	14.7069	0.0000	14.707
44	14400.0	16.3652	0.0001	16.365
45	14500.0	18.0613	0.0001	18.061
46	14600.0	19.7702	0.0002	19.770
47	14700.0	21.4650	0.0002	21.464
48	14800.0	23.1113	0.0004	23.112
49	14900.0	24.6819	0.0000	24.682
50	15000.0	26.1434	0.0008	26.144
51	15100.0	27.4648	0.0012	27.466
52	15200.0	28.6169	0.0018	28.619
53	15300.0	29.5733	0.0026	29.576
54	15400.0	30.3115	0.0038	30.315
55	15500.0	30.8138	0.0055	30.819
56	15600.0	31.0681	0.0079	31.076
57	15700.0	31.0681	0.0111	31.079
58	15800.0	30.8138	0.0157	30.829
59	15900.0	30.3115	0.0219	30.333
60	16000.0	29.5733	0.0303	29.604
61	16100.0	28.6169	0.0417	28.659
62	16200.0	27.4648	0.0570	27.522
63	16300.0	26.1434	0.0772	26.221
64	16400.0	24.6819	0.1039	24.786
65	16500.0	23.1113	0.1387	23.250
66	16600.0	21.4636	0.1839	21.647
67	16700.0	19.7702	0.2419	20.012
68	16800.0	18.0613	0.3159	18.377
69	16900.0	16.3652	0.4095	16.775
70	17000.0	14.7069	0.5270	15.234
71	17100.0	13.1085	0.6730	13.782
72	17200.0	11.5882	0.8532	12.441
73	17300.0	10.1604	1.0736	11.234
74	17400.0	8.8356	1.3409	10.176
75	17500.0	7.6206	1.6624	9.283
76	17600.0	6.5189	2.0456	8.565
77	17700.0	5.5309	2.4986	8.029
78	17800.0	4.6541	3.0293	7.683
79	17900.0	3.8844	3.6455	7.530
80	18000.0	3.2153	4.3545	7.570
81	18100.0	2.6398	5.1629	7.803
82	18200.0	2.1495	6.0760	8.226
83	18300.0	1.7360	7.0977	8.834
84	18400.0	1.3905	8.2296	9.620
85	18500.0	1.1047	9.4715	10.576
86	18600.0	0.8704	10.8199	11.690
87	18700.0	0.6802	12.2688	12.949
88	18800.0	0.5272	13.8086	14.336
89	18900.0	0.4053	15.4265	15.832
90	19000.0	0.3090	17.1063	17.415

#	NU(CM-1)	BAND(1)	BAND(2)	TOTAL
91	19100.0	0.2337	18.8285	19.062
92	19200.0	0.1753	20.5706	20.746
93	19300.0	0.1304	22.3073	22.438
94	19400.0	0.0962	24.0115	24.108
95	19500.0	0.0704	25.6544	25.725
96	19600.0	0.0511	27.2066	27.258
97	19700.0	0.0368	28.6389	28.676
98	19800.0	0.0263	29.9234	29.950
99	19900.0	0.0186	31.0338	31.052
100	20000.0	0.0131	31.9470	31.960
101	20100.0	0.0091	32.6434	32.652
102	20200.0	0.0063	33.1079	33.114
103	20300.0	0.0043	33.3302	33.334
104	20400.0	0.0029	33.3054	33.308
105	20500.0	0.0020	33.0341	33.036
106	20600.0	0.0013	32.5222	32.524
107	20700.0	0.0009	31.7811	31.782
108	20800.0	0.0006	30.8268	30.827
109	20900.0	0.0004	29.6796	29.680
110	21000.0	0.0002	28.3634	28.364
111	21100.0	0.0002	26.9048	26.905
112	21200.0	0.0001	25.3521	25.352
113	21300.0	0.0001	23.6746	23.675
114	21400.0	0.0000	21.9617	21.962
115	21500.0	0.0000	20.2218	20.222
116	21600.0	0.0000	18.4818	18.482
117	21700.0	0.0000	16.7663	16.766
118	21800.0	0.0000	15.0974	15.097
119	21900.0	0.0000	13.4939	13.494
120	22000.0	0.0000	11.9714	11.971
121	22100.0	0.0000	10.5419	10.542
122	22200.0	0.0000	9.2144	9.214
123	22300.0	0.0000	7.9944	7.994
124	22400.0	0.0000	6.8845	6.885
125	22500.0	0.0000	5.8848	5.885
126	22600.0	0.0000	4.9930	4.993
127	22700.0	0.0000	4.2050	4.205
128	22800.0	0.0000	3.5151	3.515
129	22900.0	0.0000	2.9166	2.917
130	23000.0	0.0000	2.4021	2.402
131	23100.0	0.0000	1.9637	1.964
132	23200.0	0.0000	1.5934	1.593
133	23300.0	0.0000	1.2853	1.283
134	23400.0	0.0000	1.0260	1.026
135	23500.0	0.0000	0.8141	0.814
136	23600.0	0.0000	0.6413	0.641
137	23700.0	0.0000	0.5014	0.501
138	23800.0	0.0000	0.3891	0.389
139	23900.0	0.0000	0.2997	0.300
140	24000.0	0.0000	0.2291	0.229
141	24100.0	0.0000	0.1759	0.174
142	24200.0	0.0000	0.1510	0.151
143	24300.0	0.0000	0.0979	0.098
144	24400.0	0.0000	0.0727	0.073
145	24500.0	0.0000	0.0535	0.054
146	24600.0	0.0000	0.0392	0.039
147	24700.0	0.0000	0.0284	0.028
148	24800.0	0.0000	0.0205	0.020
149	24900.0	0.0000	0.0146	0.015
150	25000.0	0.0000	0.0104	0.010
151	25100.0	0.0000	0.0073	0.007

TOTAL	POSITION
0.0	10100.00
0.0	10200.00
0.0	10300.00
0.0	10400.00
0.0	10500.00
0.0	10600.00
0.0	10700.00
0.0	10800.00
0.0	10900.00
0.0	11000.00
0.0	11100.00
0.0	11200.00
0.0	11300.00
0.0	11400.00
0.0	11500.00
0.0	11600.00
0.1	11700.00
0.1	11800.00
0.1	11900.00
0.2	12000.00
0.2	12100.00
0.3	12200.00
0.4	12300.00
0.5	12400.00
0.5	12500.00
0.7	12600.00
0.9	12700.00
1.1	12800.00
1.4	12900.00
1.7	13000.00
2.1	13100.00
2.6	13200.00
3.2	13300.00
3.9	13400.00
4.7	13500.00
5.5	13600.00
6.5	13700.00
7.6	13800.00
8.8	13900.00
10.2	14000.00
11.6	14100.00
13.1	14200.00
14.7	14300.00
16.4	14400.00
18.1	14500.00
19.8	14600.00
21.5	14700.00
23.1	14800.00
24.7	14900.00
26.1	15000.00

27,5	15100,00
28,6	15200,00
29,6	15300,00
30,3	15400,00
30,8	15500,00
31,1	15600,00
31,1	15700,00
30,8	15800,00
30,3	15900,00
29,6	16000,00
28,7	16100,00
27,5	16200,00
26,2	16300,00
24,8	16400,00
23,3	16500,00
21,6	16600,00
20,0	16700,00
18,4	16800,00
16,8	16900,00
15,2	17000,00
13,8	17100,00
12,4	17200,00
11,2	17300,00
10,2	17400,00
9,3	17500,00
8,6	17600,00
8,0	17700,00
7,7	17800,00
7,5	17900,00
7,6	18000,00
7,8	18100,00
8,2	18200,00
8,8	18300,00
9,6	18400,00
10,6	18500,00
11,7	18600,00
12,9	18700,00
14,3	18800,00
15,8	18900,00
17,4	19000,00
19,1	19100,00
20,7	19200,00
22,4	19300,00
24,1	19400,00
25,7	19500,00
27,3	19600,00
28,7	19700,00
29,9	19800,00
31,1	19900,00

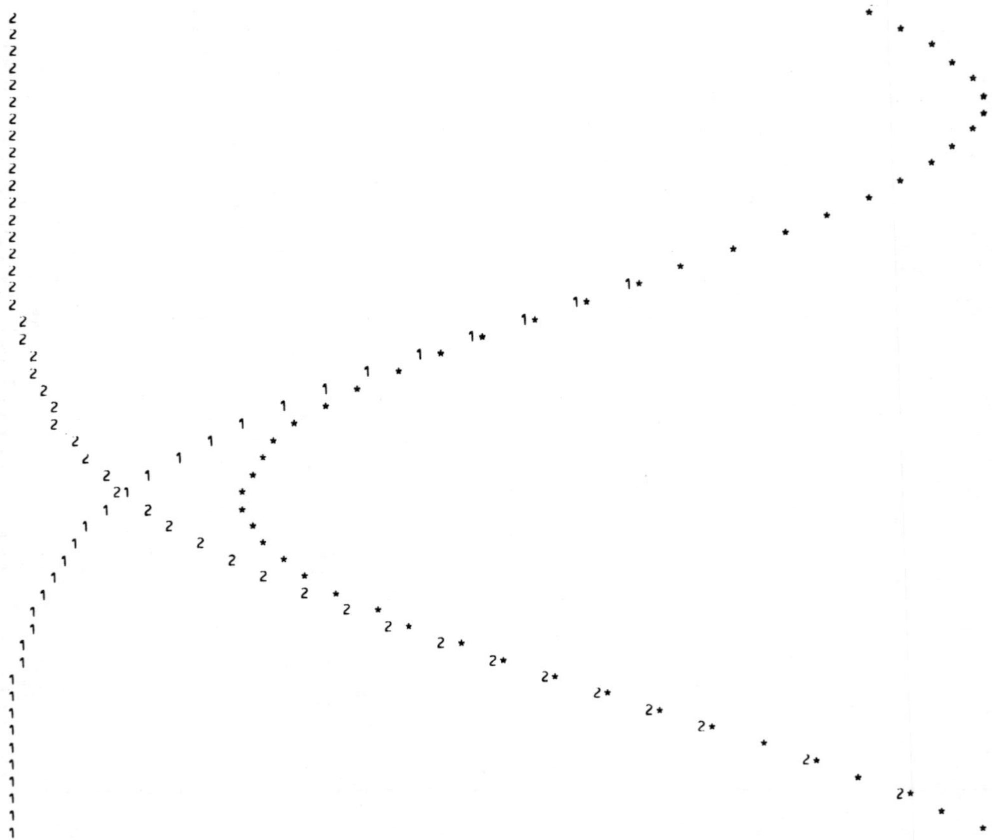

32.0	20000.00	1
32.7	20100.00	1
33.1	20200.00	1
33.3	20300.00	1
33.3	20400.00	1
33.0	20500.00	1
32.5	20600.00	1
31.8	20700.00	1
30.8	20800.00	1
29.7	20900.00	1
28.4	21000.00	1
26.9	21100.00	1
25.3	21200.00	1
23.7	21300.00	1
22.0	21400.00	1
20.2	21500.00	1
18.5	21600.00	1
16.8	21700.00	1
15.1	21800.00	1
13.5	21900.00	1
12.0	22000.00	1
10.5	22100.00	1
9.2	22200.00	1
8.0	22300.00	1
6.9	22400.00	1
5.9	22500.00	1
5.0	22600.00	1
4.2	22700.00	1
3.5	22800.00	1
2.9	22900.00	1
2.4	23000.00	1
2.0	23100.00	1
1.6	23200.00	1
1.3	23300.00	1
1.0	23400.00	1
0.8	23500.00	1
0.6	23600.00	1
0.5	23700.00	1
0.4	23800.00	1
0.3	23900.00	
0.2	24000.00	
0.2	24100.00	
0.1	24200.00	
0.1	24300.00	
0.1	24400.00	
0.1	24500.00	
0.0	24600.00	
0.0	24700.00	
0.0	24800.00	
0.0	24900.00	
0.0	25000.00	
0.0	25100.00	

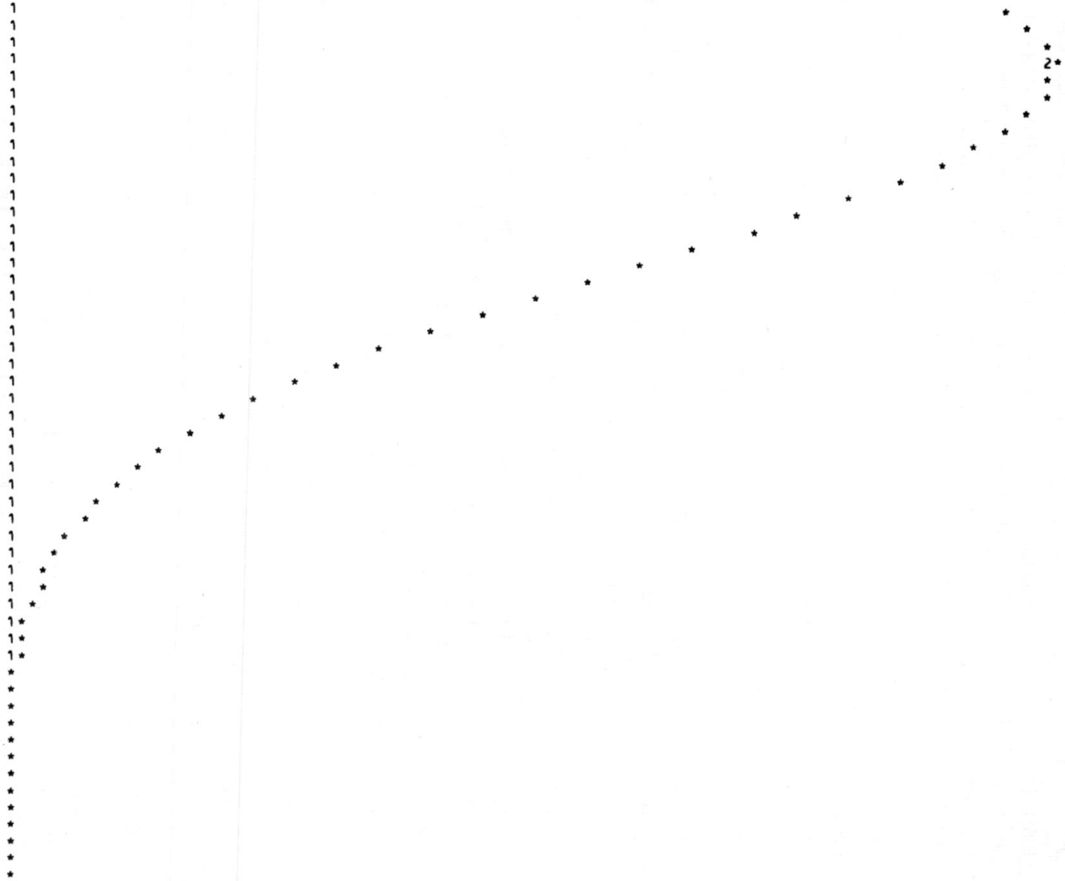

calculates the oscillator strength for each band, from equation (4.27), in lines 43 and 44. We then begin to compute the composite spectrum, from the defined $\bar{\nu}$ origin, labelled STINC. This is presumed to be in *wavenumbers*, cm^{-1}. Equation (4.28) is then used to compute ϵ_{calc} over the range STINC to (STINC + RANGE) cm^{-1} at increments of VINC; the DO loop (lines 52—65) allows up to 250 such calculations and, in this loop, the component values of ϵ are calculated. After printing the numerical values of ϵ_{calc} and of the component contributions, the section from lines 107—128 is used to generate a line printer graphical display. This makes use of a line buffer (see Section 1.3.2) to store the characters 1, 2, 3, 4, 5 or *. The numerals represent the component contributions and the * represents the total value of ϵ_{calc}. The operation of this section should be clear to you after referring to Section 1.3.2. In each line of printout, the total intensity, the position (cm^{-1}) and appropriate characters are printed. Note that the use of a line buffer is more economical than a page buffer and may be essential, in terms of available core storage, when large pictures are to be drawn. Input to SPECTRUM is as follows:

Card 1: Name (Format 18A4); number of components (I8).
Card 2: STINC, the origin value of $\bar{\nu}$ (Format F10.2).
RANGE, the total range of $\bar{\nu}$ (Format F10.2).
VINC, the increment in $\bar{\nu}$ (Format F10.2).
Card 3: Band identification (Format 5A4)

$\left.\begin{array}{l} \nu_{max} \text{ of first band} \\ \epsilon_{max} \text{ of first band} \\ \delta \text{ of first band} \end{array}\right\}$ Format 3F10.2.

Note that δ is the total line width at half-height ($\epsilon = \frac{1}{2}\epsilon_{max}$). If there is more than one component, a further card in identical format to card 3 is added for each component.

Final card: This must be a blank card to stop execution.

Suitable test data for the program are included in Table 4.9. For clarity, we repeat the data here (without regard to exact format).

Card 1: CR(DI-BENZYL-THIOCARBAMATE) IN DMF.
Card 2: 10100   5000   100
Card 3: BAND 1   15650   31.10   3120
Card 4: BAND 2   20340   33.35   3280

This program is extremely useful in demonstrating the influence of the various parameters on the final spectrum.

4.3.2 Curve Resolution
An alternative approach is to start with an experimental spectrum and attempt to resolve it into a sum of component bands described by a

function such as equation (4.28). In this case, we could start with estimates of the values of each A_j, B_j and C_j and then refine these by some least squares procedure until the experimental and calculated values of ϵ were the best possible fit. The refined values of A_j, B_j and C_j would then correspond to the 'correct' values for each component band. There is a straightforward approach to this problem, using a Taylor series approximation in much the same manner as we did in Section 3.2.4. We shall let ϵ_i represent the calculated value of the absorptivity. If we start with an initial, guessed, set of parameters A_j^0; B_j^0; C_j^0, then the next approximation will be:

$$A_j' = A_j^\circ + a_j; B_j' = B_j^\circ + b_j; C_j' = C_j^\circ + c_j \qquad (4.29)$$

the corrections a_j, b_j and c_j are obtained by expanding ϵ_i in a Taylor's series (to the first term only):

$$\epsilon_i = \sum_j f_{i,j}^\circ + \sum_j a_j \left(\frac{\partial f_{i,j}^\circ}{\partial A_j} \right) + \sum_j b_j \left(\frac{\partial f_{i,j}}{\partial B_j} \right) + \sum_j c_j \left(\frac{\partial f_{i,j}}{\partial C_j} \right) \qquad (4.30)$$

$f_{i,j}^\circ$ is strictly a function f_i $(A_j^\circ, B_j^\circ, C_j^\circ)$; the function refers to the ith data point and could be, for example, the function defined in equation (4.28). Clearly, it is necessary for the function to be differentiable, as (4.28) is. The derivation now follows traditional lines by performing a least squares fit of equation (4.30) to the experimental values of ϵ_i represented by ϵ_i'. The condition will be that the sum, S, is a minimum:

$$S = \sum_i (\epsilon_i - \epsilon_i')^2 \qquad (4.31)$$

S is then minimized by taking the first derivative of S with respect to each of the corrections a_j, b_j and c_j. For example, taking $\partial S / \partial a_k$ and then setting this equal to zero, as required for a minimum in S, we obtain:

$$\sum_i \left(\epsilon_i' - \sum_j f_{i,j}^\circ \right) \frac{\partial f_{i,j}}{\partial A_k} = \sum_i \sum_j a_j \frac{\partial f_{i,j}}{\partial A_k} \frac{\partial f_{i,j}}{\partial A_j} + \sum_i \sum_j b_j \frac{\partial f_{i,j}}{\partial A_k} \frac{\partial f_{i,j}}{\partial B_j}$$

$$+ \sum_i \sum_j c_j \frac{\partial f_{i,j}}{\partial A_k} \frac{\partial f_{i,j}}{\partial C_j}$$

The summations over i refer to the ith values of the partial derivatives. There will be k such equations, obtained on differentiating (4.31) with respect to each of the a_j variables for the k different bands. Similarly, k equations for the b_j variables are obtained and k more equations for the c_j variables, making a total of $3k$ simultaneous equations — on solving these equations for the $3k$ unknowns, we obtain the first set of corrections which, on insertion into (4.29), give a refined set of parameters. Although these parameters are 'better' estimates, it is unlikely that they are the 'best' since our procedure is based on a truncated Taylor's Series. Therefore, we repeat the whole procedure until, hopefully, a situation is reached in which the changes in the parameters are acceptably small. The only tricky aspect to the programming of this problem is the construction of the sets of simultaneous equations. The coefficient of each variable is a sum of the product of two first derivatives such as

$$\sum_i \frac{\partial f_{i,j}}{\partial A_k} \cdot \frac{\partial f_{i,j}}{\partial B_j}$$

These products are cumbersome to generate, but a considerable simplification results when we recognize that the matrix of coefficients is symmetric about its leading diagonal. Also, the matrix is conveniently divided off into nine submatrices which facilitate the construction of the composite matrix. We illustrate this feature in Table 4.10 in which we also give the form of the array of dependent variables. All of the coefficients have been set out in a manner suitable for input to our subroutine S1ML (Section 2.3) for the solution of simultaneous equations. The corrections to the parameters are stored in the array elements $X(I, 3k + 1)$, on output from S1ML. Bearing in mind the foregoing, we can inspect the program RESOLVE in Table 4.11. We begin by declaring a function $F0(I,J)$ in line 17. This function can be seen to be equivalent to

$$\epsilon_i = \epsilon_{max} \exp - \left(\frac{\bar{v} - \bar{v}_{max}}{\theta} \right)^2$$

In the program we let

$$\left. \begin{array}{l} A1(J) = \epsilon_{max} \\ B1(J) = \theta \\ C1(J) = \nu_{max} \end{array} \right\} \quad \text{for compound J}$$

The other functions $FA(I, J)$, $FB(I, J)$, and $FC(I, J)$ correspond to the first derivatives with respect to $A1(J)$, $B1(J)$ and $C1(J)$ respectively. Whilst we were able, (using an ICL computer) to declare each of these functions within the main body of the program, it may be necessary for them to be declared in external function subroutines with other compilers. Be careful to note the difference between these functions and arrays. Such functions must appear before any executable statement otherwise they will be treated as arrays. The estimates of the values of $A1(J)$, $B1(J)$ and $C1(J)$ for each component are then read in up to a maximum of NZ. We then read in the values of the wavenumbers (array $V(J)$) and the absorptivities (array $E(J)$). From line 56 to 126 the array elements as defined in Table 4.10 are

Table 4.10

A 'matrix map' showing the way in which the coefficients $x_{l,m}$ are constructed for k overlapping bands

	$m = 1$	$m = k$	$m = 2k$	$m = 3k$ $\quad m = 3k + 1$	
$l = 1$		$m =$			$l = 1$
	$\sum \dfrac{\partial f_{i,j}}{\partial A_1} \dfrac{\partial f_{i,j}}{\partial A_m}$	$\sum \dfrac{\partial f_{i,j}}{\partial A_1} \dfrac{\partial f_{i,j}}{\partial B_{m'}}$ $(m' = m - k)$	$\sum \dfrac{\partial f_{i,j}}{\partial A_1} \dfrac{\partial f_{i,j}}{\partial C_{m''}}$ $(m'' = m - 2k)$	$\Sigma z \dfrac{\partial f_{i,j}}{\partial A_1}$	
$l = k$					$l = k$
	$\sum \dfrac{\partial f_{i,j}}{\partial B_{l'}} \dfrac{\partial f_{i,j}}{\partial A_m}$ $(l' = l - k)$	$\sum \dfrac{\partial f_{i,j}}{\partial B_{l'}} \dfrac{\partial f_{i,j}}{\partial B_{m'}}$ $(l' = l - k$ $m' = m - k)$	$\sum \dfrac{\partial f_{i,j}}{\partial B_{l'}} \dfrac{\partial f_{i,j}}{\partial C_{m''}}$ $(l' = l - k$ $m'' = m - 2k)$	$\Sigma z \dfrac{\partial f_{i,j}}{\partial B_{l'}}$ $(l' = l - k)$	
$l = 2k$					$l = 2k$
	$\sum \dfrac{\partial f_{i,j}}{\partial C_{l''}} \dfrac{\partial f_{i,j}}{\partial A_m}$ $(l'' = l - 2k)$	$\sum \dfrac{\partial f_{i,j}}{\partial C_{l''}} \dfrac{\partial f_{i,j}}{\partial B_{m'}}$ $(l'' = l - 2k$ $m' = m - k)$	$\sum \dfrac{\partial f_{i,j}}{\partial C_{l''}} \dfrac{\partial f_{i,j}}{\partial C_{m''}}$ $(l'' = l - 2k$ $m'' = m - 2k)$	$\Sigma z \dfrac{\partial f_{i,j}}{\partial C_{l''}}$ $(l'' = l - 2k)$	
$l = 3k$					$l = 3k$

Notes:

(i) The summations are over the n data points.

(ii) In the right-hand, dependent vector, column Z is equal to

$$\epsilon_i - \sum_{j=1}^{k} f_{i,j}^{\circ}$$

(where $f_{i,j}^{\circ}$ is the value of the function f at data point i for component j).

computed from the various function statements for the first derivatives. Before solving the simultaneous equations which result, the matrix of coefficients is inspected and, if any very small elements are found, the whole array is scaled. The subroutine SIML is called and the corrections to the parameters are calculated (lines 139–144). The error-squared sum ERSQ(IC) defined by (4.31) is computed in lines 146–153. The complete process is repeated until the change in ERSQ(IC) is less than 1% compared with the previous iteration; the program also terminates if and when IC (the iteration count) reaches 100. Note the use of the MINO routine in output statements for controlling layout of the data. This standard routine selects the lower of two (or more) integers.

The best possible test for this program should be apply it to a spectrum synthesized by the program SPECTRUM, discussed previously in Section 4.3.1. You will recall that SPECTRUM also used Gaussian

Table 4.11

RESOLVE — program for the resolution of overlapping peaks into their components

```
0008            MASTER RESOLVE
0009            DIMENSIONA1(10),B1(10),C1(10),V(100),E(100),F(100),ERSQ(100)
0010            DIMENSIONTITLE(10)
0011            COMMON/XVAR/N,X(20,20)
0012      C     THESE ARE DEFINED AT THIS STAGE SO THAT THE USER CAN  EASILY ALTER
0013      C     THE FUNCTIONS
0014      C     THESE FUNCTIONS MAY NEED TO BE DECLARED EXTERNAL TO THE MAIN
0015      C     PROGRAM FOR NON-ICL COMPILERS
0016      C     DON'T FORGET THAT B IS THE HALF WIDTH AT HALF HEIGHT
0017            FO(I,J)=A1(J)*EXP(-((C1(J)-V(I))/B1(J))**2)
0018            FA(I,J)=EXP(-((C1(J)-V(I))/B1(J))**2)
0019            FB(I,J)=FO(I,J)*2*((C1(J)-V(I))**2)/(B1(J)**3)
0020            FC(I,J)=FO(I,J)*(-2*(C1(J)-V(I))/(B1(J)**2))
0021          1 FORMAT(2I0)
0022          2 FORMAT(3F0.0)
0023          3 FORMAT(8F10.0)
0024          4 FORMAT(10A8)
0025          5 FORMAT(1H1,120(1H*),//,10A8,//,120(1H*),//,'
0026          1          INPUT DATA',/,1X,120(1H-))
0027          6 FORMAT(1X,' FREQUENCY/CM-1',8(F10.1,2X))
0028          7 FORMAT(1X,'   ABSORPTIVITY',8(F10.3,2X),//)
0029         25 FORMAT(1X,' EXPERIMENTAL',8(2X,E10.4))
0030         26 FORMAT(1X,'  CALCULATED ',8(2X,E10.4),//)
0031         31 FORMAT(7X,I2,6X,E10.4,3(5X,E10.4))
0032         32 FORMAT((25X,3(5X,E10.4)))
0033         60 FORMAT(1X,120(1H*))
0034         55 FORMAT(1X,120(1H*),//,1X,' ITERATION    ERROR SQUARE      A
0035          1      B          C',/,1X,120(1H-))
0036            IC=0
0037            READ(1,4)(TITLE(I),I=1,10)
0038            WRITE(2,5)(TITLE(I),I=1,10)
0039      C     READ NO OF BANDS,NO OF POINTS,ESTIMATES OF PARAMETERS,EXP, DATA
0040            READ(1,1)NZ,NP
0041            READ(1,2)(A1(I),B1(I),C1(I),I=1,NZ)
0042            READ(1,3)(V(J),J=1,NP)
0043            READ(1,3)(E(J),J=1,NP)
0044            IS=1
0045            IE=8
0046        777 WRITE(2,6)(V(J),J=IS,IE)
0047            WRITE(2,7)(E(J),J=IS,IE)
0048            IF(IE.GE.NP)GO TO 77
0049            IS=MINO((IS+8),NP)
0050            IE=MINO((IE+8),NP)
0051            GO TO 777
0052         77 N=3*NZ
0053            WRITE(2,55)
0054            GO TO 225
0055      C     SET UP ARRAYS
0056         99 DO 9 K=1,NZ
0057            DO 9 L=1,NZ
0058            X(K,L)=0
0059            DO 9 I=1,NP
0060            A=FA(I,K)
0061            B=FA(I,L)
0062          9 X(K,L)=X(K,L)+A*B
0063            DO 8 K=1,NZ
0064            DO 8 L=1,NZ
0065            X(NZ+K,L)=0
0066            DO 8 I=1,NP
0067            A=FA(I,L)
0068            B=FB(I,K)
0069          8 X(NZ+K,L)=X(NZ+K,L)+A*B
0070            DO 10 K=1,NZ
0071            DO 10 L=1,NZ
0072            X(2*NZ+K,L)=0
0073            DO 10 I=1,NP
0074            A=FA(I,L)
0075            B=FC(I,K)
0076         10 X(2*NZ+K,L)=X(2*NZ+K,L)+A*B
0077            DO 11 K=1,NZ
0078            DO 11 L=1,NZ
0079            X(NZ+K,NZ+L)=0
0080            DO 11 I=1,NP
0081            A=FB(I,L)
0082            B=FB(I,K)
0083         11 X(NZ+K,NZ+L)=X(NZ+K,NZ+L)+A*B
0084            DO 40 K=1,NZ
0085            DO 40 L=1,NZ
0086            X(2*NZ+K,NZ+L)=0
0087            DO 40 I=1,NP
```

```
0088            A=FC(I,K)
0089            B=FB(I,L)
0090         40 X(2*NZ+K,NZ+L)=X(2*NZ+K,NZ+L)+A*B
0091            DO 50 K=1,NZ
0092            DO 50 L=1,NZ
0093            X(2*NZ+K,2*NZ+L)=0
0094            DO 50 I=1,NP
0095            A=FC(I,L)
0096            B=FC(I,K)
0097         50 X(2*NZ+K,2*NZ+L)=X(2*NZ+K,2*NZ+L)+A*B
0098     C      SYMMETRISE MATRIX
0099            DO 12 I=1,N
0100            DO 12 J=1,N
0101         12 X(I,J)=X(J,I)
0102     C      COMPUTE DEPENDENT-VARIABLE VECTOR
0103            DO 13 K=1,NZ
0104            X(K,N+1)=0
0105            DO 13 I=1,NP
0106            Z=E(I)
0107            DO 14 J=1,NZ
0108            A=FO(I,J)
0109         14 Z=Z-A
0110         13 X(K,(N+1))=X(K,(N+1))+Z*FA(I,K)
0111            DO 15 K=1,NZ
0112            X(K+NZ,N+1)=0
0113            DO 15 I=1,NP
0114            Z=E(I)
0115            DO 16 J=1,NZ
0116            A=FO(I,J)
0117         16 Z=Z-A
0118         15 X(K+NZ,N+1)=X(K+NZ,N+1)+Z*FB(I,K)
0119            DO 17 K=1,NZ
0120            X(K+2*NZ,N+1)=0
0121            DO 17 I=1,NP
0122            Z=E(I)
0123            DO 18 J=1,NZ
0124            A=FO(I,J)
0125         18 Z=Z-A
0126         17 X(K+2*NZ,N+1)=X(K+2*NZ,N+1)+Z*FC(I,K)
0127     C      INSPECT MATRIX FOR SMALL ELEMENTS
0128            XMIN=1
0129            DO 225 I=1,N
0130            DO 225 J=I,N
0131            IF(ABS(X(I,J)),LT,0,001)XMIN=X(I,J)
0132        225 CONTINUE
0133            IF(XMIN,GE,0,001)GO TO 224
0134            SCA=1,/XMIN
0135            DO 226 I=1,N
0136            DO 226 J=1,N+1
0137        226 X(I,J)=(X(I,J))*SCA
0138        224 CALL SIML
0139            DO 21 I=1,NZ
0140         21 A1(I)=A1(I)+X(I,N+1)
0141            DO 22 J=1,NZ
0142         22 B1(J)=B1(J)+X(NZ+J,N+1)
0143            DO 23 K=1,NZ
0144         23 C1(K)=C1(K)+X(2*NZ+K,N+1)
0145     C      COMPUTE SPECTRUM
0146        225 DO 24 I=1,NP
0147            F(I)=0
0148            DO 24 J=1,NZ
0149         24 F(I)=F(I)+FO(I,J)
0150            IC=IC+1
0151            ERSQ(IC)=0
0152            DO 30 I=1,NP
0153         30 ERSQ(IC)=ERSQ(IC)+(F(I)-E(I))*(F(I)-E(I))
0154            WRITE(2,51)IC,ERSQ(IC),A1(1),B1(1),C1(1)
0155            IF(NZ,GT,1)WRITE(2,32)(A1(I),B1(I),C1(I),I=2,NZ)
0156            IF(IC,EQ,1)GO TO 99
0157            PC=ABS((ERSQ(IC)-ERSQ(IC-1))*100/ERSQ(IC-1))
0158            IF(PC,GT,1,0,AND,IC,LT,100)GO TO 99
0159            WRITE(2,60)
0160            IS=1
0161            IE=8
0162         28 WRITE(2,25)(E(K),K=IS,IE)
0163            WRITE(2,26)(F(L),L=IS,IE)
0164            IF(IE,GE,NP)GO TO 27
0165            IS=MINO((IS+8),NP)
0166            IE=MINO((IE+8),NP)
0167            GO TO 28
0168         27 CONTINUE
0169            WRITE(2,60)
0170            STOP
0171            END
```

OUTPUT FROM SPECTRUM USED AS INPUT

INPUT DATA

FREQUENCY/CM-1	12800.0	13500.0	14000.0	14700.0	15200.0	15800.0	16400.0	17000.0
ABSORPTIVITY	1.100	4.700	10.200	21.500	28.600	50.800	24.800	15.200

FREQUENCY/CM-1	17500.0	17900.0	18100.0	18700.0	19300.0	19900.0	20400.0	20900.0
ABSORPTIVITY	9.300	7.500	7.800	12.900	22.400	51.100	53.300	29.700

FREQUENCY/CM-1	21400.0	21900.0	22300.0	22700.0
ABSORPTIVITY	22.000	13.500	8.000	4.200

ITERATION	ERROR SQUARE	A	B	C
1	0.1099E 04	0.3000E 02	0.1400E 04	0.1500E 05
		0.3800E 02	0.1200E 04	0.2000E 05
2	0.1479E 03	0.2547E 02	0.1801E 04	0.1558E 05
		0.3014E 02	0.1645E 04	0.2022E 05
3	0.4626E 01	0.3082E 02	0.1510E 04	0.1567E 05
		0.3323E 02	0.1650E 04	0.2035E 05
4	0.8845E-02	0.3106E 02	0.1563E 04	0.1565E 05
		0.3337E 02	0.1638E 04	0.2034E 05
5	0.7920E-02	0.3108E 02	0.1562E 04	0.1565E 05
		0.3337E 02	0.1638E 04	0.2034E 05
6	0.7920E-02	0.3108E 02	0.1562E 04	0.1565E 05
		0.3337E 02	0.1638E 04	0.2034E 05

EXPERIMENTAL	0.1100E 01	0.4700E 01	0.1020E 02	0.2150E 02	0.2860E 02	0.5080E 02	0.2480E 02	0.1520E 02
CALCULATED	0.1117E 01	0.4685E 01	0.1020E 02	0.2149E 02	0.2862E 02	0.5081E 02	0.2477E 02	0.1523E 02

EXPERIMENTAL	0.9300E 01	0.7500E 01	0.7800E 01	0.1290E 02	0.2240E 02	0.5110E 02	0.5330E 02	0.2970E 02
CALCULATED	0.9285E 01	0.7519E 01	0.7785E 01	0.1291E 02	0.2241E 02	0.5105E 02	0.5335E 02	0.2971E 02

EXPERIMENTAL	0.2200E 02	0.1350E 02	0.8000E 01	0.4200E 01
CALCULATED	0.2198E 02	0.1550E 02	0.7994E 01	0.4201E 01

```
*************************************************************************
O-PHENYLENEDIAMINE (3 COMPONENTS) IN ETHANOL
*************************************************************************
```

INPUT DATA

FREQUENCY/CM-1	50000.0	49000.0	48000.0	47000.0	46000.0	45000.0	44000.0	43000.0
ABSORPTIVITY	20000.000	29000.000	38000.000	32000.000	19000.000	9000.000	6000.000	6200.000

FREQUENCY/CM-1	42000.0	41000.0	40000.0	39000.0	38000.0	37000.0	36000.0	35000.0
ABSORPTIVITY	6500.000	6000.000	3800.000	1800.000	880.000	950.000	1800.000	2700.000

FREQUENCY/CM-1	34000.0	33000.0	32000.0	31000.0
ABSORPTIVITY	3200.000	2500.000	850.000	170.000

ITERATION	ERROR SQUARE	A	B	C
1	0.3142E 08	0.3750E 05	0.2500E 04	0.4800E 05
		0.6800E 04	0.2600E 04	0.4150E 05
		0.3250E 04	0.2500E 04	0.3590E 05
2	0.1254E 08	0.3701E 05	0.2435E 04	0.4798E 05
		0.6542E 04	0.2371E 04	0.4197E 05
		0.1892E 04	0.3560E 04	0.3485E 05
3	0.1792E 08	0.3706E 05	0.2421E 04	0.4799E 05
		0.6853E 04	0.2611E 04	0.4176E 05
		0.2938E 04	0.1020E 04	0.3372E 05
4	0.1045E 08	0.3706E 05	0.2419E 04	0.4799E 05
		0.6585E 04	0.2594E 04	0.4185E 05
		0.2536E 04	0.1999E 04	0.3403E 05
5	0.8362E 07	0.3704E 05	0.2430E 04	0.4798E 05
		0.6704E 04	0.2465E 04	0.4185E 05
		0.3201E 04	0.2199E 04	0.3433E 05
6	0.8350E 07	0.3704E 05	0.2429E 04	0.4798E 05
		0.6715E 04	0.2462E 04	0.4186E 05
		0.3208E 04	0.2188E 04	0.3429E 05

```
*************************************************************************
```

EXPERIMENTAL	0.2000E 05	0.2900E 05	0.3800E 05	0.3200E 05	0.1900E 05	0.9000E 04	0.6000E 04	0.6200E 04
CALCULATED	0.1857E 05	0.3107E 05	0.3705E 05	0.3154E 05	0.1943E 05	0.9525E 04	0.5674E 04	0.5970E 04
EXPERIMENTAL	0.6500E 04	0.6000E 04	0.3800E 04	0.1800E 04	0.8800E 03	0.9500E 03	0.1800E 04	0.2700E 04
CALCULATED	0.6779E 04	0.5954E 04	0.3800E 04	0.1774E 04	0.7568E 03	0.8304E 03	0.1768E 04	0.2893E 04
EXPERIMENTAL	0.3200E 04	0.2500E 04	0.8500E 03	0.1700E 03
CALCULATED	0.3152E 04	0.2263E 04	0.1070E 04	0.3329E 03

```
*************************************************************************
```

146

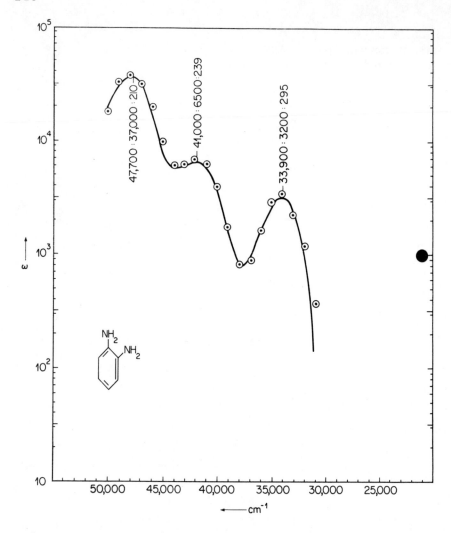

Figure 4.7 Electronic absorption spectrum of *o*-phenylenediamine. Solid line is the experimental spectrum; circles, O, are values computed by RESOLVE

functions to construct composite spectra so that, if our program is satisfactory, we should obtain almost exact agreement between the input data (for SPECTRUM) and the output data (computed from RESOLVE). As you can see (Table 4.11) the agreement is, indeed, very good. In Table 4.11 we also show an example based on some experimental data for the 3 overlapping bands of o-phenylenediamine. In Figure 4.7, the agreement between the experimental and computed data is seen to be good.

It is worthwhile mentioning at this stage that if poor starting estimates of the parameters are provided, program RESOLVE can go severely haywire, particularly in the case of complex spectra. This is characterized by the value of ERSQ diverging, rather than converging, and by the computation of ridiculous values of the parameters. If it is desired to make no changes to the program, the user can avoid this problem since it is usually possible to make good starting estimates of the parameters. The problem can also be circumvented by modifying the prógram so that the changes in the parameters are reduced by a factor λ (<1);

$$A'_j = A^\circ_j + \lambda a_j$$
$$B'_j = B^\circ_j + \lambda b_j \qquad\qquad (4.32)$$
$$C'_j = C^\circ_j + \lambda c_j$$

The corrections to the parameters in (4.32) are clearly less severe than in (4.29) and reduce the likelihood of divergence. The program can be further modified by *only* accepting refined values of the parameters that cause convergence.

Input to RESOLVE is as follows:

Card 1: Number of components (NZ); number of data points (NP) (Format 2I0).

Card 2: Estimates of A, B and C for the first component (3F0.0).

If NZ > 1,

Cards 3 to (NZ + 1): As for Card 2, one card per component.

Cards (NZ + 2) + N continuation cards if necessary: The NP Observed wavenumbers (8F10.0).

Cards (NZ + 3 + N) + N continuation cards if necessary: The NP observed absorptivities (8F10.0).

Some obvious modifications are possible: how about trying different fitting functions such as (4.26) and comparing the goodness-of-fit (Section 2.4.3)? The program could also be converted into a subroutine to service other spectroscopy programs. The program lends itself well to a data acquisition unit with which the tedium of data collection is removed (see Chapter 6).

Although we have restricted our attention to the problem of overlapping bands in electronic spectra, similar approaches to those employed in SPECTRUM and RESOLVE can be applied to overlapping bands in other circumstances. For example, infrared or NMR spectra could be fitted with suitable functions as discussed in References 36 and 37.

In the context of NMR spectroscopy, Petrakis[37] discusses the use of Gaussian and Lorentzian functions, equations (4.33) and (4.34), respectively:

$$F(H - H_0) = \frac{\sqrt{(\ln 2)}}{\sqrt{(\pi\delta)}} \exp[-(H - H_0)^2 \ln 2/\delta^2] \qquad (4.33)$$

$$F(H - H_0) = \frac{\delta}{\pi} \cdot \frac{1}{\delta^2 + (H - H_0)^2} \qquad (4.34)$$

where $F(H - H_0)$ is the signal amplitude at a magnetic field strength H (H_0 being the field strength at the peak maximum) and δ is the half-width of the peak at half-maximum. Petrakis discusses the use of the two functions and their effect on the calculation of peak areas, derivatives and statistical functions.

Another important area of curve resolution is in gas chromatography since it is desirable that the position, area and height of each peak in a composite curve should be identified. Curve resolution can be achieved by a similar program to RESOLVE with suitable modifications. These centre about the type of function which describes a GLC peak; one common function is,[38]

$$y_k = h_k \exp - (t - t_k)^2/2\sigma_k^2$$

The chromatogram is a series of deviations, y_k, from the baseline recorded as a function of time, t. Also we have:

$$\left.\begin{array}{l} h_k = \text{peak height} \\ t_k = \text{retention time (t at the kth peak maximum)} \\ \sigma_k = \text{standard deviation} \end{array}\right\} \begin{array}{l} \text{for the} \\ k\text{th} \\ \text{component} \end{array}$$

Knowing σ_k and h_k, the product of the two gives the peak area. Improved results were quoted in Reference 38 by use of the non-symmetric function:

$$y_k = h_k \left(\frac{t_k}{t}\right)^{1/2} e^{-2t_k(t_k^{1/2} - t_k)^2/\sigma_k^2}$$

Difficulties are, however, encountered in this area when the component peaks of a composite envelope are defined by *different* functions. Since it is very difficult to ascertain which function is to be associated with which peak, the problem can become very hard to solve precisely.

4.4 Spectroscopy II – Examples of Applications with Resolved Peaks

The previous section was intended as a precursor to this section in that spectroscopists usually perform calculations of, for example, oscillator strengths, band intensities, transition frequencies and the like on resolved bands. Sometimes the line shape itself is important[37] but, more frequently, the peak positions and areas are used. In this section we describe three typical applications of spectroscopic measurements which are of importance in theoretical chemistry.

4.4.1 Thermodynamic Properties of Gaseous Molecules from Infrafed Data

The application of statistical mechanics to the infrared spectra of gaseous molecules permits, in suitable cases, the calculation of thermodynamic properties such as absolute entropies and heat capacities. The methods are most easily applied to diatomic molecules but are also straightforward for linear triatomics. Since these latter display more interesting behaviour and are more worthy of computer study, we shall only discuss these cases and refer the reader to the literature[39] for further information.

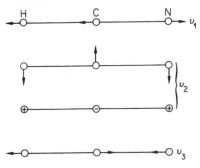

Figure 4.8   Normal modes of vibration of a triatomic linear molecule (+ and − represent motion out of and into the plane of the paper)

The infrared spectrum of a gaseous molecule may consist of vibrational absorption lines, each with rotational fine structure. For example, HCN has three normal modes of vibration (Figure 4.8) of which only v_2 and v_3 are directly observed at 712 cm^{-1} and 3317 cm^{-1}. An indirect calculation of v_1 (from the combination band $v_1 + v_2$ at 2801 cm^{-1}) yields a value of 2089 cm^{-1} for this band. An illustration of the rotational fine structure is shown in Figure 4.9 for the v_3 band of HCN.

In the case of a linear triatomic molecule, theory predicts[39] that each vibrational absorption band should be split into three branches with individual lines dependent on J, the rotational quantum number in the ground vibrational state

P branch: $\bar{v}_P = \bar{v}_{class} - 2BJ$
Q branch: $\bar{v}_Q = \bar{v}_{class}$ (forbidden for stretching vibrations of linear molecules)
R branch: $\bar{v}_R = \bar{v}_{class} + 2B(J + 1)$

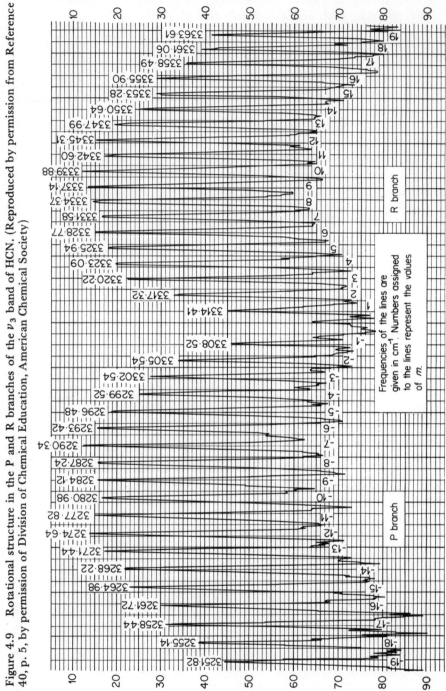

Figure 4.9   Rotational structure in the P and R branches of the ν_3 band of HCN. (Reproduced by permission from Reference 40, p. 5, by permission of Division of Chemical Education, American Chemical Society)

Frequencies of the lines are given in cm⁻¹. Numbers assigned to the lines represent the values of m.

R branch

P branch

where each $\bar{\nu}$ is in cm^{-1}, $\bar{\nu}_{\text{class}}$ is the classical vibrational frequency, B is the rotational constant (assumed equal in both ground and excited vibrational states) given by

$$B = h/8\pi^2 cI$$

where h is Planck's constant, c is the velocity of light and I is the moment of inertia of the molecule.

An average value of B can be obtained by simply averaging the spacings between the adjacent rotational line positions. (Remember that for the stretching vibration the Q branch is absent which produces effectively a double spacing at the band centre.)

We now have sufficient information to calculate the rotational and vibrational contributions to the absolute entropy and the heat capacity at constant volume. The required equations[40] are as follows:

Rotational contributions:

$$(C_v)_{\text{rot}} = R \quad \text{J mol}^{-1}\text{K}^{-1} \quad S_{\text{rot}} = R(1 + \ln[kT/(hcB)]) \quad \text{J mol}^{-1}\text{ K}^{-1}$$

Vibrational contributions:

$$(C_v)_{\text{vib}} = R \left[\left(\frac{\theta_1}{T}\right)^2 \frac{e^{\theta_1/T}}{(e^{\theta_1/T}-1)^2} + 2\left(\frac{\theta_2}{T}\right)^2 \frac{e^{\theta_2/T}}{(e^{\theta_2/T}-1)^2} \right.$$

$$\left. + \left(\frac{\theta_3}{T}\right)^2 \frac{e^{\theta_3/T}}{(e^{\theta_3/T}-1)^2} \right] \text{J mol}^{-1}\text{ K}^{-1}$$

$$S_{\text{vib}} = R \left[\frac{\theta_1}{T} \cdot 2\left(\frac{\theta_2}{T}\right)\frac{1}{(e^{\theta_2/T}-1)} + \left(\frac{\theta_3}{T}\right)\frac{1}{(e^{\theta_3/T}-1)} \right]$$

$$- R \ln \left[(1-e^{-\theta_1/T})(1-e^{-\theta_1/T})(1-e^{-\theta_3/T})\right] \quad \text{J mol}^{-1}\text{ K}^{-1}$$

where

$\theta_i = h\nu_i/k$ and ν is in hertz ($= c\bar{\nu}$)

Translational contributions:

$$(C_v)_{\text{trans}} = {}^3/_2 R \quad \text{J mol}^{-1}\text{ K}^{-1}$$
$$S_{\text{trans}} = [R({}^3/_2 \ln M + {}^5/_2 \ln T - 2.315] \quad \text{J mol}^{-1}\text{ K}^{-1}$$

where M = molar mass.

These complex equations are handled in the program STAT 2 (Table 4.12) which is based on the FORTRAN II program supplied by Little.[40] Input to the program is on a single data card:

Input card: Temperature interval; values of ν_1, ν_2 and ν_3/cm^{-1}; value of the rotational constant B; molar mass; lower (TL) and upper (TU) temperature limits (Format 8F0.0).

Table 4.12

Program STAT2 — calculation of thermodynamic parameter from infrared data

```
0008              MASTER STAT2
0009              DIMENSION XXZ(10)
0010              READ(1,1000)(XXZ(I),I=1,10)
0011         1000 FORMAT(10A8)
0012              WRITE(2,1003)(XXZ(I),I=1,10)
0013              WRITE(2,1001)
0014         1001 FORMAT(1X,'STATISTICAL CALCULATION OF THERMODYNAMIC FUNCTIONS'
0015            1,/,1X,110(1H*),/)
0016              READ(1,1002)TINT,ANU1,ANU2,ANU3,B,AMW,TL,TU
0017              WRITE(2,1004)TINT,ANU1,ANU2,ANU3,B,AMW
0018         1004 FORMAT(1X,'TINT=',1F6.1,5X,  'NU1,NU2,NU3 ARE ',3F7.1,5X,'BROT=',
0019            1F4.1,5X,'ATOMIC WT=',F5.1)
0020         1002 FORMAT(8F0.0)
0021         1003 FORMAT(1H1,10A8)
0022         1006 FORMAT(1X,'  T/K     ZROT      ZVIB      CVVIB     CVTOT    SROT     SVIB
0023            1  STR      STOT   /J K-1 MOL-1')
0024              WRITE(2,1006)
0025              FAC=6.625E-27*2.9979E10/1.3805E-16
0026              TEMP=TL-TINT
0027            6 TEMP=TEMP+TINT
0028              EX1P=EXP(-ANU1*FAC/TEMP)
0029              EX2P=EXP(-ANU2*FAC/TEMP)
0030              EX3P=EXP(-ANU3*FAC/TEMP)
0031              BEX1P=1.0/(1.0-EX1P)
0032              BEX2P=1.0/(1.0-EX2P)
0033              BEX3P=1.0/(1.0-EX3P)
0034              CEX1P=(ANU1**2)*EX1P*BEX1P**2
0035              CEX2P=2.*(ANU2**2)*EX2P*BEX2P**2
0036              CEX3P=(ANU3**2)*EX3P*BEX3P**2
0037              DEX1P=ANU1*EX1P*BEX1P
0038              DEX2P=2.*ANU2*EX2P*BEX2P
0039              DEX3P=ANU3*EX3P*BEX3P
0040          C   CALCULATE PARTITION FUNCTIONS
0041              ZROT=0.6951*TEMP/B
0042              ZVIB=BEX1P*(BEX2P**2)*BEX3P
0043          C   CALCULATE ENTROPY CONTRIBUTIONS
0044          C   SPECIFIC HEAT CONTRIBUTIONS
0045              CVVIB=((8.314*FAC**2)/TEMP**2)*(CEX1P+CEX2P+CEX3P)
0046              CVTOT=8.314+CVVIB+12.384
0047              SROT=8.314*(1.0+ALOG(ZROT))
0048              SVIB=(8.314*FAC/TEMP)*(DEX1P+DEX2P+DEX3P)+8.314*ALOG(ZVIB)
0049              STR =8.314*(1.5*ALOG(AMW)+2.5*ALOG(TEMP))-2.315
0050              STOT=SROT+SVIB+STR
0051          C   OUTPUT
0052              NTEMP=NINT(TEMP)
0053              WRITE(2,1009)NTEMP,ZROT,ZVIB,CVVIB,CVTOT,SROT,SVIB,STR,STOT
0054         1009 FORMAT(1X,I5,1X,F7.1,F9.4,2F9.3,2X,F6.2,F7.2,2X,F6.2,2X,F6.2)
0055              IF(TU-TEMP)7,7,6
0056            7 CONTINUE
0057              WRITE(2,1010)
0058         1010 FORMAT(//,1H    CVROT = 8.314  CVTR = 12.471',//)
0059              STOP
0060              END
```

A TEST RUN
STATISTICAL CALCULATION OF THERMODYNAMIC FUNCTIONS
****

TINT= 20.0   NU1,NU2,NU3 ARE 2089.0 712.0 3317.0   BROT= 1.2   ATOMIC WT= 27.0

T/K	ZROT	ZVIB	CVVIB	CVTOT	SROT	SVIB	STR	STOT	/J K-1 MOL-1
100	57.6	1.0001	0.062	20.760	42.01	0.01	134.52	176.54	
120	69.1	1.0004	0.238	20.936	43.53	0.03	138.30	181.86	
140	80.6	1.0013	0.592	21.290	44.81	0.09	141.51	186.41	
160	92.1	1.0033	1.134	21.832	45.92	0.20	144.28	190.41	
180	103.7	1.0068	1.831	22.529	46.90	0.38	146.73	194.01	
200	115.2	1.0120	2.634	23.332	47.78	0.61	148.92	197.31	
220	126.7	1.0193	3.494	24.192	48.57	0.90	150.90	200.37	
240	138.2	1.0286	4.369	25.067	49.29	1.24	152.71	203.25	
260	149.7	1.0401	5.232	25.930	49.96	1.63	154.38	205.96	
280	161.2	1.0536	6.064	26.762	50.57	2.05	155.92	208.54	
300	172.8	1.0692	6.855	27.553	51.15	2.49	157.35	210.99	
320	184.3	1.0868	7.601	28.299	51.68	2.96	158.69	213.33	
340	195.8	1.1062	8.301	28.999	52.19	3.44	159.95	215.58	
360	207.3	1.1275	8.958	29.656	52.66	3.93	161.14	217.74	
380	218.8	1.1504	9.575	30.273	53.11	4.43	162.26	219.81	
400	230.4	1.1751	10.156	30.854	53.54	4.94	163.33	221.81	

152

420	241,9	1,2013	10,704	31,402	53,94	5,45	164,34	223,74
440	253,4	1,2290	11,225	31,923	54,33	5,96	165,31	225,60
460	264,9	1,2583	11,721	32,419	54,70	6,47	166,23	227,40
480	276,4	1,2890	12,195	32,893	55,05	6,98	167,12	229,15
500	287,9	1,3212	12,650	33,348	55,39	7,49	167,97	230,85
520	299,5	1,3547	13,089	33,787	55,72	7,99	168,78	232,49
540	311,0	1,3897	13,513	34,211	56,03	8,49	169,57	234,09
560	322,5	1,4260	13,924	34,622	56,34	8,99	170,32	235,65
580	334,0	1,4637	14,324	35,022	56,63	9,49	171,05	237,17
600	345,5	1,5027	14,713	35,411	56,91	9,98	171,76	238,65
620	357,1	1,5431	15,092	35,790	57,18	10,47	172,44	240,09
640	368,6	1,5849	15,463	36,161	57,45	10,95	173,10	241,50
660	380,1	1,6280	15,826	36,524	57,70	11,43	173,74	242,87
680	391,6	1,6725	16,181	36,879	57,95	11,91	174,36	244,22
700	403,1	1,7184	16,529	37,227	58,19	12,39	174,96	245,54
720	414,6	1,7656	16,870	37,568	58,43	12,86	175,55	246,83
740	426,2	1,8143	17,204	37,902	58,65	13,32	176,12	248,09
760	437,7	1,8644	17,531	38,229	58,88	13,79	176,67	249,33
780	449,2	1,9159	17,853	38,551	59,09	14,25	177,21	250,55
800	460,7	1,9688	18,168	38,866	59,30	14,70	177,74	251,74
820	472,2	2,0233	18,477	39,175	59,51	15,15	178,25	252,91
840	483,7	2,0792	18,780	39,478	59,71	15,60	178,75	254,06
860	495,3	2,1366	19,077	39,775	59,90	16,05	179,24	255,19
880	506,8	2,1956	19,368	40,066	60,09	16,49	179,72	256,30
900	518,3	2,2561	19,653	40,351	60,28	16,93	180,18	257,39
920	529,8	2,3181	19,932	40,630	60,46	17,36	180,64	258,47
940	541,3	2,3818	20,206	40,904	60,64	17,80	181,09	259,53
960	552,9	2,4471	20,474	41,172	60,82	18,22	181,53	260,57
980	564,4	2,5141	20,736	41,434	60,99	18,65	181,95	261,59
1000	575,9	2,5827	20,995	41,691	61,16	19,07	182,37	262,60

CVROT = 8,314   CVTR = 12,471

The program is quite strightforward and you should be able to identify the calculations, listed above. The output includes the input data and nine columns of data as follows:

Column 1: Temperature.
Columns 2 and 3: Rotational and vibrational partition functions.
Columns 4 and 5: Vibrational contribution to, and total value of C_v the heat capacity at constant volume.
Columns 6, 7, 8 and 9: The contributions to the total entropy of the gas.

Finally, the rotational and translational contributions to C_v are printed.

4.4.2 Electronic Spectra of Transition Metal Complexes
Many properties of such transition metal complexes as $[Co(NH_3)_6]^{2+}$, $[Fe(H_2O)_6]^{2+}$, $[CoCl_4]^{2-}$ can be explained satisfactorily by consideration of the relative energies of the inner d-orbitals. We will concern ourselves solely with complexes of the first transition series so that the orbitals are those of the $3d$ level. The two theories that have been used most effectively in this area are the crystal field theory (CFT) and the molecular orbital (M0) theory; in many cases both lead to similar predictions, but from different standpoints. M0 theory has been used quite extensively[41] but, in this particular area, it tends to get rather complicated. CFT is simpler but necessarily more restricted in its applications since it ignores the covalent bonding between the central

metal ion and the surrounding ligands in a complex. Ligand Field Theory (LFT) allows for some covalency whilst preserving the simplicity of CFT. It is not appropriate to discuss these theories at great length here and we refer the reader to Reference 41 for suitable background material. The most important results for our purposes concern the splitting of the D or F ground electronic states of the transition metal ion when placed in the electrostatic field of the ligands. For a D ground state (d^1 or d^9 electronic configurations) the splitting is as shown in Figure 4.10(a) in which the new energy levels are identified by the symbols e, t_2 etc. Note that Δ_0 and Δ_t are used to indicate the extent of the splitting of the levels but $10Dq$ is a quantity usually used in theoretical treatments. For this simple case, a single band is observed in the electronic absorption spectrum with an energy equal to $10Dq$ (Δ_0 or Δ_t). For configurations other than d^1 or d^9, the $3d$ transition metal ions have S or F ground states in the absence of a crystal field. An S state is unaffected by the electrostatic field (crystal field) due to the ligands. An F term is split into three components   (Figure 4.10(b)) and,

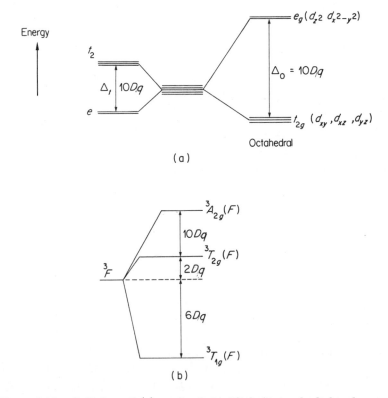

Figure 4.10   Splitting of (a) a set of d-orbitals in octahedral and tetrahedral symmetries and (b) the ^3F state of a d^8 octahedral configuration

Table 4.13
Assignments of the first three electronic transitions of metal
complexes of 3d metal ions for the d^2, d^3, d^7 and d^8
configurations

	Configuration	Transition	Assignment
or	d^2 (Tetrahedral)	ν_1	$^3T_{2g}(F) \leftarrow {}^3A_{2g}(F)$
	d^8 (Octahedral)	ν_2	$^3T_{1g}(F) \leftarrow {}^3A_{2g}(F)$
		ν_3	$^3T_{1g}(P) \leftarrow {}^3A_{2g}(F)$
or	d^3 (Tetrahedral)	ν_1	$^4T_{2g}(F) \leftarrow {}^4T_{1g}(F)$
	d^7 (Octahedral)	ν_2	$^4A_{2g}(F) \leftarrow {}^4T_{1g}(F)$
		ν_3	$^4T_{1g}(P) \leftarrow {}^4T_{1g}(F)$
or	d^3 (Octahedral)	ν_1	$^4T_{2g}(F) \leftarrow {}^4A_{2g}(F)$
	d^7 (Tetrahedral)	ν_2	$^4T_{1g}(F) \leftarrow {}^4A_{2g}(F)$
		ν_3	$^4T_{1g}(P) \leftarrow {}^4A_{2g}(F)$
or	d^2 (Octahedral)	ν_1	$^3T_{2g}(F) \leftarrow {}^3T_{1g}(F)$
	d^8 (Tetrahedral)	ν_2	$^3T_{1g}(P) \leftarrow {}^3T_{1g}(F)$
		ν_3	$^3A_{2g}(F) \leftarrow {}^3T_{1g}(F)$

Notes:
(i) ν_2 and ν_3 assignments may be reversed in some cases.
(ii) ν_1 is not always observed if of particularly low energy.
(iii) The g subscripts should be omitted for tetrahedral symmetries.

in general, both ground and excited terms may be split into components by a crystal field. Usually, only the transitions from the ground state to the next two or three higher energy levels are important in visible/ultraviolet spectra of metal complexes. The more important transitions for d^2, d^3, d^7 and d^8 ions are listed in Table 4.13.

The energy levels, and transitions between them are defined in terms of the two parameters Dq and B. The former has been mentioned already whilst the latter, called the Racah Parameter can be regarded as a measure of the interelectronic repulsions in the ion. (For example, the $^3T_{1g}(P)$ term lies at $15B$ higher than a $^3T_{1g}(F)$.) It is smaller in a complex than in a free ion and the amount of lowering of B is taken as a measure of the degree of covalency of the metal–ligand bonds.

The energetics of splitting of the various terms is more complicated than the simple case of a D term. This is because the T_{1g} terms from the P and F terms have the same multiplicity and, because of this, they interact with each other. We can show this interaction graphically as in Figure 4.11.

Notice that the result of the interaction is to introduce curvature into the T_{1g} terms; the energies of the T_{2g} and A_{2g} terms are not subject to curvature and still have precisely the same energies (denoted in Figure 4.11) regardless of the degree of interaction.

The energies of the T_{1g} terms can be expressed numerically by equations (4.35) and (4.36).

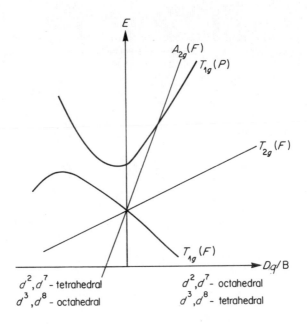

Figure 4.11   The effect of increasing Dq (in units of B) on the energies of components derived from the ^3F and ^3P terms of d^2, d^3, d^7 and d^8 configurations

Case 1: d^2, d^7 (octahedral; d^3, d^8 (tetrahedral)

$$E^2 + (6Dq - 15B)E - 16Dq^2 - 90Dq \cdot B = 0 \qquad (4.35)$$

Case 2: d^2, d^7 (tetrahedral); d^3, d^8 (octahedral)

$$E^2 - (15B + 6Dq)E + 16Dq^2 + 90Dq \cdot B = 0 \qquad (4.36)$$

Each of these equations has two solutions; these are the upper and lower values of E which refer to the energies of $T_{1g}(P)$ and $T_{1g}(F)$ respectively.

We are now in a position to appreciate the simple simulation program CRYSTAL,[42] in Table 4.14. Into this program we input ranges of Dq and of B; it then computes the energies of the $T_{1g}(F)$, T_{2g}, $T_{1g}(P)$ and A_{2g} terms and the energies of the transitions from the ground state to each of the excited states. The program is a simple one but it has a number of interesting features such as the computed GOTO which transfers control to the appropriate part of the program. You could easily use this program with a graph-plotting routine and generate graphs similar to those in Figure 4.11 with your own personalized

157

Table 4.14

A program to simulate the effect of Dq and B on the energy levels and absorption
maxima of a metal complex

```
0008                MASTER
0009                DIMENSION TITLE(10)
0010        C       THIS IS A PROGRAM TO SIMULATE THE EFFECT OF CHANGES IN DQ AND/OR B
0011        C       ON THE ENERGY LEVELS AND ABSORPTION MAXIMA OF A METAL COMPLEX,
0012        C       NS DEFINES SYMMETRY,(6 FOR OCTAHEDRAL, 4 FOR TETRAHEDRAL)
0013        C       ND DEFINES NO OF ELECTRONS (2,3,7, OR 8)
0014        C       WE SPECIFY RANGES OF DQ AND B WHICH ARE VARIED SYSTEMATICALLY
0015        C       IF DQ OR B IS TO BE FIXED,IDENTICAL VALUES MUST BE ENCODED ONTO
0016        C       THE APPROPRIATE CARD
0017        C
0018        C       ALL INPUT IS IN REAL FORMAT
0019        99 READ(1,10)(TITLE(I),I=1,10)
0020           WRITE(2,20)(TITLE(I),I=1,10)
0021        10 FORMAT(10A8)
0022        20 FORMAT(1H1,10A8,///)
0023        C       NOW READ DQ START AND STEP
0024           READ(1,1)DQ1,DQS
0025        C       NOW B RANGE AND STEP
0026           READ(1,1)B1,BS
0027         1 FORMAT(2F0,0)
0028           READ(1,2)NS,ND,NREP
0029         2 FORMAT(3I0)
0030           WRITE(2,8)NS,ND
0031         8 FORMAT(1X,'        CRYSTAL FIELD SIMULATION',////,5X,'NS=',I3,'
0032         1    ND=',I3,//)
0033           WRITE(2,5)
0034         5 FORMAT( 4X,'   DQ        B        E1        E2        E3        T1G
0035         1F     T2GF     A2GF     T1GP',//)
0036           K=NS+ND-5
0037           N1=5
0038           N2=5
0039           IF((ABS(DQS)),LT,0,0001)N1=1
0040           IF((ABS(BS)),LT,0,0001)N2=1
0041           DO 9 I=1,N1
0042           DQ=DQ1+(I-1)*DQS
0043           DO 9 J=1,N2
0044           B=B1+(J-1)*BS
0045           GO TO (3,4,4,3,3,3,4,4,3,3)K
0046        C
0047        C   CASE 1
0048        C
0049         4 DEL=10*DQ
0050           PU=(15*B-6*DQ)/2
0051           XU=(SQRT(225*B*B+18*B*DEL+DEL*DEL))/2
0052           E1=2*DQ-PU+XU
0053           E2=12*DQ-PU+XU
0054           E3=2*XU
0055           T1GF=PU-XU
0056           T2GF=2*DQ
0057           A2GF=12*DQ
0058           T1GP=PU+XU
0059           GO TO 7
0060        C
0061        C   SECOND CASE
0062        C
0063         3 DEL=10*DQ
0064           PU=(15*B+6*DQ)/2
0065           XU=(SQRT(225*B*B-18*B*DEL+DEL*DEL))/2
0066           E2=PU-XU+12*DQ
0067           E3=PU+XU+12*DQ
0068           E1=DEL
0069           A2GF=-12*DQ
0070           T2GF=-2*DQ
0071           T1GF=PU-XU
0072           T1GP=PU+XU
0073         7 WRITE(2,6)DQ,B,E1,E2,E3,T1GF,T2GF,A2GF,T1GP
0074         6 FORMAT(1X,9(3X,F7,0))
0075         9 CONTINUE
0076           IF(NREP,GT,0)GO TO 99
0077           STOP
0078           END
```

CASE 1   AQUEOUS CO(2+)   D7   OCTAHEDRAL

CRYSTAL FIELD SIMULATION

NS= 6   ND= 7

DQ	B	E1	E2	E3	T1GF	T2GF	A2GF	T1GP
800,	775,	7001,	15001,	17628,	-5401,	1600,	9600,	12226,
800,	800,	6989,	14989,	17978,	-5389,	1600,	9600,	12589,
800,	825,	6977,	14977,	18329,	-5377,	1600,	9600,	12952,
800,	850,	6965,	14965,	18681,	-5365,	1600,	9600,	13315,
800,	875,	6954,	14954,	19033,	-5354,	1600,	9600,	13679,
860,	775,	7558,	16158,	18140,	-5838,	1720,	10320,	12303,
860,	800,	7544,	16144,	18488,	-5824,	1720,	10320,	12664,
860,	825,	7531,	16131,	18836,	-5811,	1720,	10320,	13026,
860,	850,	7518,	16118,	19186,	-5798,	1720,	10320,	13388,
860,	875,	7506,	16106,	19537,	-5786,	1720,	10320,	13751,
920,	775,	8116,	17316,	18658,	-6276,	1840,	11040,	12361,
920,	800,	8102,	17302,	19003,	-6262,	1840,	11040,	12742,
920,	825,	8087,	17287,	19349,	-6247,	1840,	11040,	13102,
920,	850,	8073,	17273,	19697,	-6233,	1840,	11040,	13463,
920,	875,	8060,	17260,	20045,	-6220,	1840,	11040,	13825,
980,	775,	8678,	18478,	19180,	-6718,	1960,	11760,	12463,
980,	800,	8662,	18462,	19523,	-6702,	1960,	11760,	12822,
980,	825,	8646,	18446,	19867,	-6686,	1960,	11760,	13181,
980,	850,	8651,	18431,	20212,	-6671,	1960,	11760,	13541,
980,	875,	8617,	18417,	20559,	-6657,	1960,	11760,	13902,
1040,	775,	9241,	19641,	19707,	-7161,	2080,	12480,	12546,
1040,	800,	9224,	19624,	20048,	-7144,	2080,	12480,	12904,
1040,	825,	9207,	19607,	20390,	-7127,	2080,	12480,	13262,
1040,	850,	9191,	19591,	20733,	-7111,	2080,	12480,	13621,
1040,	875,	9176,	19576,	21077,	-7096,	2080,	12480,	13981,

values of Dq and B. Input to the program is as follows:

Card 1: Title — Format 10A8.

Card 2: DQ1, DQ2 — lower limit and step size for DQ (usually in units of cm^{-1}) — Format 2F0.0.

Card 3: B1, B2 — lower limit and step size for B (in the same units as DQ) — Format 2F0.0

Card 4: ND (number of d-electrons), NS (6 for octahedral, 4 for tetrahedral), NREP (greater than zero if another set of data follows) 3I0.

Illustrative input/output follow the program in Table 4.14.

The most generally applicable method for calculating Dq and B from experimental data is from the Tanabe–Sugano diagrams.[43] Analytical expressions have, however, been derived[44] for certain configurations, as follows:

(i) d^3, d^8 (octahedral) and d^2, d^7 (tetrahedral)

$$10Dq = v_2 - v_1 \qquad (4.37)$$

$$B = [(2v_1 - v_2)(v_1 - v_2)]/[3(5v_2 - 9v_1)] \qquad (4.38)$$

$$340Dq^2 - 18(v_2 + v_3)Dq + v_2 v_3 = 0 \qquad (4.39)$$

$$B = (v_2 + v_3 - 30Dq)/15 \qquad (4.40)$$

(ii) d^3, d^8 *(tetrahedral) and* d^2, d^7 *(octahedral)*

$$10Dq = v_2 - v \tag{4.41}$$

$$B = [v_1(v_2 - 2v_1)/3]/[9v_1 - 4v_2] \tag{4.42}$$

$$340Dq^2 + 18(v_3 - 2v)Dq + v_2^2 - v_2 v_3 = 0 \tag{4.43}$$

$$B = (v_3 - 2v_2 + 30Dq)/15 \tag{4.44}$$

The symbols v_1, v_2 and v_3 refer to the energies of the transitions from the ground state to the first, second or third excited state. Sometimes we will know v_1, v_2 and v_3 but frequently one absorption band (almost always v_1 or v_3) will be absent or obscured. This is why we quote some alternative equations in (i) and (ii) above. In the case where the solution of quadratic equations is required it will always be obvious which of the two solutions for Dq is correct (in fact, the more positive solution is the one). We can write a program to solve these equations and also we can:

(a) Set v_1 or v_3 to zero (when v_1, v_2 and v_3 are known), recalculate $10Dq$ and B and estimate v_1 and v_3.

(b) Interchange v_2 and v_3 in case (i) above in case of incorrect assignment.

By these means our data (or that from the literature) can be checked for internal consistency — this is very helpful when we are not sure of the correctness of some of our assignments.

Program BXTL (Table 4.15) uses the above principles and the reader should be able to identify the appropriate calculation steps which are quite simple and straightforward. Input to the program is as follows:

Card 1: Title — Format 10A8.

Card 2: Band positions v_1, v_2 and v_3 in wavenumbers, cm^{-1}. If v_1 or v_3 are unknown, input them as zero — Format 3I0.

Card 3: ND, NSYM, NREP — exactly as for Card 4 in program CRYSTAL.

The sum (NSYM + NREP) causes control to be passed to the appropriate section of the program. Specimen input and output are given for the aqueous Cr^{3+} ion. The assignments of the bands are:

$^4T_{2g}(F) \longleftarrow {}^4A_{2g}(F) (= 10Dq)$ 17000 cm^{-1}

$^4T_{1g}(F) \longleftarrow {}^4A_{2g}(F)$ 24000 cm^{-1}

$^4T_{1g}(P) \longleftarrow {}^4A_{2g}(F)$ (weak shoulder) 37000 cm^{-1}

Examination of the output reveals that, in 'checking routine A' (where v_1 is set to zero) an error message is printed. The calculation is, however, continued since the quantity A (line 72) is often subject to error as it is formed by the difference between two large quantities each subject to experimental error and might become negative even if the

Table 4.15

BXTL — a program to calculate $10Dq$ and B of a metal complex

```
0008        MASTER BXTL
0009      C THIS IS A PROGRAM TO EVALUATE TENDQ AND B FOR METAL IONS IN OH OR
0010      C TD SYMMETRIES WITH 2,3,7,8, D ELECTRONS,.
0011      C
0012      C FIVE DATA CARDS ARE NEEDED,THESE SPECIFY THE BAND MAXIMA, V1,V2,V3
0013      C THE NUMBER OF D ELECTRONS(ND),THE SYMMETRY(NSYM).
0014      C
0015      C
0016      C IMPORTANT
0017      C THIS PROGRAMME DOES NOT ALLOW FOR INTERCHANGE OF V1 AND V2
0018      C
0019        DIMENSIONTITLE(20)
0020      1 FORMAT(3I0)
0021    101 FORMAT(3OHOTHE VALUES OF TENDQ AND B ARE)
0022    102 FORMAT(1H ,6HTENDQ=,I7)
0023    103 FORMAT(1H ,2HB=,I7)
0024    104 FORMAT(19HOCHECKING ROUTINE A)
0025    105 FORMAT(19HOCHECKING ROUTINE B)
0026    106 FORMAT(25HOESTIMATED POSITION OF V1)
0027    107 FORMAT(1H ,4HEV1=,I7)
0028    108 FORMAT(25HOESTIMATED POSITION OF V3)
0029    109 FORMAT(1H ,4HEV3=,I7)
0030    110 FORMAT(22HOINTERCHANGE V2 AND V3)
0031    111 FORMAT(1H ,3HV2=,I7)
0032    112 FORMAT(1H ,3HND=,I2)
0033    113 FORMAT(1H ,5HNSYM=,I2)
0034    114 FORMAT(1H ,4HNDS=,I2)
0035    701 FORMAT(1H ,3HV1=,I7)
0036    901 FORMAT(1H ,3HV3=,I7)
0037     99 FORMAT(10A8)
0038    999 FORMAT(1H1,10A8)
0039      C INPUT COMPOUND NAME
0040     13 READ(1,99)(TITLE(I),I=1,10)
0041        WRITE(2,999)(TITLE(I),I=1,10)
0042        READ(1,1)N1,N2,N3
0043        READ(1,1)ND,NSYM,NREP
0044      C NREP IS REPETITION INTEGER
0045        WRITE(2,112)ND
0046        WRITE(2,113)NSYM
0047        NDS=ND+NSYM
0048        J=NDS-5
0049        N=0
0050        M1=N1
0051        M2=N2
0052        M3=N3
0053        WRITE(2,701)N1
0054        WRITE(2,111)N2
0055        WRITE(2,901)N3
0056        IF(N1)8,8,9
0057      8 N=2
0058      9 CONTINUE
0059        GO TO (7,6,6,7,7,7,6,6,7,7),J
0060      6 CONTINUE
0061      C EVALUATION OF TENDQ AND RACAH PARAMETER B
0062      C FOR D/ ORD2(OH) AND D3 ORD8(TD)
0063      C V1=T1G(F)-T2G,V2=T1G(F+-A2G,V3=T1G(F)-T1G(P),
0064      2 IF(N1)21,22,21
0065     21 IF (N3) 24,23,24
0066     24 NENDQ=(N2 -N1)
0067        K=(N3+N2-(3 *N1))/15
0068        GO TO 55
0069     22 V2=FLOAT(N2)
0070        V3=FLOAT(N3)
0071        A=324*V3*V3-64*V2*V2+64*V2*V3
0072        IF(A,GT,0)GO TO 60
0073        WRITE(2,61)A
0074     61 FORMAT(1X,' A IS EQUAL TO',E10,4,' NOW SET TO ZERO')
0075        A=0
0076     60 NENDQ=(36 *V2-18 *V3+SQRT(A))/68
0077        K=(V3-(2 *V2)+(3 *NENDQ))/15
0078        NV1=N2-NENDQ
0079        WRITE(2,106)
0080        WRITE(2,107)NV1
0081        GO TO 55
0082     23 NENDQ=(N2-N1)
0083        K=N1*(N2-(2 *N1))/3 /((9 *N1)-(4 *N2))
0084        NV3=15 *K+2 *N2-3 *NENDQ
0085        WRITE(2,108)
```

```
0086            WRITE(2,109)NV3
0087        55 CONTINUE
0088            WRITE(2,101)
0089            WRITE(2,102)NENDQ
0090            WRITE(2,103)K
0091            N=N+1
0092            GO TO(25,26,27,29)N
0093        25 IF(N1)29,29,225
0094       225 IF(N5)29,29,2225
0095      2225 N1=0
0096            WRITE(2,104)
0097            GO TO 22
0098        26 N1=M1
0099            N3=0
0100            WRITE(2,105)
0101            GO TO 23
0102        27 WRITE(2,110)
0103            N2=M5
0104            N3=M2
0105            GO TO 2
0106
0107       C
0108       C    EVALUATION OF TENDQ AND RACAH PARAMETER B
0109       C    FOR D2 OR D2(TD) AND D3 OR D8(OH)
0110       C    V1=A2G-T2G,V2=A2G-T1G(F),V3=A2G-T1G(P),
0111       C
0112         7 CONTINUE
0113         4 IF(N1)42,44,42
0114        42 IF(N3)48,46,48
0115        48 NENDQ=N1
0116            K=(N3+N2-(3 *N1))/15
0117            GO TO 5
0118        44 V2=FLOAT(N2)
0119            V3=FLOAT(N3)
0120            A=324*((V3+V2)*(V3+V2))-1360*V2*V3
0121            IF(A.GT.0)GO TO 71
0122            WRITE(2,61)A
0123            A=0
0124        71 NENDQ=(16 *(V3+V2)-SQRT(A))/68
0125            K=(N3+N2-3 *NENDQ)/15
0126            WRITE(2,106)
0127            NV1=NENDQ
0128            WRITE(2,107)NV1
0129            GO TO 5
0130        46 NENDQ=N1
0131            V1=FLOAT(N1)
0132            V2=FLOAT(N2)
0133            K=((2 *V1-V2)*(V1-V2))/(3 *(5 *V2-9 *V1))
0134            WRITE(2,108)
0135            NV3=15 *K-N2+3 *N1
0136            WRITE(2,109)NV3
0137         5 CONTINUE
0138            WRITE(2,101)
0139            WRITE(2,102)NENDQ
0140            WRITE(2,103)K
0141            N=N+1
0142            GO TO(50,52,29)N
0143        50 IF(N1)29,29,250
0144       250 IF(N3)29,29,2250
0145      2250 N1=0
0146            WRITE(2,104)
0147            GO TO 4
0148        52 N1=M1
0149            N3=0
0150            WRITE(2,105)
0151            GO TO 42
0152        29 IF(NREP.GT.0)GO TO 13
0153            STOP
0154            END
```

```
TEST DATA FOR CR(3+) ,D3 OCTAHEDRAL
ND= 3
NSYM= 6
V1=  17000
V2=  24000
V3=  37000
THE VALUES OF TENDQ AND B ARE
TENDQ=  17000
B=    666
```

```
CHECKING ROUTINE A
   A IS EQUAL TO-,2076E 10 NOW SET TO ZERO

ESTIMATED POSITION OF V1
EV1=  16147
THE VALUES OF TENDQ AND B ARE
TENDQ=  16147
B=      837

CHECKING ROUTINE B

ESTIMATED POSITION OF V3
EV3=  37605
THE VALUES OF TENDQ AND B ARE
TENDQ=  17000
B=      707
```

assignments were correct. The safest procedure in this case is to set A equal to zero and examine the results. In the specific case quoted in Table 4.15 the internal consistency is only moderate and we should therefore re-examine our experimental data.

Much interesting computer work can be done in more advanced areas of this field and References 45, 46 and 47 will be helpful in this connection.

4.4.3 High-resolution NMR Spectra

The theory of Nuclear Magnetic Resonance (NMR) spectroscopy is well-documented in excellent textbooks.[48,49] Of importance to us is the observation that, when a nucleus with a spin quantum number of ½ (e.g. ^1H, ^{19}F and ^{32}P) is placed in a magnetic field of strength H, then the nuclei become aligned or opposed to the applied field, H, with energy levels of $\pm\frac{1}{2}\gamma\hbar H$, where γ is the so-called magnetogyric ratio. Resonant transitions between the two levels can then occur when the sample is irradiated with radiation of frequency ν_0.

$$\nu_0 = \frac{\gamma H}{2\pi} \text{ Hz}$$

The resonant frequency ν_0 would be the expected value for the unshielded nucleus.

In general however, the field at a nucleus, H, does not equal that applied, H_0, because of shielding of the nucleus by its electronic environment. Because of this, it is usual to measure the positions at which resonance occurs with respect to the signal for a reference substance. Therefore, we usually measure δ, the 'chemical shift' defined by:

$$\delta = \frac{H_S - H_R}{H_0} \times 10^6 \text{ ppm (variable applied field)}$$

or

$$\delta = \frac{\nu_S - \nu_R}{\nu_0} \times 10^6 \text{ ppm (variable applied frequency)}$$

In most instruments, the field is varied to obtain the resonance condition; H_S and H_R are, therefore, the applied fields required for sample and reference nuclei respectively; ν_S and ν_R are the corresponding frequencies for a variable frequency instrument. For our purposes later in this section we note that $\nu_0 \delta$ can be calculated from a knowledge of:

(i) δ obtained *via* a variable field instrument and
(ii) the operating frequency, ν_0.

For ^{1}H NMR, δ usually covers the range 0 to 10 using tetramethylsilane $(\text{Si}(\text{CH}_3)_4; \delta = 0)$ as reference; ν_0 is typically 60 MHz on routine instruments, giving a range of 60 Hz for $\nu_0 \delta$.

For a compound containing one type of proton (keeping to proton resonances for simplicity), we find that its NMR spectrum consists of one line. If the compound contains two different protons (i.e. in different chemical environments) we expect two lines in the spectrum, as summarized in Figure 4.12. Allowed transitions are those for which the change in total spin, ΔM, is ± 1. Figure 4.12 illustrates the situations:

(a) $\nu_0 \delta/J \gg 1$, known as the AX case; J is a coupling constant measuring the strength of the spin—spin interactions between the nuclei.

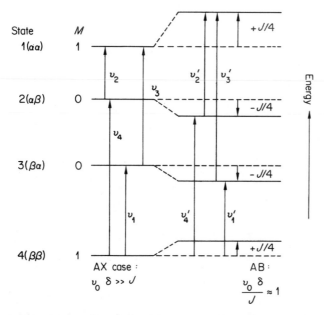

Figure 4.12 Approximate energy level diagram for 2-spin AX and AB systems Nuclear spins are denoted by α and β ($+\frac{1}{2}$ and $-\frac{1}{2}$ respectively) to give a total spin M

(b) $\nu_0 \delta/J \sim 1$, known as the AB case. Note that the similarity of the two nuclei is indicated by the nearness of letters in the alphabet.

In the AX case, $\nu_1 = \nu_2$ and $\nu_3 = \nu_4$, thus accounting for the observed two lines. The effect of coupling, for the AB case, is seen to have the effects:

$$\nu_1 < \nu_2 \quad \text{and} \quad \nu_3 > \nu_4$$

Therefore, we now see four lines in the spectrum.

Exact calculations of the transition energies[51] give slightly different results from those given in Figure 4.12. The calculations also give the relative transition energies and these are quoted in Table 4.16 for the AB system.

Table 4.16
Band energies and relative
intensities for an AB NMR
pattern

Band	Energy[a]/Hz	Intensity
1	$C + \tfrac{1}{2}J$	$1 - J/2C$
2	$C - \tfrac{1}{2}J$	$1 + J/2C$
3	$-C + \tfrac{1}{2}J$	$1 + J/2C$
4	$-C - \tfrac{1}{2}J$	$1 - J/2C$

$$C = \tfrac{1}{2}(\nu_0^2 \delta^2 + J^2)^{\frac{1}{2}}$$

[a] Relative to the centre of the band system at $(\nu_A + \nu_B)/2$.

We shall later use a computer program to calculate the transition energies and intensities for *any* values of $\nu_0 \delta$ and J; meanwhile some sample results are shown in Figure 4.13.

A slightly more complex case, and one more suitable for computer analysis is the situation for 3-spin ABX spectra.[52] This implies that two of the three nuclei (A and B) are similar, whilst the X nucleus is dissimilar. Several ABX systems can arise, such as

$$-CH=CH-CH-$$

(B) (A) (X)

We now have the possibility of varying the coupling constants J_{AB}, J_{AX} and J_{BX}; also, we can vary the resonant frequencies ν_A, ν_B and ν_X subject to the conditions that $\nu_A \approx \nu_B$ and ν_A, ν_B are significantly different from ν_X. As can be seen from Table 4.17, there are 12 possible transitions for the ABX case although less than 12 lines are sometimes observed. A full discussion of the theoretical aspects of this

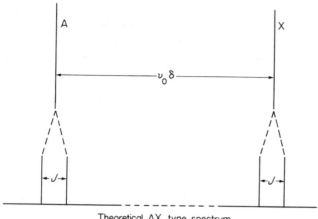

Theoretical AX type spectrum

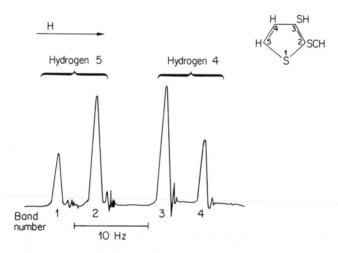

Figure 4.13 Example of an AB spectrum: Ring ^{2}H resonance at 40 MHz of liquid 2-methylthio-3-thiophenethiol (Reproduced by permission from Hoffman, R. A. and Gronowitz, S., *Arkiv. Kemi.*, 16, 503 (1960).)

and more complex spin systems is given in Reference 52, to which the reader is strongly recommended for useful background material presented in a stimulating manner.

With a knowledge of the data in Tables 4.16 and 4.17 we could either:

(i) Deduce values of chemical shifts and coupling constants from experimental data, or

(ii) Predict the appearance of spectra from input data of chemical shifts and coupling constants.

Table 4.17

Exact transition energies and relative transition intensities for the ABX system. (Reproduced from Reference 52, p. 407, by permission of Division of Chemical Education, American Chemical Society)

No.	Nucleus	Energy	Intensity
1	A	$\nu_A + \frac{1}{2}(J_{AB} + J_{AX}) + O_+$	$(C_{44} + C_{34})^2 = 1 - J_{AB}/2D_+$
2	A	$\nu_A + \frac{1}{2}(J_{AB} - J_{AX}) + O_-$	$(C_{66} + C_{56})^2 = 1 - J_{AB}/2D_-$
3	A	$\nu_A + \frac{1}{2}(J_{AX} - J_{AB}) + O_+$	$(C_{33} + C_{43})^2 = 1 + J_{AB}/2D_+$
4	A	$\nu_A - \frac{1}{2}(J_{AB} - J_{AX}) + O_-$	$(C_{55} + C_{65})^2 = 1 + J_{AB}/2D_-$
5	B	$\nu_B + \frac{1}{2}(J_{AB} + J_{BX}) - O_+$	$(C_{33} + C_{43})^2 = 1 + J_{AB}/2D_+$
6	B	$\nu_B + \frac{1}{2}(J_{AB} - J_{BX}) - O_-$	$(C_{55} + C_{65})^2 = 1 + J_{AB}/2D_-$
7	B	$\nu_B + \frac{1}{2}(J_{BX} - J_{AB}) - O_+$	$(C_{44} + C_{34})^2 = 1 - J_{AB}/2D_+$
8	B	$\nu_B - \frac{1}{2}(J_{AB} + J_{BX}) - O_-$	$(C_{66} + C_{56})^2 = 1 - J_{AB}/2D_-$
9	(Comb.)	$\nu_A + \nu_B - \nu_X$	0
10	(Comb.)	$\nu_X + \delta^\circ_{AB} + O_+ + O_-$	0
11	X	$\nu_X + \frac{1}{2}(J_{BX} + J_{AX})$	$(C_{66}C_{43} + C_{33}C_{56})^2 = J_{\hat{A}B}^2(a_- - a_+)^2/(a_-{}^2 + J_{\hat{A}B}^2)(a_+^2 + J_{\hat{A}B}^2)$
			1
12	X	$\nu_X + \frac{1}{2}(J_{AX} - J_{BX}) + O_+ - O_-$	$(C_{55}C_{33} + C_{65}C_{43})^2 = (J_{\hat{A}B}^2 + a_- a_+)^2/(a_-{}^2 + J_{\hat{A}B}^2)(a_+^2 + J_{\hat{A}B}^2)$
13	X	$\nu_X + \frac{1}{2}(J_{BX} - J_{AX}) + O_- - O_+$	$(C_{66}C_{44} + C_{56}C_{34})^2 = (J_{\hat{A}B}^2 + a_- a_+)^2/(a_-{}^2 + J_{\hat{A}B}^2)(a_+^2 + J_{\hat{A}B}^2)$
			1
14	X	$\nu_X - \frac{1}{2}(J_{BX} + J_{AX})$	$(C_{44}C_{65} + C_{55}C_{34})^2 = J_{\hat{A}B}^2(a_- - a_+)^2/(a_-{}^2 + J_{\hat{A}B}^2)(a_+^2 + J_{\hat{A}B}^2)$
15	(Comb.)	$\nu_X - \delta^\circ_{AB} - O_+ - O_-$	

$$C_{33} = C_{44} = a_+/(a_+^2 + J_{\hat{A}B}^2)^{1/2}$$
$$C_{43} = -C_{34} = J_{AB}/(a_+^2 + J_{\hat{A}B}^2)^{1/2}$$
$$C_{55} = C_{66} = a_-/(a_-^2 + J_{\hat{A}B}^2)^{1/2}$$
$$C_{65} = -C_{56} = J_{AB}/(a_-^2 + J_{\hat{A}B}^2)^{1/2}$$

$$a_\pm = [2D_\pm + \delta_{AB} \pm \frac{1}{2}(J_{AX} - J_{BX})]$$
$$O_\pm = \frac{1}{2}\{[\delta^\circ_{AB} \pm \frac{1}{2}(J_{AX} - J_{BX})]^2 + J_{\hat{A}B}^2\}^{1/2} - \frac{1}{2}[\delta_{AB} \pm \frac{1}{2}(J_{AX} - J_{BX})]$$
$$D_\pm = \frac{1}{2}\{[\delta^\circ_{AB} \pm \frac{1}{2}(J_{AX} - J_{BX})]^2 + J_{\hat{A}B}^2\}^{1/2}$$

We shall prefer (ii) as it is well suited to computer treatment and has considerable educational value.

The program NMRS will compute the appearance of an AB or ABX spectrum over a defined frequency range and output the results as a 'stick spectrum' onto the line printer. This program can be used to generate spectra for comparison with experimental data and is useful for 'getting the feel' of the influence of $\nu_0 \delta/J$ on the appearance of spectra. Some examples are given in Table 4.18.

Input to NMRS is as follows:

Card 1: Title; number of atoms A, B and C (e.g. NA = 1, NB = 1, NC = 0 for the AB case) Format 9A8, 3I2.

Card 2: (AB case): Values of $\nu_0 \delta$ and J_{AB}; frequency increment, INC, from zero; frequency range, NEND (2F0.0, 2I0). All values are in Hz.

Card 3: (ABX case): Values of J_{AB}, J_{AX}, J_{BX}, ν_A, ν_B, ν_X: frequency increment, INC, from the defined start, NS, to the end (NEND) of the computed spectrum (Format 6F0.0, 3I0).

Further sets of data will be read until a blank card terminator is read.

The program is seen to be in two sections. There is little sophisticated programming to explain and it should be possible to follow its operation by comparison with Tables 4.16 and 4.17. The stick plotting subroutine uses similar concepts to those described earlier in this chapter. This simple program could serve as an introduction to more powerful programs such as LAOCN.[53] Tabulations of data for various spin systems as a function of $J/\nu_0 \delta$ are given on pp. 625 ff. in Reference 48, which are quoted from Corio.[54] The reader can modify NMRS by using functions for other spin systems listed in the articles by Garbisch (References 51, 52 and 55).

4.5 Problems

Although some additional exercises are included here, we suggest that the problem work for this chapter should be largely devoted to the use of the programs described.

4.5.1 How would you modify program RADK so that half-lives in seconds, minutes, hours, days or years could be read in as data?

4.5.2 Compare the results obtained from BEMONTE for the system

$$^{218}\text{Po} \xrightarrow{t_{1/2} = 3\,\text{m}} {}^{214}\text{Pb} \xrightarrow{t_{1/2} = 26.8\,\text{m}} {}^{214}\text{Bi}$$

with the results produced by RADK. Investigate the effect of adjusting the ratio of N1, N2 and N3 to NLOOP.

Table 4.18

NMRS — simulation program for the AB and ABX spin systems

```
0008        MASTER NMRS
0009        DIMENSIONTITLE(9)
0010        COMMON/C/F(20),A(20),INC,NS,NEND
0011  100   READ(1,1)(TITLE(I),I=1,9),NA,NB,NC
0012    1   FORMAT(9A8,3I2)
0013        WRITE(2,2)(TITLE(I),I=1,9)
0014    2   FORMAT(1H1,9A8)
0015        NG=NA+NB+NC+1
0016        GO TO(99,99,90,91,92)NG
0017  90    READ(1,3)VOD,ABJ,INC,NEND
0018  3     FORMAT(2F0,0,2I0)
0019        VOD IS THE CHEMICAL SHIFT DIFFERENCE AND ABJ IS THE COUPLING
0020   C    CONSTANT, BOTH IN HZ.
0021   C    INC IS THE FREQUENCY INCREMENT, FROM ZERO, AND NEND IS FREQ. RANGE
0022        WRITE(2,4)VOD,ABJ,INC,NEND
0023    4   FORMAT(1X,' CHEMICAL SHIFT DIFFERENCE',F9,4,' COUPLING CONSTANT ',
0024        1F9,4,//,' SPECTRUM AT',I4,' HZ INTERVALS OVER A ',I7,' HZ RANGE'
0025        2)
0026        C=0,5*SQRT(VOD**2 + ABJ**2)
0027        F(1)=U
0028        F(2)=ABJ
0029        F(3)=2*C
0030        F(4)=2*C+ABJ
0031        A(1)=1,0
0032        A(2)=(1+ABJ/(2*C))/(1-ABJ/(2*C))
0033        A(3)=A(2)
0034        A(4)=A(1)
0035        NS=U
0036        CALL NMRPLOT(4)
0037        GO TO 10U
0038  91    CONTINUE
0039        READ(1,910)ABJ,AXJ,BXJ,VA,VB,VX,INC,NS,NEND
0040  910   FORMAT(6F0,0,3I0)
0041   C    VA,VB,VX ARE IN HZ RELATIVE TO TMS, USE VA>VB>VX
0042        WRITE(2,920)ABJ,AXJ,BXJ,VA,VB,VX,INC,NS,NEND
0043  920   FORMAT(1H1,' COUPLING CONSTANTS ARE',/,' ABJ = ',F9,4,' AXJ = ',
0044        1F9,4,' BXJ = ',F9,4,//,' RESONANT FREQUENCIES ',/,' VA, VR,
0045        2VC ARE ',3F9,1,' RESPECTIVELY',//,' SPECTRUM AT',I4,' HZ INTERVALS
0046        3 FROM ',I5,' TO',I5,' HZ')
0047        DAB=VA-VB
0048        OP=0,5*(SQRT((DAB+0,5*(AXJ-BXJ))**2+ABJ**2)-(DAB+0,5*(AXJ-BXJ)))
0049        OM=0,5*(SQRT((DAB-0,5*(AXJ-BXJ))**2+ABJ**2)-(DAB-0,5*(AXJ-BXJ)))
0050        DP=0,5*SQRT((DAB+0,5*(AXJ-BXJ))**2+ABJ**2)
0051        DM=0,5*SQRT((DAB-0,5*(AXJ-BXJ))**2+ABJ**2)
0052        AP=Z*DP+DAB+0,5*(AXJ-BXJ)
0053        AM=Z*DM+DAB-0,5*(AXJ-BXJ)
0054        F(1)=VA+U,5*(ABJ+AXJ)+OP
0055        F(2)=VA+U,5*(ABJ-AXJ)+OM
0056        F(3)=VA+U,5*(AXJ-ABJ)+OP
```

```
0057    F(4)=VA-0.5*(ABJ+AXJ)+OM
0058    F(5)=VB+0.5*(ABJ+BXJ)-OP
0059    F(6)=VB+0.5*(ABJ-BXJ)-OM
0060    F(7)=VB+0.5*(ABJ-BXJ)-OP
0061    F(8)=VB-0.5*(ABJ+BXJ)-OM
0062    F(9)=VX+DAB+OP+OM
0063    F(10)=VX+0.5*(BXJ+AXJ)
0064    F(11)=VX+0.5*(AXJ-BXJ)+OP-OM
0065    F(12)=VX+0.5*(BXJ-AXJ)+OM-OP
0066    F(13)=VX-0.5*(BXJ+AXJ)
0067    F(14)=VX-DAB-OP-OM
0068    A(1)=1-ABJ/(2*DP)
0069    A(2)=1-ABJ/(2*DM)
0070    A(3)=1+ABJ/(2*DP)
0071    A(4)=1+ABJ/(2*DM)
0072    A(5)=A(3)
0073    A(6)=A(4)
0074    A(7)=A(1)
0075    A(8)=A(2)
0076    A(9)=((ABJ+AP)**2)*(AM-AP)**2)/((AM**2)+ABJ**2)*((AP**2)+ABJ**2))
0077    A(10)=1
0078    A(11)=(((ABJ**2)+AM+AP)**2)/(((AM**2)+ABJ**2)+ABJ**2)*((AP**2)+(AP**2)+ABJ**2))
0079    A(12)=A(11)
0080    A(13)=1
0081    A(14)=A(9)
0082    CALL NMRPLOT(14)
0083 42 CONTINUE
0084    GO TO 100
0085 99 STOP
0086    END


0087    SUBROUTINE NMRPLOT(IPMAX)
0088    DIMENSION INT(20),NF(20),IP(100)
0089    COMMON/C/F(20),A(20),INC,NS,NEND
0090    DATA KAR/1H*/
0091    K1=IPMAX-1
0092    DO / K=1,K1
0093    FMIN=F(K)
0094    DO 7 I=(K+1),IPMAX
0095    IF(F(I).GT.FMIN)GO TO 7
0096    FA=F(I)
0097    FB=F(K)
0098    F(I)=FB
0099    F(K)=FA
0100    A=A(I)
0101    AB=A(K)
```

```
0102        A(I)=AB
0103        A(K)=AA
0104      7 CONTINUE
0105    C     SCALE THE INTENSITIES
0106        ALO=A(1)
0107        DO 9 I=2,IPMAX
0108        IF(A(I).LT.ALO)ALO=A(I)
0109      9 CONTINUE
0110        DO 11 I=1,IPMAX
0111     11 A(I)=A(I)/ALO
0112    C     THIS IS FOR LINE PRINTER PLOTTING A STICK SPECTRUM, IT ACCEPTS A
0113    C     LIST OF ORDERED FREQUENCIES (LOW TO HIGH) AND CONVERTS TO INTEGFK
0114    C     PRIOR TO PLOTTING
0115        DO 6 I=1,100
0116      6 IP(I)=KAR
0117        MAX=0
0118        J=1
0119        DO 1 I=1,IPMAX
0120        NF(I)=IFIX(F(I))
0121        INT(I)=IFIX(A(I))
0122      1 IF(INT(I).GT.MAX)MAX=INT(I)
0123        DO 2 I=1,500
0124        N=NS+(I-1)*INC
0125        IF(N.GT.NEND.OR.J.GT.IPMAX)GO TO 10
0126        IF(N.EQ.NF(J))GO TO 3
0127        WRITE(2,4)N
0128      4 FORMAT(1X,I7)
0129        GO TO 2
0130      3 NP=(INT(J)*100)/MAX
0131        WRITE(2,5)N,A(J),(IP(K),K=1,NP)
0132      5 FORMAT(1X,I7,3X,F7.1,3X,100A1)
0133        J=J+1
0134      2 CONTINUE
0135     10 RETURN
0136        END
```

```
        2-METHYL-3-THIOPHENOL
CHEMICAL SHIFT DIFFERENCE  14,2000  COUPLING CONSTANT    5,5000

SPECTRUM AT   1 HZ INTERVALS OVER A      50 HZ RANGE
        0      1,0    ****************************************************
        1
```

1-ACETOXY-2-NITRO-1-PHENYLETHANE

COUPLING CONSTANTS ARE
ABJ = 13.5000 AXJ = 9.7000 BXJ = 3.6000

RESONANT FREQUENCIES
VA, VB, VC ARE   280.0   262.0   382.0 RESPECTIVELY

SPECTRUM AT   1 HZ INTERVALS FROM 200 TO 500 HZ

```
       2
       3
       4
       5    *****************************************
       6
       7
       8
       9
      10
      11
      12
      13
      14
      15    2,1 *************************************
      16
      17
      18
      19
      20    1,0 ***********************************

SPECTRUM AT
  200
  201
  202
  203
  204
  205
  206
  207
  208
  209
```

```
210
211
212
213
214
215
216
217
218
219
220
221
222
223
224
225
226
227
228
229
230
231
232
233
234
235
236
237
238
239
240
241
242
243
244
245
246
247
248
249      **********
250      **********
251
252
253
254      *************
255      *************
256
257
258
```

49,0

68,6

248,3

228,9

248,3

228,9

49,U

68,4

259
260
261
262
263
264
265
266
267
268
269
270
271
272
273
274
275
276
277
278
279
280
281
282
283
284
285
286
287
288
289
290
291
292
293
294
295
296
297
298
299
300
301
302
303
304
305
306
307
308
309
310
311

```
357
358
359        1.0   *
360
361
362
363
364
365
366
367
368
369
370
371
372
373
374
375      148.7   *******************************************
376
377
378
379      147.7   *******************************************
380
381
382
383
384      147.7   *******************************************
385
386
387
388      148.7   *******************************************
389
390
391
392
393
394
395
396
397
398
399
400
401
402
403
404        1.0   *
```

4.5.3 Write a program to calculate the pH of a weak acid and its salt for various ratios of [acid] to [salt].

4.5.4 In Reference 15, the Boltzmann distribution is derived by starting with a number of particles, each of equal energy. A series of random steps is taken in which pairs of particles randomly exchange energy until they approach an equilibrium situation. Read the reference and write a program to perform the task described.

4.5.5 Take an experimental spectroscopic peak envelope produced by a chemical instrument (e.g. infrared , ultraviolet, NMR) and use either SPECTRUM or RESOLVE to find the most suitable fitting function.

References
Kinetics
1. Childs, W. C., Jr., *J. Chem. Educ.*, 50, 290 (1973).
2. Cummins, J. D., and Wartell, M. A., *J. Chem. Educ.*, 50, 544 (1973).
3. Leyden, D. E., and Morgan, W. R., *J. Chem. Educ.*, 46, 169 (1969).
4. Wilkins C. L., and Klopfenstein, C. E., *Chem. Tech.*, 681 (Nov. 1972).
5. Manock, J. J., in *Computers in Chemistry and Instrumentation* (Ed. by J. S. Maltson, H. B. Mark and H. C. MacDonald), Volume 3, Marcel Dekker, New York, 1973, p. 267.
6. Janata, J. Reference 5, p. 209.
7. Bunker, D. L., *Accounts Chem. Research*, 7, 195 (1974).
8. Householder, A. S., *Principles of Numerical Analysis*, McGraw-Hill, New York, 1953.
9. Para, A. Foglio, and Lazzarini, E., *J. Chem. Educ.*, 51, 336 (1974).
10. Rabinovitch, B., *J. Chem. Educ.*, 46, 262 (1969).
11. Schaad, L. J., *J. Amer. Chem. Soc.*, 85, 3588 (1963).

Thermodynamics
12. Neilson, W. B., *J. Chem. Educ.*, 48, 414 (1971).
13. Sabatini, A., and Vacca, A., *J. Chem. Soc. (Dalton)*, 1693 (1972).
14. Demattia, D., Gruhn, T., and Gorman, M., *J. Chem. Educ.*, 46, 398 (1969).
15. Schettler, P., *J. Chem. Educ.*, 51, 250 (1974).
16. Childs, C. W., Hallman, P. S., and Perrin, D. D., *Talanta*, 16, 1119 (1969).
17. Janis, F. T., and Kozel, T., *J. Chem. Educ.*, 60, 301 (1973).
18. Magnell, K. R., *J. Chem. Educ.*, 50, 619 (1973).
19. Groves, P. D., Huck, P. J., and Homer, J., *Chem. and Ind.*, p. 915 (1967, June 3).

20. Rossotti, F. J. C., Rossotti, H. S., and Whewell, R. J., *J. Inorg. Nucl. Chem.*, **33**, 2051 (1971) and references therein.
21. Feldberg, S., Klotz, P., Newman, L., *Inorg. Chem.*, **11**, 2860 (1972).
22. Beech, G., and Ashcroft, S. J., *Inorganic Thermodynamics*, Van Nostrand Reinhold Ltd., London, 1973, p. 15.
23. Isenhour, T. L., and Jurs, P. C., *Introduction to Computer Programming for Chemists*, Allyn and Bacon Inc., Boston, 1972, p. 127.
24. PJBT was kindly donated to this work by Mr E. J. Birch, of Oxford Polytechnic. The subroutine PPHT was modified to use the iterative method in program PHCALC.

Lattice Energies
25. Reference 22, p. 33.
26. Moore, W. J., *Physical Chemistry* (4th Edition), Longman, London, 1963, p. 693.
27. Waddington, T. C., *Advan. Inorg. Radiochem*, **1**, 157 (1959).
28. Lister, M. W., *Thermochim. Acta*, **8**, 341 (1974).
29. Hannay, N. B., *Solid State Chemistry*, Prentice Hall, Englewood Cliffs, N. J., 1967, p. 7.
30. Landolt-Bornstein, *Zahlewerte and Funktionen aus Physik Chemie U.S.W.*, 6 Aufl., II Band, I Teil, Springer-Verlag, Berlin, 1971, p. 483.
31. Mayer, J. E., *J. Chem. Phys.*, **1**, 278 (1933).

Spectroscopy (General)
32. Jorgensen, C. K., *Acta Chem. Scand.*, **8**, 1495 (1954).
33. Klabuhn, B., Spindler, D., and Goetz, H., *Spectrochimica Acta*, **29A**, 1283 (1973).
34. Ballhausen, C. J., in *Prog. Inorg. Chem.* (Ed. by F. A. Cotton), **2**, 253 (1960).
35. Program kindly contributed by Dr J. R. Wasson, University of Kentucky (some very slight changes exist between this and the original version developed at Kentucky). The program is referred to in *J. Chem. Educ.*, **50**, 177 (1973).
36. Seshadri, K. S., and Jones, R. N., *Spectrochimica Acta*, **19**, 1013 (1963).
37. Petrakis, L., *J. Chem. Educ.*, **44**, 433 (1967).
38. Littlewood, A. B. L., Gibb, T. C., and Anderson, A. H., in *Gas Chromatography 1968* (Proceedings of the 7th International Symposium on Gas Chromatography and its Exploitation, Copenhagen 25–28 June 1968. Edited by C. L. A. Harbourn, pub. by The Institute of Petroleum) p. 297.

Infrared Spectroscopy
39. Wheatley, P. J., *The Determination of Molecular Structure*, Oxford University Press, 1968.
40. Little, R., *J. Chem. Educ.*, 43, 5 (1966).

Crystal Field
41. Figgis, B. N., *Introduction to Ligand Field Theory*, Interscience, New York, 1966.
42. This program resembles that written by J. R. Wasson, *J. Chem. Educ.*, 47, 371 (1970).
43. Reference 41, p. 161.
44. Underhill, A. E., and Billing, D. E., *Nature*, 210, 834 (1966).
45. Krishnamurthy, R., and Schaap, W. B., *J. Chem. Educ.*, 47, 433.
46. Tapscott, R. E., *J. Chem. Educ.*, 49, 752 (1972).
47. Wasson, J. R., and Stoklosa, H. J., 50, 186 (1973).

NMR
48. Emsley, J. W., Feeney, J., and Sutcliffe, L. H., *High Resonance Nuclear Magnetic Resonance Spectroscopy*, Vol. 1, Pergamon Press, London 1965.
49. Roberts, J. D., *An Introduction to the Analysis of Spin–Spin Splitting in High Resolution NMR Spectra*, W. A. Benjamin Inc., New York, 1962.
50. Reference 48, p. 310.
51. Garbisch, E. W., *J. Chem. Educ.*, 45, 311 (1968).
52. Garbisch, E. W., *J. Chem. Educ.*, 45, 402 (1968).
53. Detar, D. F., *Computer Programs for Chemistry*, Vol. 1, W. A. Benjamin Inc., New York, 1968.
54. Corio, P. L., *Chem. Rev.*, 60, 363 (1960).
55. Garbisch, E. W., *J. Chem. Educ.*, 45, 480 (1968).

Chapter 5

Theoretical Chemistry

5.1 Introduction

A basic postulate of quantum mechanics[1,2] is that an electron bound
to a system of nuclei or a nucleus is associated with a stationary wave.
Classical stationary waves[3] differ from travelling waves in that the
amplitude of oscillation for a stationary wave depends on position and,
at a given point, is independent of time. Electron stationary waves also
have amplitudes which are a function of position relative to the
nucleus, or nuclei, to which the electron is bound. The mathematical
function ψ, which describes the amplitude variation of the stationary
wave with position, is referred to as the wave function. Another
postulate[1] of wave mechanics is that ψ^2 may be interpreted as a
measure of the electron probability density. (Some wave functions
involve complex numbers in which case the electron probability density
is given by $\psi \psi^*$ where ψ^* is the complex conjugate of ψ.) It follows
that if ψ is a known function then electron densities in atoms and
molecules can be obtained. The determination of ψ, and the energy, E,
of the electron in the state described by ψ, is of great significance in
chemistry and forms the basis of modern theories of valence.

In order to obtain the mathematical form of ψ, and the corresponding energy, it is necessary to solve the quantum mechanical
equivalent of a classical wave equation. For systems involving bound
electrons, in which the electron density is independent of time, ψ is
obtained by solving the time-independent Schrodinger wave equation:

$$\frac{\partial^2 \psi}{\partial x^2} + \frac{\partial^2 \psi}{\partial y^2} + \frac{\partial^2 \psi}{\partial z^2} + \frac{8\pi^2 m}{h^2}(E - V)\psi = 0 \qquad (5.1)$$

where x, y and z are Cartesian coordinates, m is the electron mass, h is
Planck's constant and V is the potential energy of the electron. The
strategy in applying this equation to a particular system is to express
the potential energy explicitly in terms of positional coordinates and
then to attempt to solve for ψ and E. Often, Cartesian coordinates are

not the most convenient, in which case the wave equation is transformed to more appropriate coordinates prior to solution. A whole series of solutions are generally obtained for a given system. For each solution ψ_i, $i = 1, 2, 3, \ldots$, there is associated a corresponding energy E_i, $i = 1, 2, 3, \ldots$. If two solutions have the same energy then the two wave functions are said to be *degenerate*. Degeneracy of solutions usually arises as a consequence of inherent symmetry of the system.

Equation (5.1) is not restricted to the description of the motion of an electron. Wave properties are associated with particles of atomic and molecular dimensions generally and so appropriate wave equations can be written for the motion of nuclei e.g. molecular vibrations and rotations.[4] For the motion of electrons in atoms, an exact solution of the Schrodinger equation is only possible for relatively simple systems such as H, He^+, Li^{2+} and H_2^+ which involve a single electron. When more than one electron is involved the potential function V is complicated by electron–electron repulsions and an analytical solution of the wave equation is not usually possible. In these cases wave functions for many electron atoms are generated in numerical form by the use of computers. We shall be more concerned in this chapter with the use of simple computer programs to give a clearer understanding of the nature of the solutions of the wave equation for selected simple systems which illustrate points of either wave mechanical or chemical significance.

5.2 Problems Involving a Square-well Potential Function

A square-well potential function may be described as a potential which is infinite at all points outside a specified region. Inside this region $V = 0$, but at the boundary of the region V rises vertically to infinity. The simplest case corresponds to the problem in one dimension, e.g. along the x coordinate. The potential function V could then be defined by equations (5.2) and is shown in Figure 5.1(a).

$$V = \infty, \qquad x \leqslant 0, x \geqslant l_x$$
$$V = 0, \qquad 0 < x < l_x \tag{5.2}$$

where the region of zero potential energy extends over a distance l_x which can be interpreted as the length of a one-dimensional enclosure or box. The problem can be extended to two or three dimensions in a perfectly straightforward way by defining quantities such as l_y and l_z as well as expressions analogous to equations (5.2) for the y- and z-coordinates. These systems are often referred to as two- and three-dimensional boxes. By making use of computer programs some of the properties of particles moving in such potential wells will be examined and their relevance to chemical problems illustrated.

5.2.1 Particle in a One-dimensional Well
The wave equation for a particle moving inside a one-dimensional well is easily obtained from equation (5.1) by deleting the terms involving

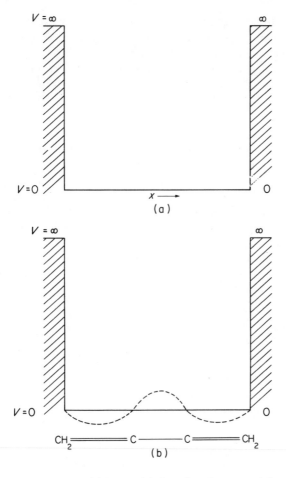

Figure 5.1 (a) Potential function for a particle in a one-dimensional well, (b) Kuhn potential function allowing for bond length alteration

the derivatives with respect to the y and z coordinates and setting $V = 0$. For a one-dimensional box with a length l, solution of the wave equation gives the allowed energies,[2] E_n, for a particle of mass m as

$$E_n = \frac{h^2 n^2}{8l^2 m} \qquad n = 1, 2, 3, \ldots \tag{5.3}$$

where n is a quantum number. Notice that n can have any integer value but cannot be zero, and the energy increases with n, as does the spacing between consecutive energy levels.

This simple model has been applied to molecules containing conjugated double bonds to predict spectral characteristics[5,6] and the extent of bond length alternation[2] along the conjugated chain. In such conjugated systems, the π-electron system can be thought of as very roughly corresponding to a one-dimensional box. Of course the potential function experienced by an electron in a conjugated π-system does not rise vertically to infinity at the ends of the chain, neither is the potential energy likely to be constant through the chain. On the other hand, there is some similarity to the one-dimensional box in that there is likely to be greater longitudinal freedom than transverse freedom for electrons in the π-system. Further, the potential energy must rise steeply at the ends of the conjugated chain.

For a conjugated system having r double bonds there are $2r$ π-electrons to be assigned to the energy levels of the box, defined by equation (5.3). Filling the levels of lowest energy first, and assigning two electrons to each level in accordance with the Pauli principle,[2] the first r levels are fully occupied in the ground electronic state of the molecule. The first excited electronic state is obtained by promoting an electron from the highest occupied level $(n = r)$ to the lowest unoccupied level $(n = r + 1)$. From equation (5.3) the transition energy, ΔE, for this excitation is

$$\Delta E = \frac{(2r + 1)}{8l^2 m} h^2 \tag{5.4}$$

The frequency, ν, of the electromagnetic radiation capable of providing this transition energy is given by $\nu = \Delta E / h$. It follows that the wavenumber $\bar{\nu}$, which is simply ν/c where c is the velocity of light in a vacuum, is given by

$$\bar{\nu} = \frac{(2r + 1)}{8l^2 mc} h \quad cm^{-1} \tag{5.5}$$

The box length l can be expressed in terms of the average bond length, R_0, for the conjugated chain and the number of double bonds r.

For the series of symmetrical carbocyanine dyes the ground state electronic structure is midway between the structure implied by the forms I and II below:

$$\tag{5.6}$$

I

$$
C_2H_5 - {}^+N \;\text{...}\; C\text{=}C \text{--} (CH\text{=}CH)_{r-4} \text{--} CH\text{=}C \;\text{...}\; N - C_2H_5
$$

(with bicyclic ring systems at each end containing CH_2-CH_2, CH_2, CH, CH groups)

(5.6)

II

The length of the conjugated chain is taken in this case as $l = (2r + 1)R_0$ since there are r double and r single bonds, in either of the structural representations I or II, between the nitrogen atoms terminating the conjugated chain. An extra R_0 is added to the length of the one-dimensional box to roughly take account of the fact that the π-electron system itself will extend slightly beyond the positions of the nitrogen atoms at both ends of the chain.

Table 5.1

Program ONEBOX for the calculation of the wavelength and wavenumber of the longest wavelength transition for symmetrical carbocydnines

```
0005                MASTER ONEBOX
0006                REAL M,NUBAR
0007                INTEGER R
0008                RO=0.14
0009                WRITE(2,100)RO
0010            100 FORMAT(1H1,1X,'PARTICLE IN A BOX MODEL',//,1X,'RO = ',F6.3,'NM')
0011              1 READ(1,1000)R
0012                IF(R.EQ.0)STOP
0013                WRITE(2,1001)R
0014           1000 FORMAT(1I0)
0015           1001 FORMAT(///,1X,'THIS CONJUGATED SYSTEM HAS ',I2,' DOUBLE BONDS',/)
0016                RR=RO*1.0E-9
0017                H=6.6/E-34
0018                M=9.107E-31
0019                C=2.997E+8
0020                XL=(2*R+1)*RR
0021                NUBAR=(2*R+1)*H/(8*XL*XL*M*C)
0022                NUBAR =NUBAR*1.0E-2
0023                WVL=1.0E+7/NUBAR
0024                WRITE(2,1002)NUBAR,WVL
0025           1002 FORMAT(1X,'FOR THE LONGEST WAVELENGTH TRANSITION',//,1X,'NUBAR =',
0026                1F8.0,' 1/CM',//,1X,'WAVELENGTH = ',F6.1,' NM',/////)
0027                GO TO 1
0028                END
```

The program ONEBOX listed in Table 5.1 calculates the wavenumber (in cm^{-1}) of the longest wavelength transition for some symmetrical carbocyanine dyes using $R_0 = 0.14$ nm. The input to the program is the number of conjugated double bonds, r, in free integer format. Successive calculations for other values of r are introduced by cards similar to the first. If a card is encountered for which $r = 0$ then the calculation is terminated. Output obtained for $r = 4$ and $r = 5$ is shown in Table 5.2. The experimental values[2] and those calculated using ONEBOX are compared in Table 5.3.

184

Table 5.2
Example of the output from ONEBOX. The corresponding input was:

Card 1: 4 (IO)
Card 2: 5 (IO)
Card 3: 0 (IO)

```
PARTICLE IN A BOX MODEL

RO =  0.140NM


THIS CONJUGATED SYSTEM HAS   4 DOUBLE BONDS

FOR THE LONGEST WAVELENGTH TRANSITION
NUBAR =   17317, 1/CM
WAVELENGTH =   577.5 NM


THIS CONJUGATED SYSTEM HAS   5 DOUBLE BONDS

FOR THE LONGEST WAVELENGTH TRANSITION
NUBAR =   14169, 1/CM
WAVELENGTH =   705.8 NM
```

Table 5.3
A comparison of experimental[7] and calculated
wavenumbers for the longest wavelength
transitions of some symmetrical carbocyanines

Number of double bonds	Experimental wavenumber/cm^{-1}	Calculated wavenumber/cm^{-1}
4	17,241	17,317
5	14,085	14,169
6	12,346	11,989
7	10,870	10,390

The agreement between calculated and experimental values is remarkable and is far better than one has a right to expect from such a simple theory. The good agreement is in fact a coincidence brought about by the mutual cancellation of errors associated with the approximate nature of the model. This cancellation is most effective when there is no bond length alternation along the conjugated chain. The extreme structures I and II (equation (5.6)) represent a given bond in the chain as being either single or double. The true description being midway between these extremes indicates a uniform bond length along the chain (i.e. no bond length alternation).

However, in the case of the linear polyenes, of general formula

$$CH_2=CH-(CH=CH)_{r-2}-CH=CH_2$$

there is considerable bond length alternation so that agreement between theory and experiment is not expected to be so good. The conjugated chain here contains r double bonds and $r-1$ single bonds. The total length is therefore $(2r - 1)R_0$. Once again adding R_0 to account approximately for the extension of the π-system beyond the centres of the terminal carbon atoms, the length of the conjugated chain is $l = 2rR_0$. Program ONEBOX is easily altered to accommodate this new definition of l by altering line 24 (Table 5.1). Table 5.4 compares the experimental wavenumbers and those calculated using the modified program ONEBOX. It is apparent that the agreement between the calculated and experimental values is much poorer than for the symmetrical carbocyanine dyes.

Table 5.4
Experimental and calculated wavenumbers for polyenes of the
type: $CH_2=CH_2-(CH_2=CH)\overline{_{r-2}}CH=CH_2$

Number of double bonds (r)	Experimental wavenumber/cm^{-1}	Calculated wavenumber/cm^{-1}	
		ONEBOX[a]	Kuhn model
2	46,080	48,704	57,805
3	37,310	30,305	40,417
4	32,890	21,917	32,535
5	29,940	17,144	28,065
6	27,470	14,070	25,193
7	25,640	11,928	23,195
8	24,390	10,350	21,726
10	22,370	8,182	19,710

[a] This program is a modified version of ONEBOX to accommodate the different definition of l for the linear polyenes compared with the carbocyonine dyes (see text).

It could be argued that a choice of $R_0 = 0.14$ nm is not appropriate for the linear polyenes, and that better agreement may be achieved by a different value for this parameter. What value of R_0 would give the best fit of the calculated to the experimental values? Program BEST, described in section 2.5.1 can be used to find the optimum value of R_0. A number of modifications to BEST are required for this application: a new function ERROR is needed to calculate the square deviation between the calculated and experimental wavenumbers; some alterations are also required in the MASTER segment all concerned with input and output. No changes are needed in subroutine BE10 which actually carries out the optimization of the parameter R_0. The modified MASTER segment and new function ERROR are listed in Table 5.5. The experimental data for the linear polyenes is given in Table 5.4. Using these data as a basis for optimization the independent variable is the number of double bonds and a suitable initial guess for the parameter R_0 to be optimized is $R_0 = 0.15$ nm.

Table 5.5

Modified MASTER and ERROR routine of program BEIO for optimization
of R_0 in calculations of wavenumbers for linear polyene spectra

```
0006                    MASTER BEST
0007                    COMMON/A/XXX(10),X(10),DX(10),DELTA(10),YFIN(100)
0008                    COMMON/B/YOBS(100),YCALC(100),XD(100)
0009                    READ(1,1)NVAR,NP
0010               1    FORMAT(2I0)
0011                    READ(1,29)(XD(I),YOBS(I),I=1,NP)
0012              29    FORMAT(2F0,0)
0013               2    FORMAT(3E10,4)
0014             222    FORMAT(1H ,E15,8,3X,E15,8,3X,E15,8)
0015            2004    FORMAT(1H1,'INPUT    ORIGIN X        STEP SIZE        LIMIT        ')
0016                    READ(1,2)(XXX(I),DX(I),DELTA(I),I=1,NVAR)
0017                    WRITE(2,2004)
0018                    WRITE(2,222)(XXX(I),DX(I),DELTA(I),I=1,NVAR)
0019                    CALL BE10(NVAR,AMIN,NP,AL)
0020                    WRITE(2,5)
0021               5    FORMAT(1H ,' FINAL   VALUES')
0022                    WRITE(2,6)AMIN
0023                    WRITE(2,9)AL
0024               9    FORMAT(1H ,'  LINEAR PARAMETER ',E10,4,/,'   NON LINEAR',/)
0025               6    FORMAT(1H ,'AMIN =',E11,4)
0026                    WRITE(2,8)(XXX(I),DX(I),I=1,NVAR)
0027               8    FORMAT(1H ,'FINAL BOND LENGTH =',F8,4,'NM',/,1X,'FINAL STEP',
0028                   1' SIZE = ',F8,6,/)
0029                    WRITE(2,31)
0030              31    FORMAT(1H1,' D-BONDS     NUBAR OBS       NUBAR CALC')
0031                    WRITE(2,50)(XD(I),YOBS(I),YFIN(I),I=1,NP)
0032              50    FORMAT(F9,2,5X,F9,1,5X,F9,1)
0033                    STOP
0034                    END


0035                    SUBROUTINE BE10(NVAR,AMIN,NP,AL)
0036                    DIMENSIONN(10),XX(10)
0037                    COMMON/A/XXX(10),X(10),DX(10),DELTA(10),YFIN(100)
0038                    COMMON/B/YOBS(100),YCALC(100),XD(100)
0039             302    FORMAT(39H JOB ABANDONED,MORE THAN 100 ITERATIONS)
0040               8    FORMAT(1H ,6H AMIN=,E15,8)
0041             101    FORMAT(1H ,25H NUMBER OF ITERATIONS WAS,I4)
0042              11    FORMAT(1H ,10X,4HSTEP,I2,2H =,E10,4)
0043             900    FORMAT(1H1,'DIRECT GRID SEARCH')
0044             555    FORMAT(1H ,20A4)
0045                    NC=0
0046                    DO 4 I=1,NVAR
0047                    X(I)=XXX(I)
0048                    XX(I)=XXX(I)
0049               4    CONTINUE
0050                    M1=1
0051                    AMIN=ERROR(NP,AL)
0052                    NC=0
0053             308    NC=NC+1
0054                    WRITE(2,8)AMIN
0055                    IF(NC,LT,100)GO TO 301
0056                    WRITE(2,302)
0057                    GO TO 313
0058             301    CONTINUE
0059                    ND=0
0060                    NL=0
0061         C          MINIMISATION LOOPS
0062                    GO TO(30,31,32,33,34,35,36,37,38,39)NVAR
0063              39    DO 29 J10=1,3
0064                    X(10)=XXX(10)+DX(10)*(J10-2)
0065              38    DO 28 J9=1,3
0066                    X(9)=XXX(9)+DX(9)*(J9-2)
0067              37    DO 27 J8=1,3
0068                    X(8)=XXX(8)+DX(8)*(J8-2)
0069              36    DO 26 J7=1,3
0070                    X(7)=XXX(7)+DX(7)*(J7-2)
0071              35    DO 25 J6=1,3
0072                    X(6)=XXX(6)+DX(6)*(J6-2)
0073              34    DO 24 J5=1,3
0074                    X(5)=XXX(5)+DX(5)*(J5-2)
0075              33    DO 23 J4=1,3
0076                    X(4)=XXX(4)+DX(4)*(J4-2)
0077              32    DO 22 J3=1,3
0078                    X(3)=XXX(3)+DX(3)*(J3-2)
```

```
0079            31 DO 21 J2=1,3
0080               X(2)=XXX(2)+DX(2)*(J2-2)
0081            30 DO 20 J1=1,3
0082               X(1)=XXX(1)+DX(1)*(J1-2)
0083           120 F=ERROR(NP,AL)
0084               IF(AMIN.LT.F)GO TO 20
0085               AMIN=F
0086               DO 66 JK=1,NVAR
0087            66 XX(JK)=X(JK)
0088               DO 67 K=1,NP
0089            67 YFIN(K)=YCALC(K)
0090            20 CONTINUE
0091               IF(NVAR.EQ.1)GO TO 40
0092            21 CONTINUE
0093               IF(NVAR.EQ.2)GO TO 40
0094            22 CONTINUE
0095               IF(NVAR.EQ.3)GO TO 40
0096            23 CONTINUF
0097               IF(NVAR.EQ.4)GO TO 40
0098            24 CONTINUE
0099               IF(NVAR.EQ.5)GO TO 40
0100            25 CONTINUE
0101               IF(NVAR.EQ.6)GO TO 40
0102            26 CONTINUE
0103               IF(NVAR.EQ.7)GO TO 40
0104            27 CONTINUE
0105               IF(NVAR.EQ.8)GO TO 40
0106            28 CONTINUE
0107               IF(NVAR.EQ.9)GO TO 40
0108            29 CONTINUE
0109            40 CONTINUE
0110               DO 377 I=1,NVAR
0111         C     STORE X FOR FINAL O/P
0112               IF((ABS(XXX(I)-XX(I))).GE.DX(I))GO TO 306
0113               IF(DX(I).GT.DELTA(I))GO TO 800
0114               GO TO 377
0115           800 DX(I)=DX(I)/2.
0116               NL=NL+1
0117           602 CONTINUE
0118               GO TO 377
0119           306 DX(I)=2.*DX(I)
0120               ND=ND+1
0121           377 XXX(I)=XX(I)
0122               IF(ND.GT.0.OR.NL.GT.0)GO TO 308
0123           313 CONTINUE
0124               WRITE(2,101)NC
0125               RETURN
0126               END
```

```
0127               FUNCTION ERROR(NP,AL)
0128               COMMON/A/XXX(10),X(10),DX(10),DELTA(10),YFIN(100)
0129               COMMON/B/YOBS(100),YCALC(100),XD(100)
0130               REAL M
0131               INTEGER R
0132               ERROR=0.0
0133               H=6.67E-54
0134               M=9.107E-31
0135               C=2.997E+10
0136               R0=X(1)*1.0E-9
0137               DO 1 I=1,NP
0138               R=NINT(XD(I))
0139               YCALC(I)=(2*R+1)*H/(32*R*R*R0*R0*M*C)
0140             1 ERROR=ERROR+(YOBS(I)-YCALC(I))**2
0141               RETURN
0142               END
```

The input data were as follows:

Card 1: 1 (number of variables); 8 (number of compounds).
Card 2: 2.0 (number of double bonds in compound 1); 46080 (experimental wavenumber for compound 1).

and similarly for cards 3 to 9 (one card for each compound) until

Card 10: 0.1500E 00 0200E 00 .0010E 00 corresponding to: initial R_0/nm, initial stepsize/nm, terminating stepsize (3E10.4).

Table 5.6
Output from program in Table 5.5 for optimization of R_0

```
INPUT    ORIGIN X       STEP SIZE       LIMIT
0.15000000E 00     0.20000000E-01    0.10000000E-02
AMIN= 0.14806460E 10
AMIN= 0.85503029E 09
AMIN= 0.85503029E 09
AMIN= 0.85503029E 09
AMIN= 0.85503029E 09
AMIN= 0.81335955E 09
AMIN= 0.81335955E 09
AMIN= 0.81335955E 09
AMIN= 0.81335955E 09
AMIN= 0.81335955E 09
NUMBER OF ITERATIONS WAS   10
FINAL   VALUES
AMIN = 0.8134E 09
   LINEAR PARAMETER 0.0000E 00
   NON LINEAR

FINAL BOND LENGTH =   0.1250NM
FINAL STEP SIZE = 0.000625
```

D-BONDS	NUBAR OBS	NUBAR CALC
2.00	46080.0	61094.7
3.00	37310.0	38014.5
4.00	32890.0	27492.6
5.00	29940.0	21505.3
6.00	27470.0	17649.6
7.00	25640.0	14962.0
8.00	24390.0	12982.6
10.00	22370.0	10263.9

The corresponding output is shown in Table 5.6. The comparison between calculated and experimental values is much poorer than for the symmetrical carbocyanine dyes in spite of the optimization. This is because the deficiencies in the model cannot be accommodated by only optimizing R_0. It is true that the fit to experimental data can always be improved by optimizing more parameters but any such improvement should not be interpreted as an indication that the model is a truer representation of reality. The poorer agreement between theory and experiment for the linear polyenes is, in fact, due to bond alternation leading to a failure of the cancellation of errors which occurred in the absence of bond alternation. It would appear that if an improved fit is to be obtained a more sophisticated model must be chosen.

In order to allow for bond alteration Kuhn[7] has used a particle in a box potential function with a sinusoidal potential superimposed inside the box, as shown in Figure 5.1b. The maxima and minima associated with the sine wave are chosen to coincide with the centres of respectively single and double bonds. The choice of the amplitude of the sinusoidal function introduces an unknown parameter. This results in the appearance of an alternative variable parameter in the expression, given below, for the wavenumber of the longest wavelength transition of the linear polyenes.

$$\bar{\nu} = \frac{h}{8ml^2}(2r + 1) + \bar{\nu}_e (1 - 1/2r) \qquad (5.7)$$

Except for $\bar{\nu}_e$ the symbols here are as defined for equation (5.4). The new parameter, $\bar{\nu}_e$, can be regarded as the limiting value of $\bar{\nu}$ as r tends to infinity. The program listed in Table 5.5 can be used to optimize the choice of $\bar{\nu}_e$ in equation (5.7) provided that FUNCTION ERROR is altered. Since $\bar{\nu}_e$ is to be optimized rather than R_0, line 136 in FUNCTION ERROR is replaced by XNUL = X(1) where XNUL = $\bar{\nu}_e$. Secondly a fixed value of R_0 must be assigned by inserting a line RO = 0.14E − 9, since 0.14 nm is a reasonable average bond length. Finally a term equivalent to $\bar{\nu}_e (1 - 1/2r)$ must be added to YCALC(I) in accordance with equation (5.7). The modified FUNCTION ERROR is shown in Table 5.7. Some simple alterations are also necessary in Master BEST to correct the headings to the output data for optimization of $\bar{\nu}_e$ rather than R_0.

Table 5.7
Modified FUNCTION ERROR for optimizing the value of ν_e

```
FUNCTION ERROR(NP,AL)
COMMON/A/XXX(10),X(10),DX(10),DELTA(10),YFIN(100)
COMMON/B/YOBS(100),YCALC(100),XD(100)
REAL M
INTEGER R
ERROR=0,0
H=6,67E-34
M=9,107E-31
C=2,997E+10
RO=0,140E-9
XNUL=X(1)
DO 1 I=1,NP
R=NINT(XD(I))
YCALC(I)=(2*R+1)*H/(32*R*R*RO*RO*M*C)
YCALC(I)=YCALC(I)+XNUL*(1,0-1,0/(2*R))
1 ERROR=ERROR+(YOBS(I)-YCALC(I))**2
RETURN
END
FINISH
```

With an initial guess for $\bar{\nu}_e$ equal to 1700 cm^{-1}, an initial step size of 100 cm^{-1} and a terminating step size of 1 cm^{-1}, 42 iterations were necessary for convergence. The final value of $\bar{\nu}_e$ obtained was 12,134 cm^{-1} and the calculated wavenumber for the polyene series are given in Table 5.4 under the heading 'Kuhn model'. It is apparent that the agreement between theory and experiment is much improved by the Kuhn modification allowing for bond alternation.

5.2.2 Particle in a Two-dimensional Well

The properties of the wave functions for a particle moving in a two-dimensional well illustrate some of the basic concepts of wave mechanics. In this section a computer program will be used to examine the wave functions which are solutions of the wave equation:

$$\frac{\partial^2 \psi}{\partial x^2} + \frac{\partial^2 \psi}{\partial y^2} + \frac{8\pi^2 m}{h^2} E\psi = 0 \tag{5.8}$$

which corresponds to a particle of mass m moving in the region $0 < x < l_x$ and $0 < y < l_y$. Within this region the potential energy is zero and it rises vertically to infinity at the boundaries. This wave equation can be solved by writing ψ as a product of a function of x only, $X(x)$, and a function of y only, $Y(y)$. On substitution for ψ in equation (5.8) two equations are obtained, each of which is equivalent to the wave equation for a one-dimensional box, one in terms of the variable x only, the other in terms of y only. The solution of the wave equation in x alone gives:

$$X(x) = \sqrt{\left(\frac{2}{l_x}\right)} \sin\left(\frac{n_x x \pi}{l_x}\right) \tag{5.9a}$$

and

$$E_{n_x} = \frac{h^2 n_x^2}{8 l_x^2 m} \tag{5.9b}$$

where n_x is a quantum number with values 1, 2, 3, etc. Similar expressions hold for the solution of the wave equation in the y variables so that another quantum number n_y is introduced where $n_y = 1,2,3, \ldots$. The total wave function is then given by

$$\psi(n_x n_y) = \frac{2}{\sqrt{(l_x l_y)}} \sin\left(\frac{n_x x \pi}{l_x}\right) \cdot \sin\left(\frac{n_y y \pi}{l_y}\right) \tag{5.10}$$

and the energy is the sum of E_{n_x} and the analogous quantity E_{n_y}.

$$E_{n_x n_y} = \frac{h^2 n_x^2}{8 l_x^2 m} + \frac{h^2 n_y^2}{8 l_y^2 m} \tag{5.11}$$

If $l_x = l_y$ then the system displays degeneracy. For example, $E_{1,2} = E_{2,1}$ and $E_{1,3} = E_{3,1}$, etc. Are there any *differences*, however, in the wave functions $\psi_{1,2}$ and $\psi_{2,1}$, etc?

Program TWOBX, shown in Table 5.8 may be used to investigate the format of the solutions given by equation (5.10) for the case $l_x = l_y$. This program is designed to produce a map of the wave function within the area of the two-dimensional box and to produce a map of a product

Table 5.8
Program TWOBX for the study of a particle in a 2-dimensional box

```
0008                    MASTER TWOBX
0009                    DIMENSION TITLE(4),PS1(51,30),PS2(51,30),PS12(51,30)
0010                    COMMON N1X,N1Y,N2X,N2Y
0011                    WRITE(2,1001)
0012                    READ(1,1002)IOP,N1X,N1Y
0013            C       IOP=1 MAP OF PS1 ONLY,LENGTH OF SIDES=1,0(ARBITRARY UNITS)
0014            C       IOP=2 MAP OF PS1,PS2 AND THEIR PRODUCT
0015         1001 FORMAT(1X,'PARTICLE IN A TWO DIMENSIONAL BOX',/,1X,107(1H*))
0016                    PI=3,1416
0017                    IF(IOP,EQ.2)READ(1,1002)N2X,N2Y
0018            9 DO 2 I=1,51
0019                    X=(I-1)/50,0
0020                    DO 2 J=1,28
0021                    Y=(J-1)/27,0
0022                    PS1(I,J)=SIN(N1X*X*PI)*SIN(N1Y*Y*PI)*2,0
0023                    IF(IOP,EQ.1)GO TO 3
0024                    PS2(I,J)=SIN(N2X*X*PI)*SIN(N2Y*Y*PI)*2,0
0025                    PS12(I,J)=PS1(I,J)*PS2(I,J)
0026            3 CONTINUE
0027            2 CONTINUE
0028                    CALL MAP(PS1,1)
0029         1002 FORMAT(4I0)
0030                    WRITE(2,103)
0031          103 FORMAT(1H1)
0032                    IF(IOP,FQ.1)GO TO 4
0033                    WRITE(2,103)
0034                    CALL MAP(PS2,2)
0035                    WRITE(2,103)
0036                    CALL MAP(PS12,3)
0037                    CALL ING(PS12)
0038                    WRITE(2,103)
0039                    READ(1,1002)N1X,N1Y,N2X,N2Y
0040                    IF(N1X,EQ.0)STOP
0041                    GO TO 9
0042            4 READ(1,1002)N1X,N1Y
0043                    IF(N1X,EQ.0)STOP
0044                    GO TO 9
0045                    END


0046                    SUBROUTINE MAP(PS,IOUT)
0047                    DIMENSION PS(51,30),IXY(51,30)
0048                    COMMON NX1,NY1,NX2,NY2
0049                    IF(IOUT,GT,1)GO TO 1
0050                    WRITE(2,100)NX1,NY1
0051                    GO TO 3
0052            1 IF(IOUT,GT,2)GO TO 9
0053                    WRITE(2,100)NX2,NY2
0054                    GO TO 3
0055            9 CONTINUE
0056                    WRITE(2,101)NX1,NY1,NX2,NY2
0057            3 CONTINUE
0058          100 FORMAT(1X,'MAP OF PSI(',I2,1X,I2,')*49,5',/)
0059          101 FORMAT(1X,'MAP OF PSI(',I2,1X,I2,')*PSI(',I2,1X,I2,')*24,75',/)
0060                    WRITE(2,105)
0061          105 FORMAT(1X,86(1H*))
0062                    IF(IOUT,EQ.3)GO TO 5
0063                    DO 6 I=1,51
0064                    DO 6 J=1,28
0065            6 IXY(I,J)=NINT(PS(I,J)*49,5)
0066                    GO TO 7
0067            5 DO 8 I=1,51
0068                    DO 8 J=1,28
0069            8 IXY(I,J)=NINT(PS(I,J)*24,75)
0070            7 CONTINUE
0071                    DO 10 I=1,51
0072           10 WRITE(2,107)(IXY(I,J),J=1,28)
0073          107 FORMAT(1X,1H*,28I3,1H*)
0074                    WRITE(2,105)
0075                    RETURN
0076                    END
0077                    SUBROUTINE ING(PS)
0078                    DIMENSION PS(51,30)
0079                    A=1,0/51
0080                    B=1,0/28
```

192

```
0081              DV=A*B
0082              SUM=0.0
0083              DO 1 I=1,51
0084              DO 1 J=1,28
0085            1 SUM=SUM+PS(I,J)
0086              XX=SUM*DV
0087              WRITE(2,100)XX
0088          100 FORMAT(1X,'THE INTEGRAL OF THIS PRODUCT OVER ALL THE BOX SPACE=',
0089              11F10.5)
0090              RETURN
0091              END
```

of any two of the wave functions. The input to the program involves an integer, IOP. If IOP = 1 then a numerical map of a wave function is produced. The input for TWOBX to generate maps of wave functions is as follows:

Card 1: IOP, n_x, n_y (3I0).
Card 2 and susequent cards: New values of n_x, n_y (2I0).
Terminator card: 0 0 (2I0).

The output shown in Figure 5.2(a) and 5.2(b) for ψ_{12} and ψ_{21} corresponds to input IOP = 1, n_x = 1, n_y = 2 for the first map and n_x = 2, n_y = 1 for the second map. The difference between these wave functions is readily apparent — one is rotated through ninety degrees relative to the other. If l_x were greater than l_y then the lobes associated with $\psi_{2,1}$ would be extended and narrowed relative to these of $\psi_{1,2}$ and degeneracy between these two solutions would lifted, as is apparent from equation (5.11). It should be noted that program TWOBX scales the calculated values of the wave function to be mapped so that the maximum value is 99. (The scaling factor used is shown at the top of the map.) This is purely a matter of convenience in producing an easily readable map. The problem of producing a square map with a line printer is not trivial because the line spacing is not equal to that associated with line printer characters. The use of this type of page buffer is discussed in Chapter 1.

For any system two different solutions of the wave equation ψ_i and ψ_j should be orthogonal. That is, the following expression holds:

$$\int \psi_i \, \psi_j \, d\tau = 0 \quad i \neq j \tag{5.12}$$

where $d\tau$ represents a volume element in general (in this case $d\tau$ actually refers to an element of area) and the integration is over all of the space available. If the wave functions are normalized to unity so that

$$\int \psi_i^2 \, d\tau = 1 \tag{5.13}$$

then the set of wave functions are called an orthonormal set (both orthogonal and normalized). (For complex wave functions once again equations (5.12) and (5.13) are modified to:

$$\int \psi_i^* \psi_i \, d\tau = 1 \text{ and } \int \psi_i^* \psi_j \, d\tau = 0)$$

PARTICLE IN A TWO DIMENSIONAL BOX
****
MAP OF PSI(1 2)*49,5

```
*************************************************************************
*  0   0   0   0   0   0   0   0   0   0   0   0   0   0   0   0   0   0   0   0   0   0   0   0   0   0   0   0  0*
*  0   1   3   4   5   6   6   6   6   5   5   3   2   1  -1  -2  -3  -5  -5  -6  -6  -6  -6  -6  -5  -4  -3  -1  0*
*  0   3   6   8  10  11  12  12  12  11   9   7   4   1  -1  -4  -7  -9 -11 -12 -12 -12 -11 -10  -8  -6  -3   0*
*  0   4   8  12  15  17  18  19  18  16  13  10   6   2  -2  -6 -10 -13 -16 -18 -19 -18 -17 -15 -12  -8  -4   0*
*  0   6  11  16  20  23  24  25  24  21  18  14   8   3  -3  -8 -14 -18 -21 -24 -25 -24 -23 -20 -16 -11  -6   0*
*  0   7  14  20  25  28  30  31  29  26  22  17  10   4  -4 -10 -17 -22 -26 -29 -31 -30 -28 -25 -20 -14  -7   0*
*  0   8  16  23  29  33  36  36  35  32  27  20  12   4 -12 -20 -27 -32 -35 -36 -36 -33 -29 -23 -16  -8   0*
*  0  10  19  27  34  39  42  42  40  37  31  23  14   5  -5 -14 -23 -31 -37 -40 -42 -42 -39 -34 -27 -19 -10   0*
*  0  11  21  31  38  44  47  48  46  41  35  26  16   6  -6 -16 -26 -35 -41 -46 -48 -47 -44 -38 -31 -21 -11   0*
*  0  12  24  34  43  49  52  53  51  46  39  29  18   6  -6 -18 -29 -39 -46 -51 -53 -52 -49 -43 -34 -24 -12   0*
*  0  13  26  37  47  53  57  58  56  50  42  32  20   7  -7 -20 -32 -42 -50 -56 -58 -57 -53 -47 -37 -26 -13   0*
*  0  15  28  41  51  58  62  63  60  55  46  35  22   7  -7 -22 -35 -46 -55 -60 -63 -62 -58 -51 -41 -28 -15   0*
*  0  16  30  44  54  62  67  68  65  59  49  37  23   8  -8 -23 -37 -49 -59 -65 -68 -67 -62 -54 -44 -30 -16   0*
*  0  17  32  46  58  66  71  72  69  62  52  40  25   8  -8 -25 -40 -52 -62 -69 -72 -71 -66 -58 -46 -32 -17   0*
*  0  18  34  49  61  70  75  76  73  66  55  42  26   9  -9 -26 -42 -55 -66 -73 -76 -75 -70 -61 -49 -34 -18   0*
*  0  18  36  51  64  74  79  80  77  69  58  44  27   9  -9 -27 -44 -58 -69 -77 -80 -79 -74 -64 -51 -36 -18   0*
*  0  19  38  54  67  77  82  83  80  72  61  46  29 -10 -10 -29 -46 -61 -72 -80 -83 -82 -77 -67 -54 -38 -19   0*
*  0  20  39  56  70  80  85  87  83  75  63  48  30 -10 -10 -30 -48 -63 -75 -83 -87 -85 -80 -70 -56 -39 -20   0*
*  0  21  40  58  72  82  88  89  86  78  65  49  31 -10 -10 -31 -49 -65 -78 -86 -89 -88 -82 -72 -58 -40 -21   0*
*  0  21  41  59  74  85  91  92  88  80  67  51  31 -11 -11 -31 -51 -67 -80 -88 -92 -91 -85 -74 -59 -41 -21   0*
*  0  22  42  61  76  86  93  94  90  82  68  52  32 -11 -11 -32 -52 -68 -82 -90 -94 -93 -86 -76 -61 -42 -22   0*
*  0  22  43  62  77  88  94  96  92  83  70  53  33 -11 -11 -33 -53 -70 -83 -92 -96 -94 -88 -77 -62 -43 -22   0*
*  0  22  44  63  78  89  96  97  93  84  71  53  33 -11 -11 -33 -53 -71 -84 -93 -97 -96 -89 -78 -63 -44 -22   0*
*  0  23  44  63  79  90  97  98  94  85  71  54  34 -11 -11 -34 -54 -71 -85 -94 -98 -97 -90 -79 -63 -44 -23   0*
*  0  23  44  64  79  91  97  99  95  86  72  54  34 -11 -11 -34 -54 -72 -86 -95 -99 -97 -91 -79 -64 -44 -23   0*
*  0  23  44  64  79  91  97  99  95  86  72  54  34 -11 -11 -34 -54 -72 -86 -95 -99 -97 -91 -79 -64 -44 -23   0*
*  0  23  44  64  79  91  97  99  95  86  72  54  34 -11 -11 -34 -54 -72 -86 -95 -99 -97 -91 -79 -64 -44 -23   0*
*  0  23  44  63  79  90  97  98  94  85  71  54  34 -11 -11 -34 -54 -71 -85 -94 -98 -97 -90 -79 -63 -44 -23   0*
*  0  22  44  63  78  89  96  97  93  84  71  53  33 -11 -11 -33 -53 -71 -84 -93 -97 -96 -89 -78 -63 -44 -22   0*
*  0  22  43  62  77  88  94  96  92  83  70  53  33 -11 -11 -33 -53 -70 -83 -92 -96 -94 -88 -77 -62 -43 -22   0*
*  0  22  42  61  76  86  93  94  90  82  68  52  32 -11 -11 -32 -52 -68 -82 -90 -94 -93 -86 -76 -61 -42 -22   0*
*  0  21  41  59  74  85  91  92  88  80  67  51  31 -11 -11 -31 -51 -67 -80 -88 -92 -91 -85 -74 -59 -41 -21   0*
*  0  21  40  58  72  82  88  89  86  78  65  49  31 -10 -10 -31 -49 -65 -78 -86 -89 -88 -82 -72 -58 -40 -21   0*
*  0  20  39  56  70  80  85  87  83  75  63  48  30 -10 -10 -30 -48 -63 -75 -83 -87 -85 -80 -70 -56 -39 -20   0*
*  0  19  38  54  67  77  82  83  80  72  61  46  29 -10 -10 -29 -46 -61 -72 -80 -83 -82 -77 -67 -54 -38 -19   0*
*  0  18  36  51  64  74  79  80  77  69  58  44  27   9  -9 -27 -44 -58 -69 -77 -80 -79 -74 -64 -51 -36 -18   0*
*  0  18  34  49  61  70  75  76  73  66  55  42  26   9  -9 -26 -42 -55 -66 -73 -76 -75 -70 -61 -49 -34 -18   0*
*  0  17  32  46  58  66  71  72  69  62  52  40  25   8  -8 -25 -40 -52 -62 -69 -72 -71 -66 -58 -46 -32 -17   0*
*  0  16  30  44  54  62  67  68  65  59  46  35  22   8  -8 -23 -37 -46 -55 -60 -63 -62 -58 -51 -41 -30 -16   0*
*  0  15  28  41  51  58  62  63  60  55  46  35  22   7  -7 -22 -35 -46 -55 -60 -63 -62 -58 -51 -41 -28 -15   0*
*  0  13  26  37  47  53  57  58  56  50  42  32  20   7  -7 -20 -32 -42 -50 -56 -58 -57 -53 -47 -37 -26 -13   0*
*  0  12  24  34  43  49  52  53  51  49  37  23   6  -6 -18 -29 -39 -46 -51 -53 -52 -49 -43 -34 -24 -12   0*
*  0  11  21  31  38  44  47  48  46  41  35  26  16   6  -6 -16 -26 -35 -41 -46 -48 -47 -44 -38 -31 -21 -11   0*
*  0  10  19  27  34  39  42  42  40  37  31  23  14   5  -5 -10 -23 -31 -37 -40 -42 -42 -39 -34 -27 -19 -10   0*
*  0   8  16  23  29  33  36  36  35  32  27  20  12   4  -4 -12 -20 -27 -32 -35 -36 -36 -33 -29 -23 -16  -8   0*
*  0   7  14  20  25  28  30  31  29  26  22  17  10   4  -4 -10 -17 -22 -26 -29 -31 -30 -28 -25 -20 -14  -7   0*
*  0   6  11  16  20  23  24  25  24  21  18  14   8   3  -3  -8 -14 -18 -21 -24 -25 -24 -23 -20 -16 -11  -6   0*
*  0   4   8  12  15  17  18  19  18  16  13  10   6   2  -2  -6 -10 -13 -16 -18 -19 -18 -17 -15 -12  -8  -4   0*
*  0   3   6   8  10  11  12  12  12  11   9   7   4   1  -1  -4  -7  -9 -11 -12 -12 -12 -11 -10  -8  -6  -3   0*
*  0   1   3   4   5   6   6   6   5   5   3   2   1  -1  -2  -3  -5  -5  -6  -6  -6  -6  -5  -4  -3  -1   0*
*  0   0   0   0   0   0   0   0   0   0   0   0   0   0   0   0   0   0   0   0   0   0   0   0   0   0   0   0  0*
*************************************************************************
```

Figure 5.2 (a) Printer plot of the wave function ψ_{12}, (b) wave function ψ_{21}, (c) printer plot of the product of $\psi_{1,2}$ and $\psi_{2,1}$

The significance of orthogonality can be appreciated using the anology of unit vectors. In n-dimensional space, any vector can be described in terms of a set of n mutually perpendicular unit vectors which are orthonormal in the sense that they are all mutually perpendicular and have unit length. In a similar sense a set of n solutions of the wave equation form a set of n mutually orthogonal and normalized functions. It follows that any mathematical function in the function space described by the wave functions can be written in terms

MAP OF PSI(2 1)*49,5

```
****************************************************************
* U  U  U  U  U  0  0  0  U  U  U  0  0  0  0  0  0  0  U  0  U  0  G  U  0  0  0  0  0  0  0  0  0  U*
* U  1  3  4  6  7  8  9 10 11 11 12 12 12 12 12 12 11 11 10  9  8  7  6  4  3  1  0*
* U  3  6  8 11 14 16 18 20 21 23 24 24 25 25 24 24 23 21 20 18 16 14 11  8  6  3  0*
* U  4  8 12 16 20 23 27 29 32 33 35 36 36 36 36 35 33 33 32 29 27 23 20 16 12  8  4  0*
* 0  6 11 16 21 26 31 35 38 41 44 46 47 48 48 47 46 44 41 38 35 31 26 21 16 11  6  0*
* U  7 13 20 26 32 37 42 47 50 53 56 57 58 58 57 56 53 50 47 42 37 32 26 20 13  7  0*
* U  8 16 23 30 37 44 49 54 59 62 65 67 68 68 67 65 62 59 54 49 44 37 30 23 16  8  0*
* U  9 18 26 34 42 49 55 61 66 70 73 75 76 76 75 73 70 66 61 55 49 42 34 26 18  9  0*
* U 10 19 29 38 46 54 61 67 72 77 80 82 83 83 82 80 77 72 67 61 54 46 38 29 19 10  0*
* U 10 21 31 40 49 58 65 72 78 82 86 88 89 89 88 86 82 78 72 65 58 49 40 31 21 10  0*
* U 11 22 32 42 52 61 68 76 82 86 90 93 94 94 93 90 86 82 76 68 61 52 42 32 22 11  0*
* U 11 22 33 44 53 63 71 78 84 89 93 96 97 97 96 93 89 84 78 71 63 53 44 33 22 11  0*
* 0 11 23 34 44 54 64 72 79 86 91 95 97 99 99 97 95 91 86 79 72 64 54 44 34 23 11  0*
* U 11 23 34 44 54 64 72 79 86 91 95 97 99 99 97 95 91 86 79 72 64 54 44 34 23 11  0*
* U 11 22 33 44 53 63 71 78 84 89 93 96 97 97 96 93 89 84 78 71 63 53 44 33 22 11  0*
* U 11 22 32 42 52 61 68 76 82 86 90 93 94 94 93 90 86 82 76 68 61 52 42 32 22 11  0*
* U 10 21 31 40 49 58 65 72 78 82 86 88 89 89 88 86 82 78 72 65 58 49 40 31 21 10  0*
* U 10 19 29 38 46 54 61 67 72 77 80 82 83 83 82 80 77 72 67 61 54 46 38 29 19 10  0*
* U  9 18 26 34 42 49 55 61 66 70 73 75 76 76 75 73 70 66 61 55 49 42 34 26 18  9  0*
* U  8 16 23 30 37 44 49 54 59 62 65 67 68 68 67 65 62 59 54 49 44 37 30 23 16  8  0*
* U  7 13 20 26 32 37 42 47 50 53 56 57 58 58 57 56 53 50 47 42 37 32 26 20 13  7  0*
* 0  6 11 16 21 26 31 35 38 41 44 46 47 48 48 47 46 44 41 38 35 31 26 21 16 11  6  0*
* U  4  8 12 16 20 23 27 29 32 33 35 36 36 36 36 35 33 32 29 27 23 20 16 12  8  4  0*
* U  3  6  8 11 14 16 18 20 21 23 24 24 25 25 24 24 23 21 20 18 16 14 11  8  6  3  0*
* U  1  3  4  6  7  8  9 10 11 11 12 12 12 12 12 12 11 11 10  9  8  7  6  4  3  1  0*
* U  0  0  0  0  0  0  0  0  0  0  0  0  0  0  0  0  0  0  0  0  0  0  0  0  0  0  0  0*
* U -1 -3 -4 -6 -7 -8 -9 -10 -11 -11 -12 -12 -12 -12 -12 -12 -11 -11 -10 -9 -8 -7 -6 -4 -3 -1  0*
* U -3 -6 -8 -11 -14 -16 -18 -20 -21 -23 -24 -24 -25 -25 -24 -24 -23 -21 -20 -18 -16 -14 -11 -8 -6 -3  0*
* U -4 -8 -12 -16 -20 -23 -27 -29 -32 -33 -35 -36 -36 -36 -36 -35 -33 -32 -29 -27 -23 -20 -16 -12 -8 -4  0*
* U -6 -11 -16 -21 -26 -31 -35 -38 -41 -44 -46 -47 -48 -48 -47 -46 -44 -41 -38 -35 -31 -26 -21 -16 -11 -6  0*
* U -7 -13 -20 -26 -32 -37 -42 -47 -50 -53 -56 -57 -58 -58 -57 -56 -53 -50 -47 -42 -37 -32 -26 -20 -13 -7  0*
* U -8 -16 -23 -30 -37 -44 -49 -54 -59 -62 -65 -67 -68 -68 -67 -65 -62 -59 -54 -49 -44 -37 -30 -23 -16 -8  0*
* U -9 -18 -26 -34 -42 -49 -55 -61 -66 -70 -73 -75 -76 -76 -75 -73 -70 -66 -61 -55 -49 -42 -34 -26 -18 -9  0*
* 0 -10 -19 -29 -38 -46 -54 -61 -67 -72 -77 -80 -82 -83 -83 -82 -80 -77 -72 -67 -61 -54 -46 -38 -29 -19 -10  0*
* U -10 -21 -31 -40 -49 -58 -65 -72 -78 -82 -86 -88 -89 -89 -88 -86 -82 -78 -72 -65 -58 -49 -40 -31 -21 -10  0*
* U -11 -22 -32 -42 -52 -61 -68 -76 -82 -86 -90 -93 -94 -94 -93 -90 -86 -82 -76 -68 -61 -52 -42 -32 -22 -11  0*
* 0 -11 -22 -33 -44 -53 -63 -71 -78 -84 -89 -93 -96 -97 -97 -96 -93 -89 -84 -78 -71 -63 -53 -44 -33 -22 -11  0*
* 0 -11 -23 -34 -44 -54 -64 -72 -79 -86 -91 -95 -97 -99 -99 -97 -95 -91 -86 -79 -72 -64 -54 -44 -34 -23 -11  0*
* 0 -11 -23 -34 -44 -54 -64 -72 -79 -86 -91 -95 -97 -99 -99 -97 -95 -91 -86 -79 -72 -64 -54 -44 -34 -23 -11  0*
* 0 -11 -22 -33 -44 -53 -63 -71 -78 -84 -89 -93 -96 -97 -97 -96 -93 -89 -84 -78 -71 -63 -53 -44 -33 -22 -11  0*
* 0 -11 -22 -32 -42 -52 -61 -68 -76 -82 -86 -90 -93 -94 -94 -93 -90 -86 -82 -76 -68 -61 -52 -42 -32 -22 -11  0*
* 0 -10 -21 -31 -40 -49 -58 -65 -72 -78 -82 -86 -88 -89 -89 -88 -86 -82 -78 -72 -65 -58 -49 -40 -31 -21 -10  0*
* 0 -10 -19 -29 -38 -46 -54 -61 -67 -72 -77 -80 -82 -83 -83 -82 -80 -77 -72 -67 -61 -54 -46 -38 -29 -19 -10  0*
* U -9 -18 -26 -34 -42 -49 -55 -61 -66 -70 -73 -75 -76 -76 -75 -73 -70 -66 -61 -55 -49 -42 -34 -26 -18 -9  0*
* U -8 -16 -23 -30 -37 -44 -49 -54 -59 -62 -65 -67 -68 -68 -67 -65 -62 -59 -54 -49 -44 -37 -30 -23 -16 -8  0*
* 0 -7 -13 -20 -26 -32 -37 -42 -47 -50 -53 -56 -57 -58 -58 -57 -56 -53 -50 -47 -42 -37 -32 -26 -20 -13 -7  0*
* 0 -6 -11 -16 -21 -26 -31 -35 -38 -41 -44 -46 -47 -48 -48 -47 -46 -44 -41 -38 -35 -31 -26 -21 -16 -11 -6  0*
* 0 -4 -8 -12 -16 -20 -23 -27 -29 -32 -33 -35 -36 -36 -36 -36 -35 -33 -32 -29 -27 -23 -20 -16 -12 -8 -4  0*
* 0 -3 -6 -8 -11 -14 -16 -18 -20 -21 -23 -24 -24 -25 -25 -24 -24 -23 -21 -20 -18 -16 -14 -11 -8 -6 -3  0*
* 0 -1 -3 -4 -6 -7 -8 -9 -10 -11 -11 -12 -12 -12 -12 -12 -12 -11 -11 -10 -9 -8 -7 -6 -4 -3 -1  0*
* 0  0  0  0  0  0  0  0  0  0  0  0  0  0  0  0  0  0  0  0  0  0  0  0  0  0  0  0*
****************************************************************
```

of a linear combination of the wave functions. This concept will be used later in considering electronic wave functions for molecular systems.

Program TWOBX can also be used to output maps of products of wave functions for a square potential well by setting IOP = 2. The subroutine ING carries out an approximate integration of the product map, over the area of the box, by summing the magnitude of the product for each array value times the area of each grid element. The integral for a product of two wave functions of different symmetry is zero (see Figure 5.2c). The integration for a product of two wave functions of the same symmetry is also zero but since the wave functions are normalized the integral of ψ^2 is unity. Both of these points, which are of general validity in wave mechanics, can be checked

MAP OF PSI(1 2)*PSI(2 1)*24.75

```
**********************************************************************
*  U  U  U  0  U  0  0  U  U  U  0  0  U  0  0  0  0  0  0  0  0  U  J  0  0  0  U  0  0  0 *
*  U  U  U  U  U  0  0  1  1  1  1  U  0  0  0  0  0 -1 -1 -1 -1  0  0  U  0  0  0  0  0  0 *
*  U  0  0  1  1  2  2  2  2  2  2  2  1  0  0 -1 -2 -2 -2 -2 -2 -2 -2 -1 -1  U  0  0 *
*  U  U  1  2  2  3  4  5  5  5  5  4  2  1 -1 -2 -4 -5 -5 -5 -5 -4 -3 -2 -2 -1  0  0 *
*  U  U  1  3  4  6  8  9  9  9  8  6  4  1 -1 -4 -6 -8 -9 -9 -9 -8 -6 -4 -3 -1  0  0 *
*  U  0  2  4  6  9 11 13 14 15 12  9  6  2 -2 -6 -9-12-13-14-13-11 -9 -6 -4 -2  0  0 *
*  U  1  3  5  9 13 16 18 19 19 17 13  8  3 -3 -8-13-17-19-19-18-16-13 -9 -5 -3 -1  0 *
*  U  1  3  7 12 16 21 24 25 24 22 17 11  4 -4-11-17-22-24-25-24-21-16-12 -7 -3 -1  0 *
*  U  1  4  9 14 20 25 29 31 30 27 21 14  5 -5-14-21-27-30-31-29-25-20-14 -9 -4 -1  0 *
*  U  1  5 11 17 24 30 35 37 36 32 25 16  6 -6-16-25-32-36-37-35-30-24-17-11 -5 -1  0 *
*  U  1  6 12 20 28 35 40 43 42 37 .29.19  6 -6-19-29-37-42-43-40-35-28-20-12 -6 -1  0 *
*  U  2  6 14 22 31 39 45 48 46 41 33 21  7 -7-21-33-41-46-48-45-39-31-22-14 -6 -2  0 *
*  U  2  7 15 24 34 43 49 52 51 45 36 23  8 -8-23-36-45-51-52-49-43-34-24-15 -7 -2  0 *
*  U  2  7 16 26 36 46 52 55 54 48 38 24  8 -8-24-38-48-54-55-52-46-36-26-16 -7 -2  0 *
*  U  2  8 16 27 38 47 54 58 56 50 39 25  9 -9-25-39-50-56-58-54-47-38-27-16 -8 -2  0 *
*  U  2  8 17 27 38 48 55 59 57 51 40 26  9 -9-26-40-51-57-59-55-49-38-27-17 -8 -2  0 *
*  U  2  8 17 27 38 48 55 58 57 51 40 25  9 -9-25-40-51-57-58-55-48-38-27-17 -8 -2  0 *
*  U  2  8 16 26 37 46 53 56 55 49 39 25  8 -8-25-39-49-55-56-53-46-37-26-16 -8 -2  0 *
*  U  2  7 15 25 35 44 50 53 52 46 36 23  8 -8-23-36-45-52-53-50-44-35-25-15 -7 -2  0 *
*  U  2  7 14 23 32 40 46 48 47 42 33 21  7 -7-21-33-42-47-48-46-40-32-23-14 -7 -2  0 *
*  U  1  6 12 20 28 35 40 43 42 37 29 19  6 -6-19-29-37-42-43-40-35-28-20-12 -6 -1  0 *
*  U  1  5 10 17 23 29 34 35 35 31 24 16  5 -5-16-24-31-35-35-34-29-23-17-10 -5 -1  0 *
*  U  1  4  8 13 18 23 26 28 27 24 19 12  4 -4-12-19-24-27-28-26-23-18-13 -8 -4 -1  0 *
*  U  1  3  5  9 12 15 18 19 18 16 13  8  3 -3 -8-13-16-18-19-18-15-12 -9 -5 -3 -1  0 *
*  U  0  1  3  4  6  8  9 10  9  8  7  4  1 -1 -4 -7 -8 -9-10 -9 -8 -6 -4 -3 -1  0  0 *
*  U  U  0  0  0  0  0  0  0  0  0  0  0  0  0  0  0  0  0  0  0  0  0  0  0  0  0  0 *
*  U  0 -1 -3 -4 -6 -8 -9-10 -9 -8 -7 -4 -1  1  4  7  8  9 10  9  8  6  4  3  1  0  0 *
*  U -1 -3 -5 -9-12-15-18-19-18-16-13 -8 -3  3  8 13 16 18 19 18 15 12  9  5  3  1  0 *
*  U -1 -4 -8-13-18-23-26-28-27-24-19-12 -4  4 12 19 24 27 28 26 23 18 13  8  4  1  0 *
*  U -1 -5-10-17-23-29-34-35-35-31-24-16 -5  5 16 24 31 35 35 34 29 23 17 10  5  1  0 *
*  U -1 -6-12-20-28-35-40-43-42-37-29-19 -6  6 19 29 37 42 43 40 35 28 20 12  6  1  0 *
*  U -2 -7-14-23-32-40-46-48-47-42-33-21 -7  7 21 33 42 47 48 46 40 32 23 14  7  2  0 *
*  U -2 -7-15-25-35-44-50-53-52-46-36-23 -8  8 23 36 46 52 53 50 44 35 25 15  7  2  0 *
*  U -2 -8-16-26-37-46-53-56-55-49-39-25 -8  8 25 39 49 55 56 53 46 37 26 16  8  2  0 *
*  U -2 -8-17-27-38-48-55-58-57-51-40-25 -9  9 25 40 51 57 58 55 48 38 27 17  8  2  0 *
*  U -2 -8-17-27-38-48-55-59-57-51-40-26 -9  9 26 40 51 57 59 55 48 38 27 17  8  2  0 *
*  U -2 -7-16-26-36-46-52-55-54-48-38-24 -8  8 24 38 48 54 55 52 46 36 26 16  7  2  0 *
*  U -2 -7-15-24-34-43-49-52-51-45-34-24 -8  8 23 36 45 51 52 49 43 34 24 15  7  2  0 *
*  U -2 -6-14-22-31-39-45-48-46-41-33-21 -7  7 21 33 41 46 48 45 39 31 22 14  6  2  0 *
*  U -1 -6-12-20-28-35-40-43-42-37-29-19 -6  6 19 29 37 42 43 40 35 28 20 12  6  1  0 *
*  U -1 -5-11-17-24-30-35-35-35-32-25-16 -6  6 16 25 32 36 37 35 30 24 17 11  5  1  0 *
*  U -1 -4 -9-14-20-25-29-31-30-27-21-14 -5  5 14 21 27 30 31 29 25 20 14  9  4  1  0 *
*  U -1 -3 -7-12-16-21-24-25-24-22-17-11 -4  4 11 17 22 24 25 24 21 16 12  7  3  1  0 *
*  U -1 -3 -5 -9-13-16-18-19-17-13 -8 -3  3  8 13 17 19 19 18 16 13  9  5  3  1  0 *
*  U  0 -2 -4 -6 -9-11-13-14-13-12 -9 -6 -2  2  6  9 12 13 14 13 11  9  6  4  2  0  0 *
*  U  0 -1 -3 -4 -6 -8 -9 -9 -9 -8 -6 -4 -1  1  4  6  8  9  9  8  6  4  3  1  0  0 *
*  U  0 -1 -2 -2 -3 -4 -5 -5 -5 -5 -4 -2 -1  1  2  4  5  5  5  5  4  3  2  2  1  0  0 *
*  U  0  0 -1 -1 -2 -2 -2 -2 -2 -2 -2 -1  0  0  1  2  2  2  2  2  2  1  1  0  0  0  0 *
*  U  0  0  0  0  0  0 -1 -1 -1 -1  0  0  0  0  0  0  1  1  1  0  0  0  0  0  0  0  0 *
*  U  0  0  0  0  0  0  U  U  0  U  0  0  0  0  0  0  0  0  0  0  0  0  0  0  0  0  0 *
**********************************************************************
THE INTEGRAL OF THIS PRODUCT OVER ALL THE BOX SPACE=    0.000
```

using ING, although one should expect some small deviation from zero for integrals of products of orbitals of the same symmetry, and from unity in the case of normalization integrals because of the approximate nature of the integration. The input for the production of maps of wave functions and their products is as follows:

Card 1: $2, n_{1x}, n_{1y}$ (310).
Card 2: n_{2x}, n_{2y} (2I0).

where n_{1x} and n_{1y} are the quantum numbers for ψ_1 and n_{2x}, n_{2y} refer to ψ_2.

Cards 1 and 2 produce maps of $\psi_{1(n_{1x}, n_{1y})}$ and $\psi_{2(n_{2x}, n_{2y})}$ and the map of their product.

Card 3: n_{1x} n_{1y} n_{2x} n_{2y} (4I$_0$) quantum numbers defining new ψ_1 and ψ_2 for the next set of maps (ψ_1, ψ_2 and ψ_1 ψ_2)

and subsequent cards

Termination card: 0 0 0 0 (4I0)

If the computer has graph-plotting facilities then the maps can be represented in terms of contours.

The detailed programming required for the use of a graph plotter depends on the software plotting package (see Appendix A). For this reason only outline details applicable to most installations will be considered here. The position of the graph plotter pen is usually specified in 'graph-plotter units', i.e. distances along the x- and y-axes of the plotter frame. In generating the graphical output the programmer must specify the following information by making use of the locally available plotter software:

(1) Define an origin and draw axes (of the required length).
(2) Annotate the axes (appropriately scaled).
(3) Generate x and y coordinates of points to be plotted (or joined etc.) in graph-plotter units. The data generated by the program therefore has to be scaled so that the calculated x and y values fill in the range of the plotter frame.
(4) Indicate how points are to be treated (e.g. specified points to be joined by straight lines or indicated by a cross or some other symbol).

Examples of contour maps drawn (using a CALCOMP ® plotter) from the output of TWOBX are shown in Figure 5.3. An example of a suitable contour program is listed in Appendix A. This program used to produce Figure 5.3 joined two points, with the same value for the function, by means of straight lines. For this reason, with the rather large grid size used in TWOBX, in regions where the function changes rapidly the true curved appearance of the contours is lost.

5.3 Calculation and Representation of Atomic Orbitals
5.3.1 Introduction
The simplest atomic system is the hydrogen atom. The single electron moves in the potential field of the nucleus, so that the potential function is given by equation (5.14)

$$V = \frac{-e^2}{4\pi\epsilon_0 r} \tag{5.14}$$

where e is the charge on the electron, ϵ_0 is the permittivity of a vacuum and r is the distance between the electron and the nucleus. The Schrodinger wave equation can be solved relatively easily for the hydrogen atom provided the position of the electron is written in terms

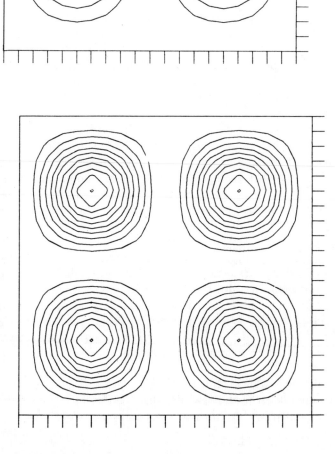

Two-dimensional box (2,2) PSISQ

Two-dimensional box (2,1) PSISQ

Figure 5.3   Output from TWOBX using a Graph Plotter

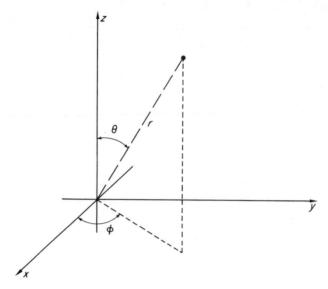

Figure 5.4   Relationship between Cartesian and spherical
polar coordinates

of the spherical polar coordinates r, θ and ϕ, defined in Figure 5.4
relative to a set of Cartesian coordinates centred on the nucleus (which
is treated here as a point charge).

Spherical polar coordinates are preferable because the wave equation,
in terms of these, is separable, as in the previous example in two
coordinates. On the other hand, no such separation is possible in terms
of the Cartesian coordinates x, y and z because the potential function
(equation (5.2)) is not a simple function of x, y and z. As in the case of
the two-dimensional box problems, the solutions of the wave equation
for the hydrogen atom can be written as a product of two terms; one
which expresses the radial dependence $R_{(r)}$ and the other the angular
dependence $\Omega(\theta, \phi)$. The different total wave functions involve the
products of different angular and radial functions. The precise form of
$R_{(r)}$ depends on two quantum numbers n and l, where n (the principal
quantum number) determines the energy associated with the wave
function and l determines the total orbital angular momentum of the
electron. The allowed values of n are 1, 2, 3, ... For a given n, l is
resticted to $l = n$, $n - 1$, . . . , 0. The angular part of the wavefunction
depends on l and a quantum number which determines the
z-component of the orbital angular momentum and for a given l value
has $2l + 1$ allowed values. That is $m = l$, $l - 1$, $l - 2$, ... The radial wave
functions depend on n and l whilst the angular functions depend on l
and m. The total wave function $|\psi_{nlm}$ has the form

$$\psi_{nlm} = R_{n,l}(r) \cdot \psi_{l,m}(\theta, \phi) \tag{5.15}$$

Table 5.9
Examples of Hydrogenic orbitals where Z = nuclear charge (units of proton charge);
$a_0 = 0.0523$ nm (the 'Bohr radius') and $\rho = 2zr/na_0$ (see Reference 1, p. 90)

Orbital type	$R_{n,l}$	$\psi_{n,l,m}$ (normalised)
s-orbitals ($l = 0$)	$R_{1,0} = 2\left(\dfrac{z}{a_0}\right)^{3/2} e^{-\rho}$	$\psi_{13} = \dfrac{R_{1,0}}{2\sqrt{\pi}}$
	$R_{2,0} = \dfrac{2}{2\sqrt{2}}\left(\dfrac{z}{a_0}\right)^{3/2}(2-\rho)e^{-\rho/2}$	$\psi_{2s} = \dfrac{R_{2,0}}{2\sqrt{\pi}}$
	$R_{3,0} = \dfrac{2}{81\sqrt{3}}\left(\dfrac{z}{a_0}\right)^{3/2}(27-18\rho+2\rho^2)e^{-\rho/2}$	$\psi_{3s} = \dfrac{R_{3,0}}{2\sqrt{\pi}}$
p-orbitals ($l = 1$)	$R_{2,1} = \dfrac{1}{2\sqrt{6}}\left(\dfrac{z}{a_0}\right)^{3/2}\rho e^{-\rho/2}$	$\psi_{2p_z} = \dfrac{R_{2,1}}{2\sqrt{(3\pi)}}\cos\theta$
		$\psi_{2p_x} = \dfrac{R_{2,1}}{2\sqrt{(3\pi)}}\sin\theta\cos\phi$
		$\psi_{2p_y} = \dfrac{R_{21}}{2\sqrt{(3\pi)}}\sin\theta\sin\phi$
	$R_{3,1} = \dfrac{4}{81\sqrt{6}}\left(\dfrac{z}{a_0}\right)^{3/2}(6\rho-\rho^2)e^{-\rho/3}$	$\psi_{3p_z} = \dfrac{\sqrt{3}}{2\sqrt{\pi}}R_{3,1}\cos\theta$
		$\psi_{3p_x} = \dfrac{\sqrt{6}}{2\sqrt{\pi}}R_{3,1}\sin\theta\cos\phi$
		$\psi_{3p_y} = \dfrac{\sqrt{6}}{2\sqrt{\pi}}R_{3,1}\sin\theta\sin\phi$

The solutions $\psi_{n\,lm}$ are referred to as hydrogenic atomic orbitals Chemists place considerable weight on an appreciation of the shapes and sizes of these orbitals. It is in these areas that a computer can help to give the correct impression of the factors which influence the different types of orbitals. Examples of the analytical form of the radial and angular functions are given in Table 5.9. In subsequent sections we shall use the computer to illustrate certain aspects of the behaviour of these mathematical functions which are relevant to the modern chemist.

5.3.2 The Radial Wave Function
The general form of the radial wave function with quantum numbers n and l is given by[9]

$$R_{n,l}(r) = -\sqrt{\left[\left(\frac{2Z}{na}\right)^3 \frac{(n-l-1)!}{2\pi\{(n+l)!\}^3}\right]} e^{-\rho/2}\rho^l L_{n+l}^{2l+1}(\rho) \qquad (5.16)$$

where Z is the atomic number, a_0 is the Bohr radius $= 0.0529$ nm, and $\rho = 2Zr/na_0$. The function $L_{n+l}^{2l+1}(\rho)$ is a polynomial in ρ known as the associated Laguerre polynomial. It is defined explicitly by equation (5.17)

$$L_{n+l}^{2l+1}(\rho) = \sum_{k=0}^{n-l-1} (-1)^{k+l} \frac{\{(n+l)!\}^2 \, \rho^k}{(n-l-1-k)!(2l+1+k)!} \qquad (5.17)$$

The program RAD1 listed in Table 5.10, computes the value of $R_{n,l}$ for a range of values of r from $r = 0$ to an upper limit of 1.225 nm in steps of 0.025 nm.

The program also calculates the radial wave function squared and the function $4\pi r^2 \cdot R_{n,l}^2(r)$, which is called the radial distribution function. The latter is significant because $4\pi r^2 \cdot R_{n,l}^2 dr$ is the probability that an electron will be found between distances r and $(r + dr)$ from the nucleus irrespective of direction. Therefore, the radial distribution function, plotted against r, can be used to indicate the most probable value of r for an electron in a specified atomic orbital. Program RAD1 (Table 5.10) involves a MASTER routine and a subroutine LAGUERRE which evaluates the polynomials $L_{n+l}^{2l+1}(\rho)$ for a given values of n, l and ρ, and a function routine which evaluates the factorial terms required by both the master routine and subroutine LAGUERRE. The input data consists of

Card 1: A title card (8A8).
Card 2: n, l, z (3I0).
Card 3: As card 2 for further n, l, z if required or

0 0 0 as a terminator card.

The output consists of tabulated values of r in cm, $R_{n,l}(r)$, $R_{n,l}^2(r)$ and the radial distribution function. Table 5.11 shows output for $2s$, $3s$ and $3p$ atomic orbitals all with unit nuclear charge. This program can be used to illustrate a number of points about the radial dependence of hydrogenic orbitals. The variation of $R_{n,l}(r)$ with r shows the presence of radial nodes and how the number of such nodes depends on the quantum numbers n and l. In the case of s-orbitals the angular wave function is a constant (see Table 5.9) so that $R^2(n, l)$ shows the variation in probability electron density with r.

It is interesting to note that the electron density for s-orbitals is non-zero at $r = 0$, i.e. at the nucleus. But why does the electron charge not neutralize the nuclear charge? The reason is that the wave function squared is a measure of the electron probability *per unit volume* not the actual probability that an electron will be in a specified point. The electron probability density at the nucleus must be multiplied by the volume of the nucleus to obtain a real probability. However, in formulating the wave equation for the hydrogen atom it is assumed that

Table 5.10
Program RADI for the calculation of radial wave functions $R_{n,l}$

```
0008              MASTER RAD1
0009              DIMENSION T(8)
0010              WRITE(2,1003)
0011              READ(1,1001)(T(I),I=1,8)
0012         1001 FORMAT(8A8)
0013              WRITE(2,1001)(T(I),I=1,8)
0014          999 CONTINUE
0015              READ(1,2001)N,L,IZ
0016              IF(N.EQ.0)STOP
0017         2001 FORMAT(3I3)
0018              WRITE(2,2002)N,L,IZ
0019              PI=3.1416
0020         1003 FORMAT(1H1,1X,'RADIAL WAVEFUNCTIONS AND RADIAL DISTRIBUTION FUNCTI
0021            1ONS FOR HYDROGEN-LIKE    ORBITALS',////)
0022         2002 FORMAT(1X,'N=',I3,2X,'L=',I3,2X,'IZ=',I3,//)
0023              WRITE(2,1005)
0024         1005 FORMAT(8X,'R/CM',11X,'RNL',12X,'RSQD',11X,'RDF',/)
0025              A0=0.529E-8
0026              DO 10 J=1,50
0027              R=(J-1)*0.25E-8
0028              F=4*PI*R**2
0029              RHO=2*IZ*R/(N*A0)
0030              FNL=IFAC(N+L)
0031              FNORM=-SQRT(((2*IZ/(N*A0))**3*IFAC(N-L-1)/(2*FNL**3*N)))
0032              CALL LAGUERRE(PLANL,N,L,RHO)
0033              RHOL=1
0034              IF(L.GT.0)RHOL=RHO**L
0035              RNL=FNORM*RHOL*PLANL*EXP(-RHO/2)
0036              RSQ=RNL*RNL
0037              RDF=F*RSQ
0038              WRITE(2,1006)R,RNL,RSQ,RDF
0039         1006 FORMAT(1X,4E15.4,/)
0040           10 CONTINUE
0041              GO TO 999
0042              END


0043              SUBROUTINE LAGUERRE(PLANL,N,L,RHO)
0044              LIM=N-L-1
0045              IF(LIM.GE.0)GO TO 1
0046              WRITE(2,1001)N,L
0047         1001 FORMAT(1X,'FAULT IN Q, NOS N=',I3,2X,'L=',I3,/)
0048              STOP
0049            1 CONTINUE
0050              SUM=0.0
0051              LUP=LIM+1
0052              DO 5 K1=1,LUP
0053              K=K1-1
0054              IF1=IFAC(N+L)
0055              RHOK=1.0
0056              IF(K.GT.0)RHOK=RHO**K
0057            3 SUM=SUM+(-1)**(K+1)*IF1*IF1*RHOK/(IFAC(N-L-1-K)*IFAC(2*L+1+K)*IFAC
0058            1(K))
0059              PLANL=SUM
0060              RETURN
0061              END


0062              FUNCTION IFAC(N)
0063              IPROD=1
0064              IF(N.EQ.0)GO TO 40
0065           20 DO 30 I=1,N
0066           30 IPROD=IPROD*I
0067           40 IFAC=IPROD
0068              RETURN
0069              END
```

Table 5.11
Output from RADI for the $2s$, $3s$, and $3p$ atomic orbitals

RADIAL WAVEFUNCTIONS AND RADIAL DISTRIBUTION FUNCTIONS FOR HYDROGEN-LIKE   ORBITALS

HYDROGENIC ORBITALS Z=1
N= 1 L= 0 IZ= 1

R/CM	RNL	RSQD	RDF
0.0000E 00	0.5198E 13	0.2702E 26	0.0000E 00
0.2500E-08	0.3240E 13	0.1050E 26	0.8247E 09
0.5000E-08	0.2020E 13	0.4081E 25	0.1282E 10
0.7500E-08	0.1259E 13	0.1586E 25	0.1121E 10
0.1000E-07	0.7850E 12	0.6162E 24	0.7744E 09
0.1250E-07	0.4894E 12	0.2395E 24	0.4702E 09
0.1500E-07	0.3051E 12	0.9306E 23	0.2631E 09
0.1750E-07	0.1902E 12	0.3617E 23	0.1392E 09
0.2000E-07	0.1185E 12	0.1405E 23	0.7064E 08
0.2250E-07	0.7390E 11	0.5462E 22	0.3475E 08
0.2500E-07	0.4607E 11	0.2122E 22	0.1667E 08
0.2750E-07	0.2872E 11	0.8248E 21	0.7838E 07
0.3000E-07	0.1790E 11	0.3205E 21	0.3625E 07
0.3250E-07	0.1116E 11	0.1246E 21	0.1653E 07
0.3500E-07	0.6957E 10	0.4840E 20	0.7451E 06
0.3750E-07	0.4337E 10	0.1881E 20	0.3324E 06
0.4000E-07	0.2704E 10	0.7310E 19	0.1470E 06
0.4250E-07	0.1685E 10	0.2841E 19	0.6448E 05
0.4500E-07	0.1051E 10	0.1104E 19	0.2809E 05
0.4750E-07	0.6550E 09	0.4290E 18	0.1216E 05
0.5000E-07	0.4083E 09	0.1667E 18	0.5237E 04
0.5250E-07	0.2545E 09	0.6479E 17	0.2244E 04
0.5500E-07	0.1587E 09	0.2518E 17	0.9570E 03
0.5750E-07	0.9891E 08	0.9784E 16	0.4065E 03
0.6000E-07	0.6166E 08	0.3802E 16	0.1720E 03
0.6250E-07	0.3844E 08	0.1478E 16	0.7253E 02
0.6500E-07	0.2396E 08	0.5742E 15	0.3049E 02
0.6750E-07	0.1494E 08	0.2231E 15	0.1278E 02
0.7000E-07	0.9312E 07	0.8671E 14	0.5339E 01
0.7250E-07	0.5805E 07	0.3370E 14	0.2226E 01
0.7500E-07	0.3619E 07	0.1309E 14	0.9256E 00
0.7750E-07	0.2256E 07	0.5089E 13	0.3841E 00
0.8000E-07	0.1406E 07	0.1978E 13	0.1590E 00
0.8250E-07	0.8766E 06	0.7685E 12	0.6573E-01

0,8500E-07	0,5465E 06	0,2986E 12	0,2711E-01
0,8750E-07	0,3407E 06	0,1161E 12	0,1117E-01
0,9000E-07	0,2124E 06	0,4510E 11	0,4591E-02
0,9250E-07	0,1324E 06	0,1753E 11	0,1884E-02
0,9500E-07	0,8253E 05	0,6811E 10	0,7725E-03
0,9750E-07	0,5145E 05	0,2647E 10	0,3162E-03
0,1000E-06	0,3207E 05	0,1029E 10	0,1293E-03
0,1025E-06	0,1999E 05	0,3997E 09	0,5277E-04
0,1050E-06	0,1246E 05	0,1553E 09	0,2152E-04
0,1075E-06	0,7769E 04	0,6036E 08	0,8766E-05
0,1100E-06	0,4845E 04	0,2346E 08	0,3567E-05
0,1125E-06	0,3019E 04	0,9116E 07	0,1450E-05
0,1150E-06	0,1882E 04	0,3543E 07	0,5887E-06
0,1175E-06	0,1173E 04	0,1377E 07	0,2388E-06
0,1200E-06	0,7314E 03	0,5350E 06	0,9681E-07
0,1225E-06	0,4560E 03	0,2079E 06	0,3921E-07

N= 2 L= 0 IZ= 1

R/CM	RNL	RSQD	RDF
0,0000E 00	0,1838E 13	0,3378E 25	0,0000E 00
0,2500E-08	0,1108E 13	0,1228E 25	0,9645E 08
0,5000E-08	0,6042E 12	0,3651E 24	0,1147E 09
0,7500E-08	0,2655E 12	0,6934E 23	0,4902E 08
0,1000E-07	0,3915E 11	0,1533E 22	0,1926E 07
0,1250E-07	-0,1023E 12	0,1047E 23	0,2056E 08
0,1500E-07	-0,1860E 12	0,3460E 23	0,9782E 08
0,1750E-07	-0,2299E 12	0,5286E 23	0,2034E 09
0,2000E-07	-0,2471E 12	0,6106E 23	0,3069E 09
0,2250E-07	-0,2469E 12	0,6095E 23	0,3878E 09
0,2500E-07	-0,2358E 12	0,5561E 23	0,4367E 09
0,2750E-07	-0,2185E 12	0,4773E 23	0,4536E 09
0,3000E-07	-0,1980E 12	0,3919E 23	0,4433E 09
0,3250E-07	-0,1764E 12	0,3113E 23	0,4132E 09
0,3500E-07	-0,1552E 12	0,2408E 23	0,3707E 09
0,3750E-07	-0,1351E 12	0,1824E 23	0,3224E 09
0,4000E-07	-0,1166E 12	0,1358E 23	0,2731E 09
0,4250E-07	-0,9984E 11	0,9968E 22	0,2263E 09
0,4500E-07	-0,8500E 11	0,7226E 22	0,1839E 09
0,4750E-07	-0,7199E 11	0,5182E 22	0,1469E 09
0,5000E-07	-0,6069E 11	0,3683E 22	0,1157E 09
0,5250E-07	-0,5095E 11	0,2596E 22	0,8993E 08

204

0.5500E-07	-0.4263E 11	0.1817E 22	0.6908E 08
0.5750E-07	-0.3555E 11	0.1264E 22	0.5252E 08
0.6000E-07	-0.2957E 11	0.8742E 21	0.3955E 08
0.6250E-07	-0.2453E 11	0.6015E 21	0.2953E 08
0.6500E-07	-0.2030E 11	0.4119E 21	0.2187E 08
0.6750E-07	-0.1676E 11	0.2809E 21	0.1608E 08
0.7000E-07	-0.1381E 11	0.1908E 21	0.1175E 08
0.7250E-07	-0.1137E 11	0.1292E 21	0.8534E 07
0.7500E-07	-0.9337E 10	0.8717E 20	0.6162E 07
0.7750E-07	-0.7658E 10	0.5864E 20	0.4426E 07
0.8000E-07	-0.6272E 10	0.3934E 20	0.3164E 07
0.8250E-07	-0.5130E 10	0.2632E 20	0.2251E 07
0.8500E-07	-0.4192E 10	0.1757E 20	0.1595E 07
0.8750E-07	-0.3421E 10	0.1170E 20	0.1126E 07
0.9000E-07	-0.2788E 10	0.7776E 19	0.7915E 06
0.9250E-07	-0.2271E 10	0.5157E 19	0.5545E 06
0.9500E-07	-0.1848E 10	0.3414E 19	0.3872E 06
0.9750E-07	-0.1502E 10	0.2256E 19	0.2695E 06
0.1000E-06	-0.1220E 10	0.1489E 19	0.1871E 06
0.1025E-06	-0.9902E 09	0.9806E 18	0.1295E 06
0.1050E-06	-0.8031E 09	0.6450E 18	0.8936E 05
0.1075E-06	-0.6509E 09	0.4236E 18	0.6152E 05
0.1100E-06	-0.5272E 09	0.2779E 18	0.4225E 05
0.1125E-06	-0.4267E 09	0.1821E 18	0.2896E 05
0.1150E-06	-0.3451E 09	0.1191E 18	0.1980E 05
0.1175E-06	-0.2790E 09	0.7786E 17	0.1351E 05
0.1200E-06	-0.2255E 09	0.5083E 17	0.9199E 04
0.1225E-06	-0.1821E 09	0.3315E 17	0.6252E 04

N= 3 L= 0 IZ= 1

R/CM	RNL	RSQD	RDF
0.0000E 00	0.1000E 13	0.1001E 25	0.0000E 00
0.2500E-08	0.5995E 12	0.3594E 24	0.2822E 08
0.5000E-08	0.3183E 12	0.1013E 24	0.3184E 08
0.7500E-08	0.1270E 12	0.1614E 23	0.1141E 08
0.1000E-07	0.2377E 10	0.5648E 19	0.7098E 04
0.1250E-07	-0.7359E 11	0.5415E 22	0.1063E 08
0.1500E-07	-0.1146E 12	0.1313E 23	0.3713E 08
0.1750E-07	-0.1311E 12	0.1719E 23	0.6615E 08
0.2000E-07	-0.1310E 12	0.1715E 23	0.8623E 08
0.2250E-07	-0.1201E 12	0.1442E 23	0.9173E 08

0,2500E-07	-0,1027E 12	0,1055E 23	0,8289E 08
0,2750E-07	-0,8203E 11	0,6730E 22	0,6395E 08
0,3000E-07	-0,6019E 11	0,3623E 22	0,4097E 08
0,3250E-07	-0,3870E 11	0,1498E 22	0,1988E 08
0,3500E-07	-0,1855E 11	0,3441E 21	0,5297E 07
0,3750E-07	-0,3340E 09	0,1115E 18	0,1971E 04
0,4000E-07	0,1563E 11	0,2442E 21	0,4911E 07
0,4250E-07	0,2922E 11	0,8537E 21	0,1938E 08
0,4500E-07	0,4046E 11	0,1637E 22	0,4165E 08
0,4750E-07	0,4946E 11	0,2446E 22	0,6936E 08
0,5000E-07	0,5639E 11	0,3180E 22	0,9991E 08
0,5250E-07	0,6147E 11	0,3778E 22	0,1309E 09
0,5500E-07	0,6490E 11	0,4212E 22	0,1601E 09
0,5750E-07	0,6691E 11	0,4477E 22	0,1860E 09
0,6000E-07	0,6771E 11	0,4585E 22	0,2074E 09
0,6250E-07	0,6750E 11	0,4556E 22	0,2237E 09
0,6500E-07	0,6646E 11	0,4417E 22	0,2345E 09
0,6750E-07	0,6477E 11	0,4195E 22	0,2402E 09
0,7000E-07	0,6255E 11	0,3913E 22	0,2409E 09
0,7250E-07	0,5995E 11	0,3594E 22	0,2374E 09
0,7500E-07	0,5708E 11	0,3258E 22	0,2303E 09
0,7750E-07	0,5401E 11	0,2918E 22	0,2202E 09
0,8000E-07	0,5085E 11	0,2585E 22	0,2079E 09
0,8250E-07	0,4764E 11	0,2269E 22	0,1941E 09
0,8500E-07	0,4444E 11	0,1975E 22	0,1793E 09
0,8750E-07	0,4130E 11	0,1705E 22	0,1641E 09
0,9000E-07	0,3824E 11	0,1462E 22	0,1488E 09
0,9250E-07	0,3529E 11	0,1246E 22	0,1339E 09
0,9500E-07	0,3248E 11	0,1055E 22	0,1196E 09
0,9750E-07	0,2980E 11	0,8882E 21	0,1061E 09
0,1000E-06	0,2728E 11	0,7442E 21	0,9351E 08
0,1025F-06	0,2491E 11	0,6205E 21	0,8192E 08
0,1050E-06	0,2270E 11	0,5151E 21	0,7137E 08
0,1075E-06	0,2064E 11	0,4258E 21	0,6184E 08
0,1100E-06	0,1873E 11	0,3507F 21	0,5332E 08
0,1125E-06	0,1696E 11	0,2878E 21	0,4577E 08
0,1150E-06	0,1534E 11	0,2353E 21	0,3911E 08
0,1175E-06	0,1385E 11	0,1918E 21	0,3328E 08
0,1200E-06	0,1248E 11	0,1559E 21	0,2821E 08
0,1225E-06	0,1124E 11	0,1263E 21	0,2382E 08

206

```
N=  2  L=  1  IZ=  1
```

R/CM	RNL	RSQD	RDF
0.0000E 00	0.0000E 00	0.0000E 00	0.0000E 00
0.2500E-08	0.1980E 12	0.3919E 23	0.3078E 07
0.5000E-08	0.3126E 12	0.9772E 23	0.3070E 08
0.7500E-08	0.3702E 12	0.1371E 24	0.9688E 08
0.1000E-07	0.3897E 12	0.1519E 24	0.1909E 09
0.1250E-07	0.3846E 12	0.1479E 24	0.2905E 09
0.1500E-07	0.3644E 12	0.1328E 24	0.3755E 09
0.1750E-07	0.3357E 12	0.1127E 24	0.4337E 09
0.2000E-07	0.3029E 12	0.9175E 23	0.4612E 09
0.2250E-07	0.2691E 12	0.7239E 23	0.4605E 09
0.2500E-07	0.2360E 12	0.5571E 23	0.4376E 09
0.2750E-07	0.2050E 12	0.4202E 23	0.3994E 09
0.3000E-07	0.1766E 12	0.3118E 23	0.3526E 09
0.3250E-07	0.1510E 12	0.2281E 23	0.3028E 09
0.3500E-07	0.1284E 12	0.1649E 23	0.2539E 09
0.3750E-07	0.1086E 12	0.1180E 23	0.2085E 09
0.4000E-07	0.9149E 11	0.8370E 22	0.1683E 09
0.4250E-07	0.7675E 11	0.5891E 22	0.1337E 09
0.4500E-07	0.6416E 11	0.4117E 22	0.1048E 09
0.4750E-07	0.5347E 11	0.2859E 22	0.8107E 08
0.5000E-07	0.4444E 11	0.1975E 22	0.6205E 08
0.5250E-07	0.3684E 11	0.1357E 22	0.4702E 08
0.5500E-07	0.3047E 11	0.9287E 21	0.3550E 08
0.5750E-07	0.2516E 11	0.6328E 21	0.2629E 08
0.6000E-07	0.2072E 11	0.4295E 21	0.1943E 08
0.6250E-07	0.1704E 11	0.2905E 21	0.1426E 08
0.6500E-07	0.1400E 11	0.1959E 21	0.1040E 08
0.6750E-07	0.1148E 11	0.1317E 21	0.7540E 07
0.7000E-07	0.9396E 10	0.8829E 20	0.5436E 07
0.7250E-07	0.7684E 10	0.5904E 20	0.3900E 07
0.7500E-07	0.6276E 10	0.3939E 20	0.2784E 07
0.7750E-07	0.5120E 10	0.2622E 20	0.1979E 07
0.8000E-07	0.4175E 10	0.1741E 20	0.1401E 07
0.8250E-07	0.3398E 10	0.1155E 20	0.9874E 06
0.8500E-07	0.2764E 10	0.7640E 19	0.6936E 06
0.8750E-07	0.2247E 10	0.5047E 19	0.4856E 06
0.9000E-07	0.1824E 10	0.3328E 19	0.3388E 06
0.9250E-07	0.1480E 10	0.2192E 19	0.2357E 06

0,9500E-07	0,1200E 10	0,1441E 19	0,1634E 06
0,9750E-07	0,9728E 09	0,9463E 18	0,1130E 06
0,1000E-06	0,7878E 09	0,6206E 18	0,7798E 05
0,1025E-06	0,6375E 09	0,4064E 18	0,5366E 05
0,1050E-06	0,5156E 09	0,2659E 18	0,3684E 05
0,1075E-06	0,4166E 09	0,1737E 18	0,2523E 05
0,1100E-06	0,3367E 09	0,1134E 18	0,1724E 05
0,1125E-06	0,2719E 09	0,7394E 17	0,1176E 05
0,1150E-06	0,2195E 09	0,4816E 17	0,8004E 04
0,1175E-06	0,1770E 09	0,3134E 17	0,5438E 04
0,1200E-06	0,1428E 09	0,2038E 17	0,3688E 04
0,1225E-06	0,1151E 09	0,1324E 17	0,2497E 04

N= 3 L= 1 IZ= 1

R/CM	PNL	RSQD	RDF
0,0000E 00	0,0000E 00	0,0000E 00	0,0000E 00
0,2500E-08	0,1169E 12	0,1367E 23	0,1074E 07
0,5000E-08	0,1827E 12	0,3337E 23	0,1048E 08
0,7500E-08	0,2122E 12	0,4503E 23	0,3183E 08
0,1000E-07	0,2168E 12	0,4699E 23	0,5905E 08
0,1250E-07	0,2049E 12	0,4197E 23	0,8240E 08
0,1500E-07	0,1827E 12	0,3338E 23	0,9439E 08
0,1750E-07	0,1549E 12	0,2399E 23	0,9234E 08
0,2000E-07	0,1247E 12	0,1554E 23	0,7814E 08
0,2250E-07	0,9430E 11	0,8893E 22	0,5658E 08
0,2500E-07	0,6529E 11	0,4263E 22	0,3348E 08
0,2750E-07	0,3860E 11	0,1490E 22	0,1416E 08
0,3000E-07	0,1476E 11	0,2179E 21	0,2464E 07
0,3250E-07	-0,5966E 10	0,3560E 20	0,4725E 06
0,3500E-07	-0,2354E 11	0,5543E 21	0,8533E 07
0,3750E-07	-0,3808E 11	0,1450E 22	0,2562E 08
0,4000E-07	-0,4975E 11	0,2475E 22	0,4977E 08
0,4250E-07	-0,5882E 11	0,3460E 22	0,7854E 08
0,4500E-07	-0,6557E 11	0,4299E 22	0,1094E 09
0,4750E-07	-0,7027E 11	0,4938E 22	0,1400E 09
0,5000E-07	-0,7321E 11	0,5360E 22	0,1684E 09
0,5250E-07	-0,7466E 11	0,5574E 22	0,1931E 09
0,5500E-07	-0,7486E 11	0,5604E 22	0,2130E 09
0,5750E-07	-0,7404E 11	0,5482E 22	0,2278E 09
0,6000E-07	-0,7241E 11	0,5243E 22	0,2372E 09
0,6250E-07	-0,7013E 11	0,4918E 22	0,2414E 09

0,6500E-07	-0,6737E 11	0,4539E 22	0,2410E 09
0,6750E-07	-0,6426E 11	0,4129E 22	0,2364E 09
0,7000E-07	-0,6090E 11	0,3709E 22	0,2284E 09
0,7250E-07	-0,5741E 11	0,3295E 22	0,2177E 09
0,7500E-07	-0,5384E 11	0,2899E 22	0,2049E 09
0,7750E-07	-0,5027E 11	0,2528E 22	0,1908E 09
0,8000E-07	-0,4675E 11	0,2186E 22	0,1758E 09
0,8250E-07	-0,4332E 11	0,1877E 22	0,1605E 09
0,8500E-07	-0,4001E 11	0,1601E 22	0,1453E 09
0,8750E-07	-0,3685E 11	0,1357E 22	0,1305E 09
0,9000E-07	-0,3381E 11	0,1143E 22	0,1164E 09
0,9250E-07	-0,3096E 11	0,9587E 21	0,1031E 09
0,9500E-07	-0,2828E 11	0,7999E 21	0,9072E 08
0,9750E-07	-0,2578E 11	0,6644E 21	0,7937E 08
0,1000E-06	-0,2344E 11	0,5495E 21	0,6906E 08
0,1025E-06	-0,2128E 11	0,4527E 21	0,5977E 08
0,1050E-06	-0,1928E 11	0,3716E 21	0,5149E 08
0,1075E-06	-0,1744E 11	0,3040E 21	0,4415E 08
0,1100E-06	-0,1574E 11	0,2479E 21	0,3769E 08
0,1125E-06	-0,1419E 11	0,2015E 21	0,3204E 08
0,1150E-06	-0,1278E 11	0,1633E 21	0,2714E 08
0,1175E-06	-0,1149E 11	0,1320E 21	0,2290E 08
0,1200E-06	-0,1031E 11	0,1064E 21	0,1925E 08
0,1225E-06	-0,9250E 10	0,8556E 20	0,1613E 08

the nucleus is a point charge which has zero volume. Therefore the probability of an electron being at the nucleus is zero even for s-orbitals. The output from RAD1 can also be used to examine the behaviour of the radial distribution function. Notice that even for s-orbitals. it has a zero value at $r = 0$ (see Table 5.11). The most probable value of r can be determined roughly from the output for the radial distribution function for any choice of n and l. A more accurate estimate can be obtained by plotting the computed points against r (either manually or using the computer graph-plotting facilities if available).

5.3.3 The Total Wave Function
A number of ways exist for representing the form of atomic orbitals in both two and three dimensions. Each method has its weakness and strength. For example, the three-dimensional model which represents an s-orbital as a sphere and p-orbitals as having dumb-bell-type shapes does give the correct impression concerning the angular dependence of

the wave function. These models represent a surface inside of which there is a high probability of electron containment. However they give little impression of the radial dependence of the orbitals. Alternative methods using two-dimensional representations involve the use of electron density maps evaluated for cross-sections through the atom. The charge-cloud representation can be regarded as a map in which the density of printing is a measure of the electron density (See Section 1.3.2).

Numerical plots of either ψ or ψ^2 for atomic orbitals in a plane, say perpendicular to the z-axis, are easily generated. Program CONTOUR, listed in Table 5.12 is designed for this purpose. It will output ψ or ψ^2 maps in any chosen plane perpendicular to the z-axis for the following hydrogenic orbitals: $1s, 2s, 2p_x, 2p_y, 2p_z, 3s, 3p_x, 3p_y, 3p_z, 3d_{z^2}, 3d_{xz}, 3d_{yz}, 3d_{x^2-y^2}$ 2 and $3d_{xy}$.

The output is in either scaled numeric form or is symbolic to represent ranges of ψ^2 and takes the form of a square of dimension chosen by the user as in the case of TWOBX. This square is divided into a number of grid elements for the purpose of computation. For each grid the value of ψ or ψ^2 is computed and printed out after scaling to a maximum absolute value of 99.

The form of the output is determined by a control integer OUTOP. If OUTOP is 0 then ψ^2 is printed in the grid elements, scaled to a maximum value of 99. To facilitate a comparison with other orbitals the scale of electron density used in any plot is related to the plot for ψ_{1s}^2 by the factor appearing at the top of the map. If OUTOP is 1 then maps of ψ are printed. As for OUTOP = 0, the scale used to indicate the magnitude of ψ is related to that used for ψ_{1s} (see Figure 5.5a). If OUTOP is 2 then the printing of numbers is replaced by symbols which differ for different ranges of ψ^2 as shown in Figure 5.5(b).

The card input for CONTOUR is as follows:

Card 1: IOP, OUTOP, NZ (nuclear charge), WDTH (plot width/nm), Z (value of z-coordinate/nm for chosen plane (format 3I0, 2F0.0).

IOP selects the orbital to be mapped in accordance with the key:

orbital	$1s$,	$2s$,	$2p_z$,	$2p_x$,	$2p_y$,	$3s$,	$3p_z$,	$3p_x$
IOP	1	2	3	4	5	6	7	8

orbital	$3p_y$,	$3d_{z^2}$,	$3d_{xy}$,	$3d_{yz}$,	$3d_{x^2-y^2}$	$3d_{xy}$
IOP	9	10	11	12	13	14

Card 2 and subsequent cards: As card 1, one card for each plot.
Terminator card: As card 1 with a zero value for IOP

By varying NZ, the nuclear charge, the program can be used to demonstrate orbital contraction. In addition, an accurate three-

Table 5.12

Program CONTOUR — for generating numeric or printer plots of cross-sections of
atomic orbitals

```
0008            MASTER CONTOUR
0009            DIMENSION IA(51,84),HOL(20),KTOR(10),XAA(51,84)
0010            DATA HOL(1)/4H  1S/,HOL(2)/4H   2S/,HOL(3)/4H 2PZ/,HOL(4)/4H 2PX/,
0011           1HOL(5)/4H 2PY/,HOL(6)/4H   3S/,HOL(7)/4H 3PZ/,HOL(8)/4H 3PX/,
0012           2HOL(9)/4H 3PY/,HOL(10)/4H3DZ2/,HOL(11)/4H3DXZ/,HOL(12)/4H3DYZ/,
0013           3HOL(13)/4HDXMY/ ,HOL(14)/4H3DXY/
0014            DATA KTOK(1)/1H /,KTOR(2)/1H./,KTOR(3)/1H,/,KTOR(4)/1H-/,KTOR(5)
0015           1/1H+/,KTOR(6)/1H=/,KTOR(7)/1H*/,KTOR(8)/1HO/,KTOR(9)/1HZ/,KTOR(10)
0016           2/1HX/
0017            INTEGER OUTOP
0018      C     OPTION1   SELECTS ORBITAL AS INDICATED IN DATA STATEMENT
0019      C     OUTOP=0 FOR NUMERIC DENSITY MAPS
0020      C     OUTOP=1 FOR WAVE FUNCTION MAP
0021      C     OUTPUT=2 FOR SYMBOLIC DENSITY MAPS
0022            A0=0.0529
0023            RTPI=1.7725
0024            AAIS=SQRT(1/(A0*A0*A0))/RTPI
0025      990   CONTINUE
0026            READ(1,1901)IOP,OUTOP,NZ,WDTH,Z
0027      1901  FORMAT(3I0,2F0.0)
0028            IF(IOP.EQ.0)STOP
0029            IF(OUTOP.EQ.0)WRITE(2,1073)
0030            IF(OUTOP.EQ.1)WRITE(2,1074)
0031            IF(OUTOP.EQ.2)WRITE(2,2075)
0032      2075  FORMAT(1H1,1X, 77(1H*),/,30X,'ELECTRON DENSITY MAP')
0033      1073  FORMAT(1H1,1X, 77(1H*),/,30X,'ELECTRON DENSITY MAPS RELATIVE TO',
0034           1' PSI IS AT R=0')
0035      1074  FORMAT(1H1,1X, 77(1H*),/,30X,'MAP OF WAVE FUNCTION RELATIVE  TO',
0036           1' PSI IS AT R=0')
0037      1020  FORMAT(1X,'OPTION=',I2,2X,'NUCLEAR CHARGE=',I1,2X,'PLOT WIDTH/NM'
0038           1,' = ',1F10.4)
0039            WRITE(2,1020)IOP,NZ,WDTH
0040      1075  FORMAT(1X, 77(1H*))
0041            XX=SQRT(NZ*NZ*NZ/(A0*A0*A0))
0042            NUP=25
0043            IF(OUTOP.EQ.2)NUP=75
0044            DO 2 J=1,46
0045            Y=(23.5-J)*WDTH/45.0
0046            DO 2 K=1,NUP
0047            IF(OUTOP.EQ.2)X=(K-38.0)*WDTH/74.0
0048            IF(OUTOP.NE.2)X=(K-13.0)*WDTH/24.0
0049            R=SQRT(X*X+Y*Y+Z*Z)
0050            XXZ=NZ*R/A0
0051            AA=0.0
0052            IF(R.LT.0.00001)GO TO 701
0053            COST=Z/R
0054            SINT=SQRT(1.-COST**2)
0055            IF(IOP.EQ.3)AA=0.17677*XX/RTPI*COST*EXP(-XXZ/2)*XXZ
0056            IF(IOP.EQ.7)AA=0.017459*XX/RTPI*(6.0-XXZ)*XXZ*COST*EXP(-XXZ/3)
0057            IF(IOP.EQ.10)AA=0.005040*XX/RTPI*XXZ*XXZ*(3+COST*COST-1)*EXP(-XXZ/
0058           13)
0059            IF(SINT.LT.0.00001)GO TO701
0060            D=R*SINT
0061            COSP=X/D
0062            SINP=Y/D
0063            IF(IOP.EQ.4)AA=0.176776*XX/RTPI*COSP*SINT*XXZ*EXP(-XXZ/2)
0064            IF(IOP.EQ.5)AA=0.176776*XX/RTPI*SINP*SINT*XXZ*EXP(-XXZ/2)
0065            IF(IOP.EQ.8)AA=0.017459*XX/RTPI*(6.0-XXZ)*XXZ*SINT*COSP*EXP(-XXZ/
0066           13)
0067            IF(IOP.EQ.9)AA=0.017459*XX/RTPI*(6.0-XXZ)*XXZ*SINT*SINP*EXP(-XXZ/
0068           13)
0069            IF(IOP.GE.11)AFAC=0.017459*XX/RTPI*XXZ*XXZ*EXP(-XXZ/3)
0070            IF(IOP.EQ.11)AA=AFAC*SINT*COST*COSP
0071            IF(IOP.EQ.12)AA=AFAC*SINT*COST*SINP
0072            IF(IOP.EQ.13)AA=AFAC*SINT*(COSP*COSP-SINP*SINP)/2
0073            IF(IOP.EQ.14)AA=AFAC*SINT*SINT*SINP*COSP
0074      701   IF(IOP.EQ.1)AA=XX*EXP(-XXZ)/RTPI
0075            IF(IOP.EQ.2)AA=XX/RTPI*(2.0-XXZ)*EXP(-XXZ/2)*0.176776
0076            IF(IOP.EQ.6)AA=0.007127*XX/RTPI*(27.0-18*XXZ+2*XXZ*XXZ)*EXP(-XXZ/
0077           13)
0078            IF(OUTOP.EQ.0.OR.OUTOP.EQ.2)XAA(J,K)=AA**2
0079            IF(OUTOP.EQ.1)XAA(J,K)=AA
0080      2     CONTINUE
0081      1     CONTINUE
0082            ZMAX=0.0
0083            DO 60 J=1,46
0084            DO 60 K=1,NUP
0085      60    IF(ABS(XAA(J,K)).GT.ZMAX)ZMAX=ABS(XAA(J,K))
```

```
0086                    XFAC=ZMAX/AAIS
0087                    IF(OUTOP.EQ.0)XFAC=ZMAX/AAIS**2
0088                    XFAC=1.0/XFAC
0089                    WRITE(2,1003)Z
0090               1003 FORMAT(1X,'IN THE PLANE Z=',1F10.4,'NM',/)
0091                    IF(OUTOP.EQ.2)GO TO 497
0092                    IF(OUTOP.EQ.0)GO TO 631
0093                    WRITE(2,1004)HOL(IOP),XFAC
0094               1004 FORMAT(1X,'PSI MAP FOR THE ',1A4,'-ORBITAL TIMES ',F10.3,' RELATI'
0095                   1'VE TO PSI IS AT R=0',//,1X,77(1H*))
0096                    GO TO 632
0097                651 CONTINUE
0098                    WRITE(2,1008)HOL(IOP),XFAC
0099               1008 FORMAT(1X,'PSI SQUARED MAP FOR THE ',1A4,'-ORBITAL TIMES',F10.3,
0100                   1' RELATIVE TO PSI**2 IS AT R=0',//,1X,77(1H*))
0101                632 CONTINUE
0102                    DO 62 J=1,46
0103                    DO 62 K=1,25
0104                    IA(J,K)=NINT(99*XAA(J,K)/ZMAX)
0105                 62 CONTINUE
0106                    DO 10 J=1,46
0107                 10 WRITE(2,1006)(IA(J,K),K=1,25)
0108               1006 FORMAT(1X,1H*,25I3,1H*)
0109               1010 FORMAT(1X,77(1H*),//,1H1)
0110                    WRITE(2,1010)
0111                    GO TO 61
0112                497 CONTINUE
0113                    WRITE(2,3008)HOL(IOP)
0114                    WRITE(2,1075)
0115               3008 FORMAT(1X,'MAP FOR THE ',1A4,' ORBITAL')
0116                    DO 82 J=1,46
0117                    DO 82 K=1,75
0118                    XJJK = (9.9*XAA(J,K)/ZMAX)
0119                    JK=XJJK+1
0120                 82 IA(J,K)=KTOR(JK)
0121                    DO 83 J=1,46
0122                    WRITE(2,1740)(IA(J,K),K=1,75)
0123                 83 CONTINUE
0124               1740 FORMAT(1X,1H*,75A1,1H*)
0125                    WRITE(2,1010)
0126                 61 CONTINUE
0127                    GO TO 990
0128                    STOP
0129                    END
```

dimensional picture of orbitals can be built by constructing plots for different values of the coordinate z which determines the position of the plane of the plot perpendicular to the z-axis.

As mentioned earlier, a scaling factor is applied to each plot so that the maximum computed value anywhere in the grid is equal to 99. A different scaling factor is required for each orbital. In order that two different orbitals can be compared with respect to, e.g., electron density, the ratio of the scaling factor for the plot to that for the plot of ψ_{1s}^2 (or ψ_{1s} for wave function plots) is given at the top of the map. It is therefore possible to compare the electron densities at given points in space for two different orbitals by dividing the numeric data in each plot by the scaling factor (relative to the 1s-orbital scaling) for each plot.

5.4 Molecular Orbital Calculations
5.4.1 Introduction
The time-independent wave equation can be written in terms of a differential operator $\mathcal{H}$, known as the Hamiltonian operator for the system. In terms of $\mathcal{H}$, equation (5.1) takes the form:

$$\mathcal{H}\psi = E\psi \tag{5.18}$$

```
********************************************************************
                              MAP OF WAVE FUNCTION RELATIVE  TO PSI IS AT R=0
OPTION= 6  NUCLEAR CHARGE=+1   PLOT WIDTH/NM =      0.6000
IN THE PLANE Z=    0.0000NM

PSI MAP FOR THE    3S-ORBITAL TIMES      5.909 RELATIVE TO PSI IS AT R=0
********************************************************************
*  3  2  1  0 -1 -2 -3 -4 -5 -6 -6 -7 -7 -6 -6 -5 -4 -3 -2 -1  0  1  2  3*
*  3  2  0 -1 -2 -3 -4 -6 -6 -7 -8 -8 -8 -8 -8 -7 -6 -6 -4 -3 -2 -1  0  2  3*
*  2  1  0 -2 -3 -4 -6 -7 -8 -8 -9 -9 -9 -9 -9 -8 -8 -7 -6 -4 -3 -2  0  1  2*
*  2  0 -1 -3 -4 -5 -7 -8 -9-10-10-11-11-11-10-10 -9 -8 -7 -5 -4 -3 -1  0  2*
*  1  0 -2 -4 -5 -7 -8 -9-10-11-11-12-12-12-11-11-10 -9 -8 -7 -5 -4 -2  0  1*
*  0 -1 -3 -4 -6 -8 -9-10-11-12-13-13-13-13-13-12-11-10 -9 -8 -6 -4 -3 -1  0*
*  0 -2 -4 -5 -7 -9-10-11-12-13-13-14-14-14-13-13-12-11-10 -9 -7 -5 -4 -2  0*
* -1 -3 -4 -6 -8-10-11-12-13-14-14-14-14-14-14-14-13-12-11-10 -8 -6 -4 -3 -1*
* -1 -3 -5 -7 -9-11-12-13-14-14-15-15-15-15-15-14-14-13-12-11 -9 -7 -5 -3 -1*
* -2 -4 -6 -8-10-11-13-14-15-15-15-15-15-15-15-15-15-14-13-11-10 -8 -6 -4 -2*
* -3 -5 -7 -9-11-12-14-14-15-15-15-15-14-14-15-15-15-15-14-12-11 -9 -7 -5 -3*
* -3 -5 -7 -9-11-13-14-15-15-14-13-13-14-15-15-15-15-14-13-11 -9 -7 -5 -3*
* -4 -6 -8-10-12-14-14-15-15-14-13-12-11-12-13-14-15-15-14-12-10 -8 -6 -4*
* -4 -7 -9-11-13-14-15-15-14-13-11 -9 -9 -9-11-13-14-15-14-13-11 -9 -7 -4*
* -5 -7 -9-11-13-14-15-14-13-11 -8 -6 -5 -6 -8-11-13-14-15-14-11 -9 -7 -5*
* -5 -8-10-12-14-15-15-14-12 -8 -4 -1  0 -1 -4 -8-12-14-15-15-14-12-10 -8 -5*
* -6 -8-10-12-14-15-15-15-10 -5  0  5  7  5  0 -5-10-13-15-15-14-12-10 -8 -6*
* -6 -8-11-13-14-15-14-12 -8 -2  6 13 15 13  6 -2 -8-12-14-15-15-14-11 -8 -6*
* -6 -9-11-13-14-15-14-11 -6  2 12 22 26 22 12  2 -6-11-14-15-14-13-11 -9 -6*
* -6 -9-11-13-15-15-14-10 -4  6 19 33 39 33 19  6 -4-10-14-15-15-13-11 -9 -6*
* -7 -9-11-13-15-15-13 -9 -2 10 26 45 56 45 26 10 -2 -9-13-15-15-13-11 -9 -7*
* -7 -9-11-13-15-15-13 -9 -1 13 32 57 75 57 32 13 -1 -9-13-15-15-13-11 -9 -7*
* -7 -9-12-14-15-15-13 -8  0 14 35 66 99 66 35 14  0 -8-13-15-15-14-12 -9 -7*
* -7 -9-12-14-15-15-13 -8  0 14 35 66 99 66 35 14  0 -8-13-15-15-14-12 -9 -7*
* -7 -9-11-13-15-15-13 -9 -1 13 32 57 75 57 32 13 -1 -9-13-15-15-13-11 -9 -7*
* -7 -9-11-13-15-15-13 -9 -2 10 26 45 56 45 26 10 -2 -9-13-15-15-13-11 -9 -7*
* -6 -9-11-13-15-15-14-10 -4  6 19 33 39 33 19  6 -4-10-14-15-15-13-11 -9 -6*
* -6 -9-11-13-14-15-14-11 -6  2 12 22 26 22 12  2 -6-11-14-15-14-13-11 -9 -6*
* -6 -8-11-13-14-15-14-12 -8 -2  6 13 15 13  6 -2 -8-12-14-15-14-13-11 -8 -6*
* -6 -8-10-12-14-15-15-13-10 -5  0  5  7  5  0 -5-10-13-15-15-14-12-10 -8 -6*
* -5 -8-10-12-14-15-15-14-12 -8 -4 -1  0 -1 -4 -8-12-14-15-15-14-12-10 -8 -5*
* -5 -7 -9-11-13-14-15-14-13-11 -8 -6 -5 -6 -8-11-13-14-15-14-13-11 -9 -7 -5*
* -4 -7 -9-11-13-14-15-15-14-13-11 -9 -9 -9-11-13-14-15-15-14-13-11 -9 -7 -4*
* -4 -6 -8-10-12-14-15-15-15-14-13-12-11-12-13-14-15-15-15-14-12-10 -8 -6 -4*
* -3 -5 -7 -9-11-13-14-15-15-15-14-13-13-13-14-15-15-15-15-14-13-11 -9 -7 -5 -3*
* -3 -5 -7 -9-11-12-14-14-15-15-15-14-14-15-15-15-15-14-14-12-11 -9 -7 -5 -3*
* -2 -4 -6 -8-10-11-13-14-15-15-15-15-15-15-15-15-15-14-13-11-10 -8 -6 -4 -2*
* -1 -3 -5 -7 -9-11-12-13-14-14-15-15-15-15-14-14-13-12-11 -9 -7 -5 -3 -1*
* -1 -3 -4 -6 -8-10-11-12-13-14-14-14-14-14-14-13-12-11-10 -8 -6 -4 -3 -1*
*  0 -2 -4 -5 -7 -9-10-11-12-13-13-14-14-14-13-13-12-11-10 -9 -7 -5 -4 -2  0*
*  0 -1 -3 -4 -6 -8 -9-10-11-12-13-13-13-13-13-12-11-10 -9 -8 -6 -4 -3 -1  0*
*  1  0 -2 -4 -5 -7 -8 -9-10-11-11-12-12-12-11-11-10 -9 -8 -7 -5 -4 -2  0  1*
*  2  0 -1 -3 -4 -5 -7 -8 -9-10-10-11-11-11-10-10 -9 -8 -7 -5 -4 -3 -1  0  2*
*  2  1  0 -2 -3 -4 -6 -7 -8 -8 -9 -9 -9 -9 -9 -8 -8 -7 -6 -4 -3 -2  0  1  2*
*  3  2  0 -1 -2 -3 -4 -6 -6 -7 -8 -8 -8 -8 -8 -7 -6 -6 -4 -3 -2 -1  0  2  3*
*  3  2  1  0 -1 -2 -3 -4 -5 -6 -6 -7 -7 -7 -6 -6 -5 -4 -3 -2 -1  0  1  2  3*
********************************************************************
```

Figure 5.5   (a) Digital map of 3s orbital, (b) Symbolic map of ψ^2 for the $d_{x^2-y^2}$ orbital

where $\mathcal{H}$, in terms of Cartesian coordinates is given by the expression

$$\mathcal{H} = -\left(\frac{\partial^2}{\partial x^2} + \frac{\partial^2}{\partial y^2} + \frac{\partial^2}{\partial z^2}\right)\frac{h^2}{8\pi^2}m + V \tag{5.19}$$

which can be verified by comparing equations (5.1) and (5.18). In the latter equation note that ψ cannot be deleted from both sides precisely because $|\mathcal{H}|$ is an operator in the same sense as d/dx or $\partial/\partial x^2$. The set of wave functions for a system have the property that any one of the set is modified by the application of the operator $\mathcal{H}$ to the extent of multiplication by a constant which is the energy corresponding to the

```
*********************************************************************
                            ELECTRON DENSITY MAP
OPTION=13  NUCLEAR CHARGE=+1  PLOT WIDTH/NM =      1,7000
IN THE PLANE Z=    0,0000NM

MAP FOR THE DXMY ORBITAL
*********************************************************************
*                                                                 *
*                                                                 *
*                          . . . . . . . . . .                    *
*                      . . . . . . . . . . . . . . .              *
*                    . . . . / / / / / / / / / . . .             *
*                  . . . / / / - - - - - - / / / / . .           *
*                 . . . / / - - - + + + + + + - - - / / . .       *
*                 . . / / - - - + + + + + + + + - - / / , .       *
*               . . , / - - + + = = = = = = = + + - - / , ,       *
*               , . , / - - + = * * * * * * * = = + - - , , ,     *
*               , , / - + + = * 0 0 0 0 0 0 0 * = + + - , , ,     *
*               , , - + = * 0 Z Z Z Z Z Z 0 * * = + - , , ,       *
*               , / - = * 0 Z Z X X X Z Z 0 * = - , , ,           *
*               , . - = 0 Z X X X X Z 0 * + - , ,                 *
. . . . . . .   , , - = * 0 Z X X X Z 0 * = - , ,     . . . . . . *
. . . . . . . . . . , . , - = * 0 Z Z Z 0 * = - , ,   . . . . . . . . . . *
. . . , / / / / - - - - - . . . .   , , - + = * 0 * + - , ,   . . . . / / / / / / / . . . *
. . . , / / - - - + + + = = = = = + - - , ,   . - + + = + = - ,   , , - - + + = = = + + - - - , / / / . . . . *
. . . . / / / - - + + = = = * * * 0 0 * * * = + - , ,   , , / - , , ,   , , - + = * * * 0 0 * * * = + + - - , / / . . . . *
. . . . / / / - - + = = * * 0 0 Z Z Z Z Z Z 0 * = + - , ,   . . .   , , - + = = * Z Z X X X X Z Z 0 0 * = + - - , , , . . . . *
. . . . . . / / - - + = = * * 0 0 Z Z X X X X X Z Z * = + - , ,   , , - + = * Z Z X X X X X Z Z 0 0 * = + + - - , , , . . . . *
. . . . . . / / - - + = = * * 0 0 Z Z Z Z Z Z 0 * = + - , ,   , , - + = * * 0 0 Z Z Z Z Z Z 0 0 * = + + - - , , , . . . . *
. . . . . / / / - - + + = = * * 0 0 * * * = + - , ,   . . .   , , - + = * * 0 0 * * * = + + + - - , , , . . . . *
. . . , / / / - - - + + = = = = = + - - , ,   , , / - , , ,   , , - - + + = = = = + + - - - , / / / . . . . *
. . . , / / / / / / / / / / / . , ,   , - + + = + = - - ,   , , - . . . . . . . / / / / / . . . . *
. . . . . . . . . . . . . . . .   , , - + = * 0 * + - , ,   . . . . . . . . . . . . . *
                                , , - = * 0 Z Z Z 0 * = - , ,                       *
                                , , - = 0 Z X X X Z 0 * = - , ,                      *
                                , + = 0 Z X X X X Z 0 = + - , ,                      *
                                , , - = * 0 Z X X X Z Z 0 * = - , , ,                *
                                , , - + = * 0 Z Z Z Z Z 0 * = + - , , ,             *
                                , , - + + = * 0 0 0 0 0 0 0 * = + + - , , ,         *
                                , , , / - - + + = = = = = = = + + - - , , , ,       *
                                , , , , / - - + + + + + + + + - - - , , , ,         *
                                , , , , / - - - + + + + + + - - - , , , ,           *
                                , , , / / / - - - - - - - - - / / / . . .           *
                                  . , / / / / / / / / / / , . .                     *
                                    . . . . / / / / / / / . . . .                   *
                                      . . . . . . . . . . . .                       *
                                          . . . . . . . . .                         *
*                                                                 *
*                                                                 *
*********************************************************************
```

wave function. The wave functions are referred to as *eigenfunctions* of
$\mathcal{H}$ and the corresponding energies as *eigenvalues*. The solution of the
wave equation for a system can be regarded as a search for a set of
functions each of which is multiplied by a constant when operated on
by $\mathcal{H}$.

The exact solution of the wave equation for electrons in molecules of
chemical interest is not feasible at the present time. A number of
approximate methods of soution have therefore been devised. The
molecular orbitals, which are one-electron wave functions in general,
extending over the whole molecule, are assumed to be expressable as a
linear combination of atomic orbitals (LCAO approach) as follows:

$$\psi_i = \sum_{|\nu=1}^{n} \chi_\nu c_{\nu i} \tag{5.20}$$

Here ψ_i is one of a set of n atomic orbitals (a.o.s) referred to as the basis set. The quantity $c_{\nu i}$ is the molecular orbital coefficient for the νth a.o. and the ith m.o. The atomic orbital basis set is chosen on the basis of 'chemical intuition' from the atomic orbitals centred on the various atoms in the molecule. We shall find a matrix formulation convenient in discussing the computer calculation of molecular orbitals. Equation (5.20) can be cast in matrix notation by denoting the set of atomic orbitals as a row vector χ and the set of molecular orbital coefficients for the ith molecular orbital as a column vector c_i as shown in equation (5.21).

$$\psi_i = [\chi_1, \chi_2, \ldots, \chi_n] \begin{bmatrix} c_{1i} \\ c_{2i} \\ \cdot \\ \cdot \\ \cdot \\ c_{ni} \end{bmatrix} \quad \text{or} \quad \psi_i = \chi.c_i \qquad (5.21)$$

Equation (5.21) defines a whole set of molecular orbitals, ψ_i, with correspondingly different values of i. The set of equations implied by (5.21) can be thought of as a transformation from one basis set (the a.o. set) to another basis set of orbitals (the m.o. set) rather like the transformation from one set of Cartesian coordinates to another set which is rotated relative to the first set (Figure 2.6). The dimensionality of the space in the latter case remaining at 3. Similarly the dimensions of the space spanned by the basis functions is unchanged by such a transformation. Therefore, we expect to produce by the LCAO method n m.o.s from the n a.o. basis set. The complete transformation from the a.o. set to the m.o. set can be summarized conveniently in matrix form by extending equation (5.21)

$$[\psi_1 \psi_2 \ldots \psi_n] = [\chi_1 \chi_2 \ldots \chi_n] \begin{bmatrix} c_{11} & c_{12} & \cdots & c_{1n} \\ c_{21} & c_{22} & \cdots & c_{2n} \\ \cdot & \cdot & & \cdot \\ \cdot & \cdot & & \cdot \\ \cdot & \cdot & & \cdot \\ c_{n1} & c_{n2} & \cdots & c_{nn} \end{bmatrix}$$

or

$$\psi = \chi c \qquad (5.22)$$

The problem of determining the form of the m.o.s is now reduced to one of calculating the m.o. coefficients contained in the matrix c. The

ith column of $\mathbf{c}$ is the set of m.o. coefficients for ψ_i and the νth element of column i is the coefficient $c_{\nu i}$.

Introducing equation (5.20) into equation (5.18) gives:

$$\sum_{\nu=1}^{n} \mathcal{H} \chi_\nu c_{\nu i} = E_i \sum_{\nu=1}^{n} \chi_\nu c_{\nu i} \qquad (5.23)$$

Multiplying this expression from the left by χ_μ and integrating both sides over all space gives the result.

$$\sum_{\nu=1}^{n} H_{\mu\nu}\, c_{\nu i} = E_i \sum_{\nu=1}^{n} S_{\mu\nu}\, c_{\nu i} \qquad (5.24)$$

where $H_{\mu\nu}$ is an element of the Hamiltonian matrix defined by

$$H_{\mu\nu} = \int \chi_\mu \mathcal{H} \chi_\nu d\tau$$

and $S_{\mu\nu}$ is an element of the overlap matrix defined as

$$S_{\mu\nu} = \int \chi_\mu \chi_\nu\, d\tau$$

$S_{\mu\nu}$ is known as the overlap integral between χ_μ and χ_ν.

Equation (5.24) is one of a set of n such equations (one for each choice of χ_μ). The complete set of equations are conveniently summarized in matrix form:

$$
\begin{bmatrix}
H_{11} & H_{12} & \cdots & H_{n1} \\
H_{12} & H_{22} & \cdots & H_{n2} \\
\cdot & \cdot & & \cdot \\
\cdot & \cdot & & \cdot \\
\cdot & \cdot & & \cdot \\
H_{1n} & H_{2n} & \cdots & H_{nn}
\end{bmatrix}
\begin{bmatrix}
c_{1i} \\
c_{2i} \\
\cdot \\
\cdot \\
\cdot \\
c_{ni}
\end{bmatrix}
=
\begin{bmatrix}
S_{11} & S_{12} & \cdots & S_{1n} \\
S_{21} & S_{22} & \cdots & S_{2n} \\
\cdot & \cdot & & \cdot \\
\cdot & \cdot & & \cdot \\
\cdot & \cdot & & \cdot \\
S_{n1} & S_{n2} & \cdots & S_{nn}
\end{bmatrix}
$$

$$
\begin{bmatrix}
c_{1i} \\
c_{2i} \\
\cdot \\
\cdot \\
\cdot \\
c_{ni}
\end{bmatrix}
\begin{bmatrix} E_i \end{bmatrix}
\qquad (5.25)
$$

A set of equations of the type (5.25) exist for each molecular orbital. The whole set may be combined to yield one matrix expression as follows:

$$
\left[\begin{array}{c} H_{\mu\nu} \end{array} \right] \left[\begin{array}{c} c_{\mu i} \end{array} \right] =
$$

$$
\left[\begin{array}{c} S_{\mu\nu} \end{array} \right] \left[\begin{array}{c} c_{\mu i} \end{array} \right] \left[\begin{array}{ccccc} E_{11} & 0 & \dots & 0 \\ 0 & E_{22} & \dots & 0 \\ \cdot & & & \cdot \\ \cdot & & & \cdot \\ \cdot & & & \cdot \\ 0 & \dots & & 0\,E_{nn} \end{array} \right]
$$

$$\text{(5.26)}$$

or

$$\mathbf{Hc = ScE}$$

Notice that the energies appear as the diagonal elements of a diagonal matrix (i.e. all non-diagonal elements are zero). This ensures that the product of the c and E matrices results in the multiplication of all of the elements of the ith column of c by E_{ii}.

Some simplification of equation (5.26) is possible if the basis a.o.s are assumed to be orthonormal, i.e. both orthogonal and normalized. Within this approximation we have $S_{\mu\nu} = 0$, for $\mu \neq \nu$ and if $\mu = \nu$ the $S_{\mu\mu} = 1$. S therefore becomes a unit matrix (non-diagonal elements all zero and diagonal elements all equal to unity) and can be left out of the expression giving

$$\mathbf{Hc = cE} \qquad \text{(5.27)}$$

The assumption that the χs form an orthogonal set implies that c is a transformation matrix from one orthogonal set (the χs) to another orthogonal set (the ψs). Such a transformation matrix has the special property that its inverse is obtained merely by interchanging rows and columns, i.e. by transposing the matrix. Denoting $c\dagger$ as the transpose of c we have that c^{-1}, the inverse of c, is equal to $c\dagger$. Multiplying equations (5.27) from the left by $c\dagger$ gives

$$\mathbf{c\dagger Hc = c\dagger cE} \qquad \text{(5.28)}$$

or

$$c\dagger Hc = E$$

since

$$c\dagger c = c^{-1} c = I, \text{ the unit matrix}$$

In order to obtain c, the matrix of molecular orbital coefficients, we must assign values to the elements of H and seek a matrix which reduces H to the diagonal form, E, by pre- and post-multiplication of H as implied by (5.28). Such a process is known as matrix diagonalization for which standard computer programs are available (as discussed in Section 2.5).

5.4.2 Huckel Molecular Orbital Calculations

Huckel molecular orbital theory is a very simple form of m.o. theory which has been successfully applied[8,9] to unsaturated organic molecules. The electronic wave functions for such molecules are assumed to be separable into two parts, one associated with π-electrons and the other with the σ-electrons. The latter orbitals involve σ-bonds between the atoms of the molecule and the electric field due to both the σ-electrons and the nuclei is assumed to determine the potential function experienced by the π-electrons. The form of the Hamiltonian cannot therefore be explicitly defined, so that the elements $H_{\mu\nu}$ of the Hamiltonian matrix cannot be calculated. For this reason the $H_{\mu\nu}$ are assigned values on an empirical basis. The basis set of a.o.s appropriate to the definition of π-molecular orbitals in C, N, O compounds comprises the set of $2p_z$-orbitals perpendicular to the molecular plane. One $2p_z$-orbital is contributed by each atom involved in the π-electron system.

The empirical values chosen for $H_{\mu\nu}$ or $H_{\mu\mu}$ depends on the type of atom on which the $2p_z$ orbital is centred. For carbon atoms the following rules are used:

(1) $H_{\mu\mu}$ is assumed to be the same for each carbon atom and equal to α_C which is known as the Coulomb integral for carbon. It can be interpreted as the energy of the $2p_z$ orbital in the environment of the molecule in the absence of a π-electron system.

(2) If χ_μ and χ_ν are centred on carbon atoms which are directly bonded then $H_{\mu\nu} = \beta_{CC}$, the resonance integral for carbon—carbon bonds. Otherwise (i.e. when centres μ and ν are not bonded directly) $H_{\mu\nu} = 0$.

If the π-system involves a heteroatom, X, then the appropriate Coulomb and resonance integrals are defined as follows:

$$\alpha_X = \alpha_C + h_X\,\beta_{CC}$$
$$\beta_{CX} = k_{CX}\,\beta_{CC}$$

(5.29)

where acceptable empirical parameters[10] h_X and k_{CX} are:

$$X \quad : 0-\ ,\ 0=\ ,\ N-\ ,\ N=$$
$$h_X \quad : 1 \quad ,\ 2 \quad ,\ 0.5\ ,\ 1.5$$
$$k_{CX} : 0.8\ ,\ 2 \quad ,\ 0.8\ ,\ 1$$

In order to carry out Huckel molecular orbital calculations it is therefore only necessary to specify the elements of H and then use a diagonalization routine to obtain the m.o. coefficients and orbital energies. In practice a further simplification is usually made by a suitable choice of origin and units for the energy which are to be calculated. If we regard the energy of a carbon $2p_z$ orbital in the molecular environment in the absence of a π-system as defining the energy zero then $\alpha_C = 0$. Further if we choose β_{CC} as the unit of energy then all matrix elements which involve β_{CC} are simplified. Let us consider the matrix which is to be diagonalized for 1,3-butadiene in which we number the atoms involved in the π-system as follows.

$$C_1 = C_2$$
$$\diagdown$$
$$C_3 = C_4$$

There are 4 basis orbitals (one from each carbon) so that H is a 4 x 4 matrix. Then $H_{11} = H_{22} = H_{33} = H_{44} = \alpha = 0$, and $H_{12} = H_{24} = \beta$ from rules (1) and (2) above. Adopting β as the unit of energy then the off-diagonal H_{12}, etc. become equal to 1. If we choose the origin of energy as the energy of a $2p_z$ orbital in the hypothetical state in which it is in the environment of the σ-electron core but is not perturbed by the rest of the π-system then $H_{\mu\mu} = \alpha$ (which is the energy of such an orbital) can be set to zero. Thus all the diagonal elements of H can be set to zero because of this choice of origin for the energy. The matrix to be diagonalized is shown in Table 5.13 along with the c and E

Table 5.13
Matrices involved in the Huckel m.o. calculation for
1,3 butadiene (atoms are numbered consecutively along the
carbon chain)[a]

$$H = \begin{pmatrix} 0 & 1 & 0 & 0 \\ 1 & 0 & 1 & 0 \\ 0 & 1 & 0 & 1 \\ 0 & 0 & 1 & 0 \end{pmatrix} \qquad c = \begin{pmatrix} 0.37 & 0.60 & 0.60 & 0.37 \\ 0.60 & 0.37 & -0.37 & -0.60 \\ 0.60 & -0.37 & -0.37 & 0.61 \\ 0.37 & -0.60 & 0.60 & -0.37 \end{pmatrix}$$

$$E = \begin{pmatrix} 1.62 & 0 & 0 & 0 \\ 0 & 0.62 & 0 & 0 \\ 0 & 0 & -0.62 & 0 \\ 0 & 0 & 0 & -1.62 \end{pmatrix}$$

[a](Note that c is not always a symmetrical matrix).

matrices. The reader can confirm the solutions are in accordance with equation (5.28) by manual matrix multiplication.

With reference to Table 5.13, the ith columns of c defines the coefficients for the ith m.o. and the corresponding orbital energy is the ith diagonal element of E.

The resonance integral β (which is our unit of energy) is expected on theoretical grounds to be a negative quantity (although it cannot be calculated because of the vague definition of the Hamiltonian in Huckel m.o. theory). The question which arises is whether positive or negative values of the eigenvalues correspond to lower energies. The eigenvalues computed are relative to the unperturbed $2p_z$ orbitals in units of β. Therefore if $E_1 = 1.62$ (see Table 5.13) then this means that $E_1 = 1.62\beta$ relative to the $2p_z$ orbital energy which represents a stabilization (since β is negative). Therefore, the orbital of lowest energy has eigenvalue 1.62 with the corresponding wave function.

$$\psi_1 = 0.37\chi_1 + 0.60\chi_2 + 0.60\chi_3 + 0.37\chi_4$$

which is obtained from the first column of coefficients in c. In general if E_{ii} is the eigenvalue computed the corresponding energy is

$$E_i = \alpha + E_{ii}\beta \qquad (5.30)$$

In order of increasing energy the remaining three m.o.s for butadiene have eigenvalues equal to 0.62, -0.62 and -1.62 with corresponding m.o. coefficients given by the 2nd, 3rd and 4th columns of c. Since we have assumed that the set of a.o.s are orthonormal then, as the eigenvectors obtained from a matrix diagonalization routine are normalized to unity (i.e. the sum of the squares of the coefficients in any column of c is equal to unity), then the molecular orbitals so obtained are already normalised.

Program BEE2 (which is based on the program in the text by Dickson[11]) listed in Table 5.14 calculates the Hamiltonian matrix, from input data which indicates the atoms which are directly bonded, and computes the eigenvalues and eigenvectors of H by using a diagonalization subroutine (JACB) that uses Jacobi's method as described in Chapter 2. The subroutine JACB is listed in Table 2.14.

The input for BEE2 comprises the number of atoms, the number of electrons and the non-zero elements of the Hamiltonian matrix. In general the number of atoms, N, means the number of atoms involved in the π-system and is equal to the size of the a.o. basis set. For a neutral hydrocarbon the number of electrons, NE, is equal to N.

In the presence of a heteroatom the total number of electrons depends on how many electrons are contributed by the heteroatom. For example, the nitrogen in pyridine contributes two electrons but only one in the case of pyrrole.

Table 5.14

Listing of program BEE2 with output for cyclooctatetraene

```
0008          MASTER BEE2
0009        C PROGRAM TO COMPUTE EIGENVALUES AND EIGENVECTORS BY JACOBI
0010        C METHOD
0011          DIMENSIONNATOM(50)
0012          DIMENSIONTITLE(20)
0013          COMMON/JAC01/N,A(40,40),S(40,40)
0014        C FORMAT STATEMENTS
0015        2 FORMAT(I5,I5,F10.4,25X,25X,I2)
0016        3 FORMAT(1H0,4X,I5,I5,E20.8)
0017        5 FORMAT(1H0,2X,10HEIGENVALUE,I3,3H = ,E15.8,5H BETA)
0018        6 FORMAT(1H0,2X,30HCOEFFICIENTS OF WAVE FUNCTION ,I3)
0019        7 FORMAT(18A4)
0020        8 FORMAT(1H0)
0021        9 FORMAT(1H ,E12.5)
0022       66 FORMAT(1H0,20F6.2)
0023       77 FORMAT(1H1,18A4)
0024      199 FORMAT(1H ,13HPI-BOND ORDER,2I2,5H  IS ,E12.5,4X,21HELECTRON DENSI
0025        1TY ATOM,I2,4H IS ,E12.5)
0026      204 FORMAT(1H ,38X,21HELECTRON DENSITY ATOM,I2,4H IS ,E12.5)
0027      270 FORMAT(1H ,35HNUMBER OF ATOMS IN THIS MOLECULE IS,2X,I3,2X,30HAND
0028        1THE NUMBER OF ELECTRONS IS,2X,I3)
0029      299 FORMAT(30I2)
0030      600 FORMAT(57H EIGENVALUES LISTED IN DECREASING ENERGY,IN UNITS OF BET
0031        1A)
0032        C INPUT NAME OF COMPOUND
0033      135 READ (1,7)(TITLE(I),I=1,18)
0034      170 CONTINUE
0035        C INPUT ARRAY SIZE
0036          READ (1,2)N
0037          READ (1,2)NE
0038          READ (1,299)(NATOM(I),I=1,N)
0039        C SET EIGENVALUE MATRIX TO ZERO
0040        C GENERATE IDENTITY MATRIX AS FIRST APPROXIMATION TO X
0041          DO 150 I=1,N
0042          DO 150 J=1,N
0043          A(I,J)=0.0
0044          IF(I-J)100,101,100
0045      100 S(I,J)=0.0
0046          GO TO 150
0047      101 S(I,J)=1.0
0048      150 CONTINUE
0049      152 READ(1,2)I,J,FA,IND
0050          A(I,J)=FA
0051          A(J,I)=FA
0052          IF(IND)151,152,151
0053      151 CONTINUE
0054          WRITE(2,77)(TITLE(I),I=1,18)
0055          WRITE(2,270)N,NE
0056          DO 500 I=1,N
0057      500 WRITE(2,66)(A(I,J),J=1,N)
0058          CALL JACB
0059        C FOLLOWING SUB-STRUCTURE IS TO SORT EIGENVALUES INTO ORDER OF
0060        C DESCENDING ENERGIES TOGETHER WITH ASSOCIATED EIGENVECTORS,
0061          I=1
0062          AMIN=A(1,1)
0063          IN=2
0064      300 DO 301 ID=IN,N
0065          IF(A(ID,ID)-AMIN)302,301,301
0066      302 AMAX=AMIN
0067          AMIN=A(ID,ID)
0068          A(I,I)=AMIN
0069          A(ID,ID)=AMAX
0070          DO 301 IP=1,N
0071          AS=S(IP,I)
0072          BS=S(IP,ID)
0073          S(IP,ID)=AS
0074          S(IP,I)=BS
0075      301 CONTINUE
0076          I=I+1
0077          IF(I-N)305,304,304
0078      305 AMIN=A(IN,IN)
0079          IN=IN+1
0080          GO TO 300
0081      304 CONTINUE
0082        C
0083        C END OF SUB-STRUCTURE
0084        C
0085        C OUTPUT COMPOUND NAME
```

```
0086          C     OUTPUT ROOTS TO POLYNOMIAL
0087                DO 130 I=1,N
0088                DO 130 J=1,N
0089                IF(I-J)130,131,130
0090                WRITE(2,600)
0091            151 WRITE(2,5)I,A(I,J)
0092          C     OUTPUT CORRESPONDING COEFFICIENTS
0093                WRITE(9,6)J
0094                DO 208 IP=1,N
0095            208 WRITE(9,5)IP,J,S(IP,J)
0096          C
0097            130 CONTINUE
0098                II=1
0099            201 I=NATOM(II)
0100                IF(N-II)203,203,202
0101            202 K=NATOM(II+1)
0102            203 D=0
0103                PBO=0
0104                DO 99 J=1,N
0105                IF(NE-2*J)260,261,261
0106            261 EL=2.0
0107                GO TO 265
0108            260 IF(NE+1-2*J)266,262,262
0109            262 EL=1.0
0110                GO TO 265
0111            266 EL=0.0
0112            265 CONTINUE
0113                D=D+S(I,J)**2*EL
0114                IF(II-N)599,99,99
0115            599 PBO=PBO+S(I,J)*S(K,J)*EL
0116             99 CONTINUE
0117                IF(II-N)799,699,699
0118            799 WRITE(9,199)I,K,PBO,I,D
0119                GO TO 111
0120            699 WRITE(9,204)I,D
0121            111 II=II+1
0122                IF(II-N)201,201,200
0123            200 CONTINUE
0124                WRITE(9,8)
0125                WRITE(9,8)
0126          C     BRANCH TO THE BEGINNING OF THE PROGRAM
0127                IF(IND-99)211,212,211
0128            212 GO TO 135
0129            211 STOP
0130                END
```

```
     CYCLOOCTATETRAENE    GFTH65M1
NUMBER OF ATOMS IN THIS MOLECULE IS     8  AND THE NUMBER OF ELECTRONS IS     8

 0.00  1.00  0.00  0.00  0.00  0.00  0.00  1.00

 1.00  0.00  1.00  0.00  0.00  0.00  0.00  0.00

 0.00  1.00  0.00  1.00  0.00  0.00  0.00  0.00

 0.00  0.00  1.00  0.00  1.00  0.00  0.00  0.00

 0.00  0.00  0.00  1.00  0.00  1.00  0.00  0.00

 0.00  0.00  0.00  0.00  1.00  0.00  1.00  0.00

 0.00  0.00  0.00  0.00  0.00  1.00  0.00  1.00

 1.00  0.00  0.00  0.00  0.00  0.00  1.00  0.00

EIGENVALUE   1 = -0.20009003E 01 BETA

COEFFICIENTS OF WAVE FUNCTION   1

          1    1    -0.35346502E 00

          2    1     0.35325295E 00

          3    1    -0.35325291E 00

          4    1     0.35339422E 00

          5    1    -0.35360626E 00
```

```
     6    1      0.35381856E 00

     7    1     -0.35388908E 00

     8    1      0.35374772E 00

EIGENVALUE  2 = -0.14149498F 01 BETA
CUEFFICIENTS OF WAVE FUNCTION   2

     1    2      0.19142154E 00

     2    2      0.19126904E 00

     3    2     -0.46198438E 00

     4    2      0.46223873E 00

     5    2     -0.19184489E 00

     6    2     -0.19087731E 00

     7    2      0.46165530E 00

     8    2     -0.46186292E 00

EIGENVALUE  3 = -0.14147499F 01 BETA
CUEFFICIENTS OF WAVE FUNCTION   3

     1    3      0.46196081E 00

     2    3     -0.46213204E 00

     3    3      0.19175589E 00

     4    3      0.19088057E 00

     5    3     -0.46167701E 00

     6    3      0.46196960E 00

     7    3     -0.19151038E 00

     8    3     -0.19126600E 00

EIGENVALUE  4 = -0.10000000F-03 BETA
CUEFFICIENTS OF WAVE FUNCTION   4

     1    4     -0.35357110E 00

     2    4     -0.35364171E 00

     3    4      0.35357110E 00

     4    4      0.35357101E 00

     5    4     -0.35357110E 00

     6    4     -0.35350032E 00

     7    4      0.35350044E 00

     8    4      0.35350032E 00

EIGENVALUE  5 =  0.10000001F-03 BETA
CUEFFICIENTS OF WAVE FUNCTION   5

     1    5      0.35357101E 00

     2    5     -0.35364179E 00

     3    5     -0.35357101E 00

     4    5      0.35357109E 00

     5    5      0.35357100E 00

     6    5     -0.35350042E 00
```

```
        7    5    -0.35350035E 00

        8    5     0.35350042E 00

EIGENVALUE  6 =  0.14147499E 01 BETA

COEFFICIENTS OF WAVE FUNCTION    6

        1    6    -0.46196196E 00

        2    6    -0.46213087E 00

        3    6    -0.19175310E 00

        4    6     0.19088538E 00

        5    6     0.46167819E 00

        6    6     0.46196844E 00

        7    6     0.19150757E 00

        8    6    -0.19126880E 00

EIGENVALUE  7 =  0.14149498F 01 BETA

COEFFICIENTS OF WAVE FUNCTION    7

        1    7     0.19141874E 00

        2    7    -0.19127185E 00

        3    7    -0.46198554E 00

        4    7    -0.46223758E 00

        5    7    -0.19184208E 00

        6    7     0.19088011E 00

        7    7     0.46165446E 00

        8    7     0.46186176E 00

EIGENVALUE  8 =  0.20009003E 01 BETA

COEFFICIENTS OF WAVE FUNCTION    8

        1    8     0.35346502E 00

        2    8     0.35325295E 00

        3    8     0.35325291E 00

        4    8     0.35339422E 00

        5    8     0.35360625E 00

        6    8     0.35381836E 00

        7    8     0.35388908E 00

        8    8     0.35374772E 00
```

```
PI-BOND ORDER 1 2  IS -0.35540E 00     ELECTRON DENSITY ATOM 1 IS  0.10000E 01
PI-BOND ORDER 2 3  IS -0.85361E 00     ELECTRON DENSITY ATOM 2 IS  0.10000E 01
PI-BOND ORDER 3 4  IS -0.35554E 00     ELECTRON DENSITY ATOM 3 IS  0.10000E 01
PI-BOND ORDER 4 5  IS -0.85556E 00     ELECTRON DENSITY ATOM 4 IS  0.10000E 01
PI-BOND ORDER 5 6  IS -0.35557E 00     ELECTRON DENSITY ATOM 5 IS  0.10000E 01
PI-BOND ORDER 6 7  IS -0.85553E 00     ELECTRON DENSITY ATOM 6 IS  0.10000E 01
PI-BOND ORDER 7 8  IS -0.35563E 00     ELECTRON DENSITY ATOM 7 IS  0.10000E 01
                                       ELECTRON DENSITY ATOM 8 IS  0.10000E 01
```

The details of the input format are:

Card 1: Name of compound (18A4).
Card 2: Number of atoms (basis function) (I5).
Card 3: Number of electrons (I5).
Card 4: Atom numbers (30I2) for atom identification.
    Card 5: I, J, $H_{i,j}$, Integer K (H_{ij} are in units of β so that a typical entry is 1.0). Format is I5, I5, F10.4, 50X, I2).

If K (in columns 71, 72) is zero then another card like card 5 is read. If it equals 99 another set of data is expected. The input for octatetraene is as follows:

Card 1: Octatetraene.
Card 2: 8
Card 3: 8
Card 4: 1 2 3 4 5 6 7 8
Card 5: 1 2 1.0
Card 6: 2 3 1.0
Card 7: 3 4 1.0
Card 8: 4 5 1.0
Card 9: 5 6 1.0
Card 10: 7 8 1.0   99

The corresponding output is shown in Table 5.14.

After reading the number of atoms and number of electrons program BEE2 constructs the Hamiltonian matrix by assigning the values of H_{ij} to the appropriate elements of A. Before calling the diagonalization routine an initial guess at the eigenvectors is generated by setting array S equal to the unit matrix of order N x N. After returning from JACB the eigenvalues are stored as the diagonal elements of A, the original Hamiltonian matrix being destroyed. The eigenvectors appear as the columns of array S in the program. Both the eigenvalues and eigenvectors are ordered in terms of decreasing energy (i.e. increasing eigenvalue) prior to output. The computation from line 102 onwards is concerned with the manipulation of the m.o. coefficients to give a bond order matrix to be discussed below.

5.4.3 π-electron Densities and Bond Orders

The electron configuration for the ground electronic state of the π-electron system is built up by feeding electrons into the π-molecular orbitals two at a time, filling the lowest energy orbitals first. For a conjugated system involving n carbon atoms there are n basis a.o.s and n π-electrons to be assigned to n m.o.s. Clearly only the lower half of the orbitals will be occupied in the ground state (for the singly charged positive or negative ion one electron more, or one less respectively is involved). Each m.o. ψ_i, may be assigned an occupancy number n_i for a particular electron configuration. For normalized m.o.s the electron probability density for a single electron in the ith m.o. is given by ψ_i^2. More generally, for n_i electrons occupying ψ_i the electron density is given by $n_i\psi_i^2$. The *total* molecular π-electron density, ρ_π, is then a sum of such contributions from all the molecular orbitals (unoccupied orbitals have $n_i = 0$).

$$\rho_\pi = \sum_i n_i\psi_i^2 \tag{5.31}$$

Using equation (5.21) to expand ψ_i in terms of the a.o. basis orbitals gives:

$$\rho_\pi = \sum_{\substack{\mu,\nu \\ \mu \neq \nu}}^{n} P_{\mu\nu}\chi_\mu\chi_\nu + \sum_\mu^n P_{\mu\mu}\chi_\mu^2 \tag{5.32}$$

where summations are over all values of μ and ν for the molecule. $P_{\mu\nu}$ is known as the bond order between basis orbitals χ_μ and χ_ν and is defined as

$$P_{\mu\nu} = \sum_i^n n_i c_{\mu i} c_{\nu i} \tag{5.33}$$

and $P_{\mu\mu}$ is similarly defined as

$$P_{\mu\mu} = \sum n_i c_{\mu i}^2 \tag{5.34}$$

The bond order $P_{\mu\nu}$ has been interpreted as a measure of the π-bonding between atomic centres μ and ν and has been correlated with the bond length for directly bonded atoms. A typical empirically based correlation for carbon–carbon bonds is of the type:[2]

$$R_{\mu\nu} = 0.15 - 0.015\, P_{\mu\nu} \tag{5.35}$$

where $R_{\mu\nu}$ is the bond length, between directly bonded atoms μ and ν, in nanometers.

The quantity $P_{\mu\mu}$ is interpreted[2] as the total π-electron density centred on atom μ. For many unsaturated hydrocarbons the total π-electron density, calculated in this way, turns out to be unity for all of the carbon atoms. Such systems are known as *Alternant Hydrocarbons* and may be distinguished by the simple procedure of labelling the atoms, alternately either * or 0 throughout the conjugated chain. In an Alternant system no two identical labels (* or 0) are found to be adjacent. The Huckel molecular orbital solutions for Alternant systems show many simple and interesting relationships between the calculated eigenvalues and also between the molecular orbital coefficients. The reader is referred to specialized texts[2,10] for further details.

Program BEE2 calculates the π-electron bond orders and total electron densities. The calculation is carried out by straightforward manipulation of the molecular orbital coefficients with equations (5.33) and (5.34). The reader is referred again to Table 5.14 for the format of output for these quantities.

For a system comprising only carbon atoms the charge contributed by the carbon nucleus and σ-electron core corresponds to +1 unit of electron charge. This is because only one of the six carbon electrons is involved in the π-system. The nuclear charge of carbon, equivalent to six protons, is screened by the five electrons involved in σ-bonding leaving and effective charge of +1. If $P_{\mu\mu}$ is the π-electron density on atom μ then the net charge on atom μ is:

$$q_{\mu} = 1 - P_{\mu\mu} \tag{5.36}$$

The π-electron dipole moment may be calculated from the geometry of the system usingsthe expression:

$$\mu_{\pi,x} = \sum_{\mu} q_{\mu}x_{\mu} \tag{5.37}$$

$$\mu_{\pi,y} = \sum_{\mu} q_{\mu}y_{\mu} \tag{5.38}$$

Summations are over all atoms, μ, to give the π-dipole moments $\mu_{\pi,x}$ and $\mu_{\pi,y}$ in the x and y directions and x_{μ} and y_{μ} are the coordinates of the atom μ referred to an origin in the molecular plane (x, y plane).

For neutral alternant hydrocarbons the dipole moment calculated in this way is zero. For non-alternant hydrocarbons such as:

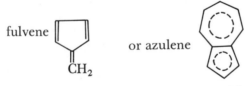

fulvene   or azulene

non-zero dipole moments are obtained.[10] In general Huckel calculations tend to overestimate the dipole moments of such molecules.

The applications of Huckel m.o. theory are numerous and include, in addition to the above calculations, studies of charge distributions in various positive and negative ions and free-radicals; prediction of unpaired spin densities in π-electron systems for rationalization of electron spin resonance spectra; studies of spectroscopic transitions in aromatic hydrocarbons; explanation of the stability of aromatic and various pseudo-aromatic systems; rationalization of substituent effects in organic reactions mechanisms. The details of these applications are beyond the scope of this text and the reader is referred to some of the many excellent texts[2,10] on the subject.

5.4.4 More Advanced Calculations
Huckel m.o. theory represents the simplest member of a series of m.o. calculations of varying degrees of sophistication. The simplest extension

of the theory is to allow carbon atoms with different electronegativities to have different values for the Coulomb integral, α (the Coulomb integral can be thought of as the energy of the a.o. basis function in the potential field of the σ-electron framework in the absence of interactions with other basis orbitals). In non-alternant hydrocarbons, Huckel calculations give predicted charge densities which in general differ from unity. The electronegativities of the C-atoms in the conjugated chain should therefore differ from one carbon atom to the next. An improved method of calculation would therefore allow the αs to vary depending on the π-electron density on the carbon atoms. This is the basis of the ω-technique in which the value of a α, for a certain atom μ, is given by[10]

$$\alpha_{\mu\mu} = \alpha_0 + (1 - q_\mu)\beta\omega \qquad (5.39)$$

Where α_0 is the value of the Coulomb integral for a neutral carbon atom, ω is an empirical parameter usually taken as equal to 1.4 and the other symbols are as defined earlier.

The Huckel m.o. values for α are taken as an initial guess of α_0 and the calculated π-densities form the basis for calculating new values. The calculation is carried out as an iterative process and is terminated when no further change is observed in the αs. This technique leads to improved calculated dipole moments for non-alternant hydrocarbons.[10]

Extended Huckel m.o. theory represents an extension of simple Huckel theory to include electrons associated with the σ-framework. This type of calculation is increased in complexity because more than one atomic orbital is centred on each atom and the mathematical form of the orbitals must be specified. Normally, approximations to atomic orbitals[2] known as *Slater orbitals* for the valence shell electrons for each atom are used to form the atomic orbital basis set. The matrix elements $H_{\mu\mu}$ are taken as a measure of the electron attracting power of particular atoms; for example, typical values for carbon $2p$ and carbon $2s$ are -11.4 eV and -21.4 eV (1 eV $= 96.4$ kJ mol^{-1}). The off-diagonal elements $H_{\mu\mu}$ are often represented by

$$H_{\mu\nu} = \frac{K}{Z}(H_{\mu\mu} + H_{\nu\nu})S_{\mu\nu} \qquad (5.40)$$

where K is an empirical parameter and $S_{\mu\nu}$ is the overlap integral, between atomic orbitals χ_μ and χ_ν which has to be known or calculated. Expressions for evaluating overlap integrals are available and have been given by Mulliken.[12] Having computed the elements of the Hamiltonian matrix, the m.o.s are computed by means of matrix diagonalization. Extended calculations have been used for calculating approximate charge distributions in σ-electron systems and estimates of molecular geometries. The latter calculations involve the computation of the molecular energy as a function of bond angles and bond length.

The predicted equilibrium geometry corresponds to an energy minimum.

Returning for the moment to π-electron calculations, the next step in sophistication is to use the so-called self-consistent-field (SCF) approach to calculate the molecular orbital coefficients defining each molecular orbital.[13] In order to understand the basis of the SCF approach one must realize that the mathematical form of a molecular orbital is determined (once the basis orbitals have been defined) by the potential function experienced by the electron in the orbital. This has contributions from the molecular core (i.e. the atomic nucleus and electrons in the sigma system) and also from the repulsion due to other electrons in the π-system. The electron repulsion terms present a problem, because if we are interested in defining the potential experienced by an electron in orbital ψ_i we need to know the mathematical form of all the other molecular orbitals which are occupied by electrons. But we do not know the m.o. coefficients for the other orbitals and we cannot apparently calculate them for, e.g., ψ_j because we do not know the mathematical form of ψ_i which contributes to the potential function defining ψ_j.

The SCF approach provides a convenient way out. Firstly a Huckel calculation is performed which gives an initial guess at the m.o. coefficients for all of the orbitals. Electrons are then fed into the energy levels and the orbitals which are occupied, for the electron configuration of interest, are defined. The detailed theory is beyond the scope of the present text, but by the inclusion of explicit electron repulsion terms in the Hamiltonian and making the assumption that the total electronic wave function can be written as a product of one-electron functions (molecular orbitals) one can obtain[22,23] a set of SCF equations of the form:

$$FC = SCE \tag{5.41}$$

where the matrices C, S and E have the same significance as in equation (5.25). The elements of matrix F involve sums of terms comprising integrals over the a.o. basis functions multiplied by the LCAO coefficients which define the occupied orbitals in our initial guess. The technique is then to generate the elements of F numerically (using the explicit equations the form of which need not concern us here but are given in many specialized texts[14,15]) and then to diagonalize the F-matrix to obtain new LCAO coefficients and new molecular orbital energies. The electrons are then fed into the energy levels once more, defining new occupied orbitals and a set of LCAO coefficients which can then be used to calculate a new F-matrix. The procedure is repeated until no further change is observed in the m.o. coefficients. The result is the best set of m.o.s which can be obtained within the limitations imposed by the chosen a.o. basis set and any approximations made in evaluating the elements of **F**. One should realize that the computation of some molecular integrals involved in SCF calculations is time consuming even on large computers and since the number of such integrals involving N basis a.o.s is of the order N^4 (in the absence of molecular symmetry) literally thousands of such calculations would be necessary for a molecule of even moderate size. This has led to the development of a number of semi-empirical approaches,[13] in which many integrals are set to zero and others are regarded as parameters chosen to give calculated molecular properties which agree with experimental observation. A multiplicity of programs have been developed to carry out SCF calculations ranging from the more rigorous *ab initio* calculations (in which every attempt is made to evaluate all the molecular integrals) to approximate calculations in which seemingly drastic approximations are made in conjunction with a great reliance on semi-empirical parameters.[14]

To add to the range of programs which have been developed there are now different choices of basis orbitals since Slater-type a.o.s give rise to integrals which, in general are difficult to compute. Gaussian type basis orbitals[15] are preferable in this context but unfortunately one usually has a larger number of such basis orbitals than with Slater-type orbitals.

The programs used to carry out these calculations are too lengthy to be listed in this text. However, a number of these programs have already been published. Pople and Beveridge[13] have listed and discussed some programs of the semi-empirical type. Further, the Quantum Chemistry Program Exchange[16] is a source of a large number of SCF programs of varying degrees of sophistication.

Readers who are interested in more details about the theory of SCF calculations and different methods of obtaining molecular wave functions (e.g. Configuration Interaction and Valence bond calculations) are referred to more specialized texts referred to earlier. Some further details are to be found in Chapter 7.

5.5 Problems
5.5.1 A particle confined to a three-dimensional potential well (the three-dimensional analogue of the potential function discussed in Section 5.2) has energy levels given by:

$$E_{n_x, n_y, n_z} = \frac{h^2}{8m}\left\{\frac{n_x^2}{l_x^2} + \frac{n_y^2}{l_y^2} + \frac{n_z^2}{l_z^2}\right\}$$

where l_x, l_y and l_z are the dimensions of the well along the x-, y- and z-coordinates and n_x, n_y and n_z are quantum numbers which have independent integer values from 1 upwards. Write a program to calculate the energy for any value of the quantum numbers and any dimensions of the well. Use the program to investigate the behaviour of the energy levels when $l_x = l_y = l_z$, $l_x = l_y \neq l_z$ and $l_x \neq l_y \neq l_z$. Find the connection between degeneracy and the symmetry of the potential function.

5.5.2   Use program TWOBX to calculate the map of $\psi_{(1,1)}$ and $\psi_{(2,2)}$ and check that these functions are orthogonal.

5.5.3   Use program RAD1 to examine the effect of nuclear charge on the size of $1s$, $2s$ and $3s$ hydrogenic orbitals. Use the radial distribution function output to obtain the most probable radial distance from the nucleus of electrons in $1s$, $2s$, $3s$, $2p$ and $3d$ orbitals.

5.5.4   Use program CONTOUR to produce, for the $4s$ orbital, a wave function map and electronic density maps in both numeric and symbolic form (you will need to choose different scales for the wave function and electron density maps).

5.5.5   Calculate, using the bond order output from the Huckel program (BEE2), the predicted C—C bond lengths for 1,3-butadiene (CH_2=CH—CH=CH_2). The experimental bond lengths are $R_{12} = R_{24} = 0.134$ nm and $R_{23} = 0.148$ nm.

5.5.6   Modify program BEE2 to calculate bond lengths in conjugated π-systems (involving only carbon atoms).

5.5.7   Modify program BEE2 for use in ω-technique calculations and compare the charge densities and bond orders for both alternant and non-alternant hydrocarbons.

References

1. Eyring, H., Walter, J., and Kimball, G. E., *Quantum Chemistry*, John Wiley, New York, 1944.
2. Murrell, J. N., Kettle, S. F. A., and Tedder, J. M., *Valence Theory*, John Wiley, London, 1965.
3. Heitler, W., *Elementary Wave Mechanics*, Clarendon Press, Oxford, 1956.
4. Schutte, C. J. H., *The Wave Mechanics of Atoms, Molecules and Ions*, Edward Arnold, 1968.
5. Murrell, J. N., *The Theory of the Electronic Spectra, of Organic Molecules*, Methuen and Co., London, 1963.
6. Jaffe, H. H. and Orchin, M., *Theory and Application of Ultraviolet Spectroscopy*, John Wiley, London, 1962.
7. Kuhn, H., *J. Chem. Phys.*, 17, 1198 (1949).
8. Schaad, L. J., and Hess, B. A., *J. Chem. Educ.*, 51, 640 (1974).
9. Duke, B. J., *J. Chem. Educ.*, 49, 703 (1972).

231

10. Streitwieser, A., *Molecular Orbital Theory for Organic Chemists*, John Wiley, London, 1961.
11. Dickson, T. R., *The Computer and Chemistry*, W. H. Freeman, San Fransisco, 1968.
12. Mulliken, R. S., Reike, C. A., Orloff, D., and Orloff, H., *J. Chem. Phys.*, 17, 1248 (1949).
13. Pople, J. A., and Beveridge, D. L., *Approximate Molecular Orbital Theory*, McGraw-Hill, London, 1970.
14. Richards, W. C., and Horsley, J. A., *Ab Initio Molecular Orbital Calculations for Chemists*, Clarendon Press, Oxford, 1970.
15. Barnett, M. P., and Coulson, C. A., *Phil. Trans. R. Soc.*, 221, 4243 (1951).
16. Quantum Chemistry Program Exchange, Indiana University, Indiana, U.S.A.

Chapter 6

Data Acquisition in Chemistry

6.1 Introduction

We shall use the phrase 'data aquisition' to describe the conversion of an electrical analogue signal from an instrument into some form of digital record. More specifically, we shall restrict our attention to paper tape as a means of storing the digital information since this is the cheapest method. The techniques described in this chapter are, however, easily applied to other media of data storage and can be extended to direct data acquisition using an on-line computer.

Data acquisition is most useful to a chemist whenever he wishes to:

(i) Record, semi-continuously, data collected at very short intervals from a high-resolution instrument for subsequent analysis;

(ii) Record, intermittently at long time intervals as, for example, during a kinetic run using some measuring instrument such as a pH meter

We shall concern ourselves solely with the former application. Instruments well-suited to this form of digital data acquistion include:

(i) Chromatography equipment;
(ii) Nuclear magnetic resonance spectrometers;
(iii) Mass spectrometers.

The first stage of data acquisition is analogue to digital conversion.[1] This is accomplished in many commercial digital voltmeters by conversion of the analogue voltage, E_i, into a series of 'clock' pulses of a set frequency. A digital indication of E_i is obtained by accumulating these clock pulses in an integrator for a set period, followed by subtraction of reference voltage pulses of opposite polarity until the contents of the integrator fall to zero. This is the basis of the 'dual-ramp' A–D converter.

Digital data is almost always converted into a binary representation of decimal digits known as 'binary coded decimal' (BCD). For example, a four-bit register can be used to represent the digits 0 to 9:

Decimal	0	1	2	3	4	5	6	7	8	9
BCD code	0000	0001	0010	0011	0100	0101	0110	0111	1000	1001

We wish now to transfer the BCD representation to an output device such as a tape punch or magnetic recorder. This is accomplished by an interface which decodes the BCD data into an appropriate character on the output medium. An example of the output on paper tape corresponding to the digits 0 to 9 is shown in Figure 6.1.

The interface is responsible for the insertion of the correct sign and 'new-line' characters and for initiation of the next reading (Figure 6.2).

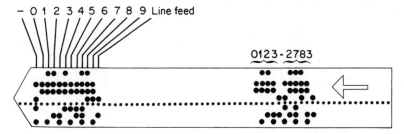

Figure 6.1   Example of punched-tape, as used for program NMRD (Table 6.4)

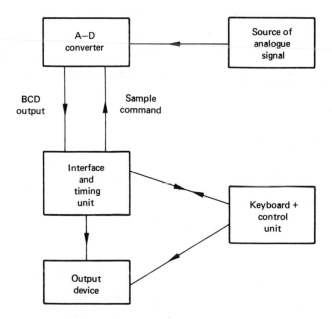

Figure 6.2   Block diagram of a simple A–D unit

The most popular forms of output media for general purpose laboratory application are paper tape or magnetic tape (seven or nine track). The prime difference between the two is that a tape punch can produce up to about 110 output characters per second whereas magnetic tape recorders can produce up to 10^5. This higher rate of data acquistion is necessary in some applications such as mass spectrometry but is more expensive (currently more than twice the cost of a tape punch) and, of course, the computer must have a suitable magnetic tape deck. Cassette recorders are somewhat cheaper but, at the time of writing, still exceed the cost of a tape punch.

As an example of a suitable collection of A–D equipment, we may cite that used in our own laboratory.[2] This consists of:

(i) Digital voltmeter: 100 mV to 10 V; reading rate, 25/sec; accuracy, 0.01% ± 1 digit in a 4-digit display.

(ii) Interface: sample intervals, 0.2 to 5000 seconds; 15-character keyboard; output, sign with four characters and line feed in ASC11 code.

(iii) Paper tape punch: speed, up to 75 characters (11 readings)/sec.

Note that it is not necessary to have a decimal point in the digital output. This has two advantages:

(i) Integer arithmetic is faster to handle in a computer (and is more economical in storage).

(ii) Maximum speed of data collection is increased by a factor of 5/4 by removing one character, assuming a 4-digit reading.

It is also not necessary to use the A/D equipment to record elapsed time since the timing interval is a reproducible constant factor. This and other constants can either be added to the paper tape by means of the keyboard or can be input separately on cards to the program.

Excluding alphameric information, the digital output will consist of coded forms of:

(i) Digits;
(ii) + or −;
(iii) Line feed (new line) or carriage return characters.

The latter character must be present to separate each reading (see Figure 6.1). In Figure 6.1 you will also observe that the sum of the number of holes in each vertical column on the tape is even. This tape is said to have even *parity* but odd parity is also allowed. The parity is fixed by a parity check hole which is present or absent in one track of the tape. Before processing the tape, it is necessary to tell the computer when a digital record is complete. This can be done by punching a specific number not normally expected from the signal source onto the tape, followed by a line feed character. (We usually punch −9999.)

6.2 Digital Techniques for Treatment of Instrumental Data

In this section some general techniques will be introduced prior to a discussion of their applications in Section 6.3. The techniques include:

(i) Correction for baseline drift;
(ii) Smoothing of noisy signals and peak identification;
(iii) Cross-correlation of digital records.

6.2.1 Correction for Linear Baseline Drift

A common situation is for an instrumental baseline to drift in either a linear or non-linear fashion. The former is easier to deal with in a text of this size and is of considerable interest in the improvement of the precision of calculations. Let us presume that we wish to calculate the precise displacement, allowing for linear drift, of a noisy signal from a baseline. The situation is illustrated in Figure 6.3. Such a situation is common in atomic absorption spectrometry and other techniques in which there is a non-periodic displacement of the signal from the baseline. The technique is as follows:[3]

(i) The blank baseline signal is recorded for T seconds. A baseline displacement occurs and after T_R seconds a sample baseline is established for a period T_S.

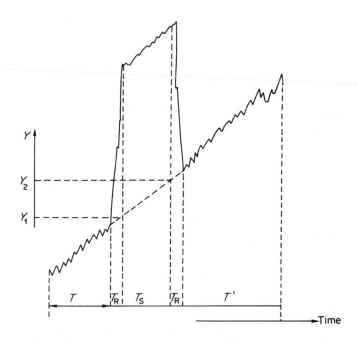

Figure 6.3   Signal amplitude as a function of time with a linearly sloping baseline

(i) The value of the *extrapolated* baseline Y_1 at the time $(T + T_R)$ is calculated from:

$$Y_1 = \bar{Y} + \frac{S2 - S1}{TI(N - Nu)Nu} \left[\frac{T + 2\,T_R}{2} \right]$$

where

TI = timing interval/seconds
N = number of data points in the period T
Nu = largest integer $\leqslant (N/3)$

$$S2 = \sum_{i=N-Nu}^{N} Y_i$$

$$S1 = \sum_{i=1}^{Nu} Y_i$$

Y_i are the signal amplitudes

$\bar{Y}$ = mean value of Y_i during T

(iii) A sample recording is made for the period TS.

(iv) A new baseline is recorded during T' and a predicted value, Y_2, of the baseline at the end of T_S is calculated:

$$Y_2 = \bar{Y} - \frac{(S2 - S1)}{TI(N - Nu)Nu} \left[\frac{T' + 2T_R}{2} \right]$$

(v) A mean value of the sample signal above a baseline interpolated between Y_1 and Y_2 is calculated, together with the standard deviation.

It is also possible to eliminate non-linear baseline drift; for the background to this and the techniques described in this section, the reader is referred to Reference 3.

6.2.2 Smoothing of Digital Representations of Analogue Signals

We shall assume that the representation consists of n equally spaced measurements recorded at such a rate that any real peak has an arbitrary minimum of five measurements across the breadth of the peak. One method of increasing the signal-to-noise ratio of a noisy spectrum (S/N) is to repetitively record the same sample sequence. If m representations of the same sample sequence can be obtained and the amplitudes of the different records at the same time intervals are summed, S/N is increased by a factor of $\sqrt{m}$. This can be accomplished with a device called a CAT (computer of average transients) or with an on-line computer;[4] during each recording sequence, signal amplitudes are fled into channels, or elements of an array, using either an internal or external signal as a reference point. The drawbacks of this method are, firstly, that it is relatively slow to improve S/N and that peak distortion can be marked if signal drift occurs.

x	y
5.0	89
5.5	90
6.0	92
6.5	90
7.0	87
7.5	91
8.0	93
8.5	90
9.0	92
9.5	88
10.0	87
10.5	91
11.0	90
11.5	89
12.0	88

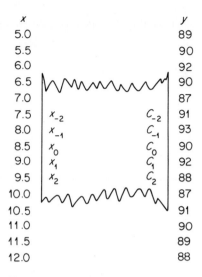

Figure 6.4   Illustration of a convolution operation for sets of 5 consecutive data points. (The data points, y, are in the far right-hand column and the convoluting integers, c, when multiplied with the corresponding y give the convolute at the abscissa value x_0)

A more attractive method which is well-suited to off-line computers using paper tape (or magnetic tape) input is to use a weighted moving average operation.[5] This method is best illustrated with an example (Figure 6.4) in which the ordinate (y) values are listed together with the corresponding x (time) values. Between the two we interpose a piece of paper identifying $2l + 1$ (l = integer) x-values and a set of numbers C_{-l} to C_l, called *convoluting integers*. The weighted average or *convolute* at the centre point is obtained by multiplying the y-values by the corresponding C_i and dividing the sum of products by a normalizer, F. In equation form, this means

$$y_0' = \sum_{-l}^{+l} \frac{C_i y_i}{F}$$

To obtain the convolute at the next point, we imagine the paper in Figure 6.4 to slide down by one line and the process is then repeated. There are many ways of obtaining the C_i values but a popular method has been described by Savitzky and Golay:[5] their method presumes that the $(2l + 1)$ x, y-values could be fitted to a polynomial such that the best curve would pass through the points. For a small number of points we could, for example, always use a quadratic as our polynomial.

If each data point is identified by the integer, i, the required fitting polynomial is

$$f_i = \sum_{k=0}^{k=n} b_{nk} i^k$$

where n is the order of the polynomial and f_i is the fitted value at data point i.

The value of i runs from $-l$ to l, for the $2l + 1$ consecutive values, so at the centre point we have

$$i = 0$$

and

$$f_0 = b_{n0}$$

By applying the least-squares method, so that the error sum

$$\sum_{i=-l}^{i=+l} (y_i - f_i)^2$$

is minimized, b_{n0} can be calculated as a function of the $(2l + 1)$ data points in the convolution operation. For details, the reader is referred to Golay and Savitzky but the result is of the form

$$f_0 = \frac{\sum_{-l}^{+l} C_i y_i}{F}$$

where the C_is are integers and F is an integer constant. Typical values are listed in Table 6.1. (See also Reference 5a for some corrections to the original Savitzky and Golay paper.)

The signal-to-noise ratio should increase approximately with the square root of the number of points used in the convolution set. This is advantageous when compared with simple signal averaging; for example, we could improve S/N by a factor of 10 either by:

(i) Simple averaging of 100 runs, or
(ii) Averaging only 4 runs (2 x improvement), followed by a 25 point least-squares smoothing treatment (further 5 x improvement).

Note that a very noisy signal would normally be accumulated at least twice to remove noise spikes which might be mistaken for peaks.

There are potential disadvantages in using many points for a least-squares smooth since peak distortion can be introduced. Best results are obtained by digitizing at high densities (short sampling times) and also by ensuring that no more than one inflexion (deviation from baseline, peak or shoulder) is present in any one convolution operation. This ensures that the polynomial will always approximate closely to the true dependence of the analytical data.

Table 6.1
Convoluting integers C_i for smoothing up to 11
consecutive points using quadratic and cubic
polynomial equations

	Number of points in set			
	5	7	9	11
x_{-5}				−36
x_{-4}			−21	9
x_{-3}		−2	14	44
x_{-2}	−3	3	39	69
x_{-1}	12	6	54	84
x_0	17	7	59	89
x_1	12	6	54	84
x_2	−3	3	39	69
x_3		−2	14	44
x_4			−21	9
x_5				−36
F (normalizing coefficient)	35	21	231	429

Table 6.2
9-Point smoothing routine using Savitzky and Golay method[a]

```
      SUBROUTINE SMOO(I1, I2, IY, IS)
      DIMENSION IY(1000), IS(1000)
C     I1 = INDEX NUMBER OF FIRST RAW DATA POINT
C     I2 = INDEX NUMBER OF LAST POINT
      I1 = I1 + 4
      I2 = I2 − 4
      DO 1 I=I1, I2
   1  IS(I) = (59*IY(I) + 54*(IY(I−1) + IY(I+1)) + 39*
              (IY(I−2) + IY(I+2))+14*(IY(I−3) + IY(I+3)) − 21*
              (IY(I−4) + IY(I+4)))/231
      RETURN
      END
```

[a]Note that this subroutine alters the values of I1 and I2 in the calling program.

An example of a possible smoothing routine is shown in Table 6.2. The array IY(1000) stores the raw data and IS(1000) stores the smoothed data. Since a 9-point smooth is used, we lose 4 points from either end of the input array and this is allowed for by the DO loop limits I1 and I2.

6.2.3 Location of Peaks in Noisy Signals
Having smoothed our data, it is still necessary to calculate peak positions and to be reasonably sure that the peaks are not simply

random noise. One simple procedure for this is as follows:

(i) Calculate the standard deviation, σ, of the first, say, 50 points in which it is known that no peak is present.

(ii) Calculate the first derivative (dy/dt) of all of the signals with respect to time. The behaviour of the first derivative is illustrated in Figure 6.5.

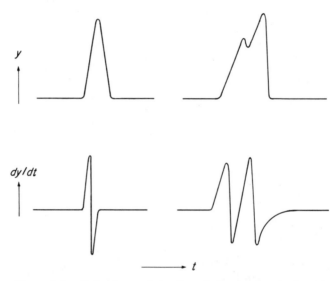

Figure 6.5   Behaviour of the first derivative for single and overlapping peaks

(iii) Only accept a peak as being detected if:

(a) dy/dt passes from + to − through zero,

(b) at the apparent peak position the peak amplitude, y exceeds 3σ (assuming a zero base line).

The approach of Savitzky and Golay[5,5a] can again be applied since sets of convoluting integers are available that calculate derivative values from input $(y$, time$)$ data. As before, sets of $(2l + 1)$ points are submitted to a convolution operation and the least-squares value of the first derivative at the centre point is obtained. The convoluting integers required are given in Table 6.3.

When derivative values are being calculated, it is necessary to incorporate the timing interval Δx, in the convolution formula:

$$\left(\frac{dy}{dx}\right)_{i=0} = \frac{\sum\limits_{i=-m}^{i=+m} c_i y_i}{F \cdot \Delta x}$$

Table 6.3
Convoluting integers, C_i, for first derivative (dy/dx)
at the midpoint of up to 11 consecutive points fitted
to a quadratic

	Number of points in set			
	5	7	9	11
x_{-5}				−5
x_{-4}			−4	−4
x_{-3}		−3	−3	−3
x_{-2}	−2	−2	−2	−2
x_{-1}	−1	−1	−1	−1
x_0	0	0	0	0
x_1	1	1	1	1
x_2	2	2	2	2
x_3		3	3	3
x_4			4	4
x_5				5
F (normalizing coefficient)	10	28	60	110

The value of dy/dx obtained refers of course, to the centre point value, $i = 0$. For example, for a 5-point operation, we would have

$$\left(\frac{dy}{dx}\right)_{i=0} = \frac{(-2y_{-2} - y_{-1} + y_{+1} + 2y_{+2})}{\Delta x \cdot 10}$$

6.2.4 Integration of Single and Overlapping Peaks

In the case of a single peak, located by the above method, one of the many methods used for calculating the area is as follows:

(i) Step backwards from the peak maximum until the amplitude is less than one standard deviation of the random noise; at this point the beginning, x_i, of the peak is located.

(ii) Step forwards from the maximum as in (i) to locate the end of the peak x_f.

The peak area, A_j, is then given by:

$$A_j = \Delta x \sum_{k=i}^{k=f} y_k \tag{6.1}$$

Storage requirements can be minimized by continually storing the apparent beginning of a peak; this can be achieved by, for example, noting when the amplitude first exceeds the standard deviation.

For a multiple peak, the most straightforward method of area apportionment is to drop perpendiculars as in Figure 6.6. This

242

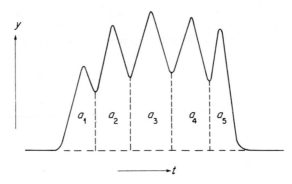

Figure 6.6   The 'perpendicular drop' method for
peak area allocation

procedure increases in accuracy with high peak/valley ratios. Other
methods of higher accuracy for severely overlapping peaks resolve the
peaks into separate Gaussian or Lorentzian components. For our simple
method the procedure would be:

(i) Locate the peak maximum as before.

(ii) For the first component, find the beginning of the peak as
previously.

(iii) The first valley is located at the x-coordinate where dy/dx
passes from negative through zero to positive. A convenient means of
identifying this point is to obtain an estimate of the standard deviation,
σ' of dy/dx. The valley is then the point at which dy/dx is increasing
and is within $\pm\sigma'$.

(iv) The valley is the end of the first peak and the summation
formula (6.1) can be applied for the first integration.

(v) The procedure is repeated for each peak in the group until the
amplitude drops to within $\pm\sigma$, denoting the end of the group of peaks.

6.2.5 Cross-Correlation of Digitized Records
If we were to examine a very noisy spectrum with the naked eye, or
with a computer, it is possible that we would be unable to ascertain
which peaks in the spectrum were due to the sample being examined
and which were due to random fluctuations. In this case, the techniques
that we have discussed would be of little value. We would, therefore, be
interested to know if the spectrum contained purely random noise or if
genuine signals were being obscured. There are two ways in which this
can be ascertained:

(i) To accumulate several digitized records, using the signal averaging method with either a small dedicated computer (CAT) or the more
efficient storage available in a large general-purpose computer, as
described previously.

(ii) To use a cross-correlation test, and only if the result is positive to proceed with (i).

The signal-averaging method has been described previously (Section 6.2.2). The main limitation of this is the extremely long time which is required in order to obtain a significant signal-to-noise improvement. The other drawback, loss of resolution, is largely overcome by the use of larger computers with many more storage locations than the small dedicated type. Nevertheless, it is an inefficient diagnostic test for the presence of a signal.

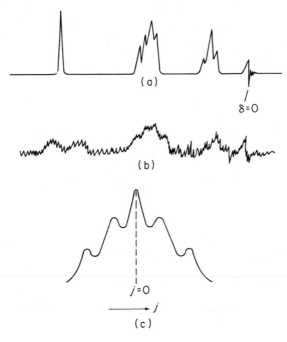

Figure 6.7   Cross-correlation of a noise-free NMR spectrum (a), with a noisy test spectrum of the same compound (b), to give a cross-correlogram (c)

Cross-correlation is a method of data treatment that permits the rapid identification of signals buried in noise, prior to possible signal averaging. Consider the pair of signal records $x(t)$ and $y(t)$ in Figure 6.7. (a) and (b) respectively. These could be two recordings of the same spectrum. The cross-correlation function, $R_{xy}(\tau)$ of the two records is defined as:

$$R_{xy}(\tau) = \lim_{T \to \infty} \frac{1}{T} \int_0^T x(t)y(t + \tau)dt$$

where T is the recording time.

The quantity τ is the amount of time lag between the two records and it is usual to display $R_{xy}(\tau)$ as a function of τ. If the two records were identical, $R_{xy}(\tau)$ would have a maximum at $\tau = 0$; if they were identical except for a time displacement τ', the maximum would occur at $\tau = \tau'$. Therefore, this function is a measure of how similar, or well-correlated, the records are.

The cross-correlation function can also be used in a discrete form suitable for use in a computer program:

$$R_{xy}(j\Delta t) = \frac{1}{N} \sum_{i=1}^{N-j} x_i y_{i+j}$$

in which Δt is the sampling interval between consecutive data points and j is the lag (in terms of data points). The usefulness of this function for our purposes is that if one has a noise-free signal replica, the presence of which one wishes to test for in a noisy record, then a strong maximum in the cross-correlation of the two records will tentatively confirm the presence of the signal. An example of this is shown in the NMR signals in Figure 6.7 in which the expected properties of the function are clearly displayed: the symmetry of the correlogram and the maximum at $j = 0$. A further point of interest is that smaller peaks also occur in the correlogram which are caused by the multiplet structure of the signals in each record.

The appearance of the correlogram for two different signals is quite dissimilar from the regular symmetric behaviour noted for records of the same signal; therefore, it can be seen that it is easy to confirm that two signals (e.g. spectra) have different origins. For example, the infrared spectra of methane and formaldehyde would not give a symmetric cross-correlogram, whereas two spectra of either compound would give the predicted correlation behaviour. It should be emphasized however, that two different compounds having very similar spectra may also yield a symmetric cross-correlogram, so that any conclusion should be regarded as tentative.

6.3 Practical Applications

In order to illustrate the discussion in the previous sections, we will now examine the operation of one typical data acquisition program for the analysis of NMR spectra. We will also indicate the application of the methods to other areas of instrumental analysis, but with less-detailed descriptions. The NMR program assumes that the data is presented as a series of integers, proportional to signal amplitudes and that each integer number is separated from the next by an appropriate character, e.g. line-feed in the case of paper tape input. Digitized data is accepted on the peripheral (e.g. paper tape reader) identified in format statements as device 3. The card reader (device 1) is used to read ancillary data. It is presumed that the termination integer is -9999.

6.3.1 NMR Data Acquisition Program

We presume that the reader is familiar with the basic concepts of NMR.[9] An example of an NMR spectrum, recorded at 60 MHz and using tetramethyl silane as internal reference was illustrated in Figure 6.7(a). The position of each peak relative to the reference is denoted by its 'chemical shift' (δ). This is defined as:

$$\delta = \frac{H_r - H_s}{H_r} \times 10^6 \text{ ppm} \tag{6.2}$$

where H_r and H_s are the applied magnetic fields required to bring the reference and sample nuclei, respectively, into resonance (sometimes δ is quoted in Hz rather than ppm). For most organic compounds δ ranges from 0 to 9 ppm (0 to 540 Hz for a 60 Mz spectrometer), such that higher values of δ correspond (equation (6.2)) to lower applied fields. The appearance of the spectrum can be altered by changing the scale expansion; the scan time from low- to high-field is also selectable depending on the resolution required.

Let us assume that a digital recording of an NMR spectrum of an organic compound, scanned from low- to high-field, is available. We wish to write a program that will:

(i) Smooth the spectrum to remove unwanted noise;

(ii) Record peak positions in parts per million, with reference to an internal standard for which $\delta = 0$, and

(iii) Calculate the area of each peak (with the exception of the reference peak).

A suitable program for this purpose is shown in Table 6.4. Input to the program consists of one card and a paper tape representation of the digitized spectrum recorded from low field to high field. The data card contains the following information:

(i) TI, the timing interval is seconds between sampling points;

(ii) EXP, the scale expansion (1 for 0 to 9 ppm; 2 for 0 to 18 and so on);

(iii) ST, the scan time for the spectrum in minutes;

(iv) NS, the number of times that the spectrum is to be smoothed if at all (using a 5-point smooth).

The format for this card is 3F0.0, I0.

The paper tape recording of the spectrum contains a series of integers with a plus or minus sign, each one being separated by a 'new line' character. Each integer represents the mean signal amplitude during a particular sampling interval, as described in Section 6.1. The end of the tape is denoted by the integer -9999, which can be added either from a keyboard attached to the A/D equipment or from a normal data preparation punch. Note that we prefer to use card input for the

Table 6.4
Program for the analysis of digital paper tape recording of NMR spectra

```
0008              MASTER NMRD                                                    100
0009              COMMON/B/IY(3000),X(3000),Z(3000)                             300
0010              WRITE(2,700)
0011          700 FORMAT(1H1,////,'   ANALYSIS OF DIGITIZED SPECTRAL DATA BY THE SA
0012         1VITZKY AND GOLAY METHOD',////)
0013              READ(1,1)TI,EXP,ST,NS
0014              WRITE(2,2)TI,EXP,ST
0015        C     TI=TIMING INTERVAL IN SECONDS
0016        C     EXP=SCALE EXPANSION OF NMR SPECTROMETER (X1=9PPM,X2=18PPM,ETC)   200
0017        C     ST=SCAN TIME IN MINUTES
0018            1 FORMAT(3F0,0,I0)
0019            2 FORMAT(1H ,' TI= ',F7,3,' SEC ',' SCALE EXPANSION ',F4,2,' SCAN
0020         1TIME ',F5,2,' MINS')
0021              N=0
0022              J=1
0023           10 READ(5,3)K
0024            3 FORMAT(I5)
0025              IF(K,EQ,-9999)GO TO 25
0026              N=N+1
0027              IY(N)=K                                                         400
0028              T=T+TI
0029              X(N)=T
0030              GO TO 10
0031           25 CALL REFR(ST,EXP,N)
0032              CALL SAVAY(J,N,NS)
0033           99 STOP
0034              END


0035              SUBROUTINE SAVAY(J,N,NS)
0036              COMMON/B/IY(3000),X(3000),Z(3000)
0037              COMMON/C/AREA(100),PEAK(100),K
0038              WRITE(2,9)NS
0039            9 FORMAT(10X,' SPECTRUM TO BE SMOOTHED ',I2,' TIMES')
0040            4 M=-1
0041           50 M=M+1
0042              IF(M,GT,NS)GO TO 2
0043           15 IYMAX=0                                                         500
0044              IYSUM=0
0045              DEV=0                                                           600
0046              DO 500 I=J,59+J
0047          500 IYSUM=IYSUM+IY(I)
0048              IYM=IYSUM/60
0049              DO 600 I=J,59+J
0050          600 DEV=DEV+(IYM-IY(I))**2                                          700
0051              STD=SQRT(DEV/59)
0052              DO 16 I=J,N
0053              IY(I)=IY(I)-IYM
0054              IF(IY(I),LE,IYMAX)GO TO 16
0055              IYMAX=IY(I)
0056              IMAX=I
0057           16 CONTINUE
0058              SN=IY(IMAX)/(2*STD)
0059              WRITE(2,800)M,SN,STD,IY(IMAX),X(IMAX),IYM
0060          800 FORMAT(1H ,' M= ',I1,' S/N= ',F9,2,' SIGMA= ',F9,4,' PEAK HEIGHT '
0061         1,I6,' AT ',F9,2,10X,'MEAN VALUE',I6)
0062            1 DO 3 I=J+2,N-2
0063            3 IY(I-2)=(-3*(IY(I-2)+IY(I+2))+12*(IY(I-1)+IY(I+1))+17*IY(I))/35
0064              DO 90 I=J+2,N-2
0065              K=N-I+1
0066           90 IY(K)=IY(K-2)
0067              J=J+2
0068              N=N-2
0069              GO TO 50
0070            2 DO 5 I=J+2,N-2
0071            5 Z(I)=(-2 *(IY(I-2)-IY(I+2))-IY(I-1)+IY(I+1))/10
0072              CALL SPRINT(J,N)
0073              WRITE(2,30)
0074           30 FORMAT(1H,///)
0075              WRITE(2,31)(IP,PEAK(IP),AREA(IP),IP=1,K)
0076           31 FORMAT(1H ,'     PEAK NUMBER',I3,'     POSITION = ',F6,2,'
0077         1     PEAK AREA = ',F9,1,' UNITS')
0078            6 RETURN
0079              END


0080              SUBROUTINE REFR(ST,EXP,N)
0081              COMMON/B/IY(3000),X(3000),Z(3000)
0082              REF=0
0083              DO 1 I=(N-(N/18)),N
0084              IF(IY(I),LT,REF)GO TO 1
0085              REF=IY(I)
0086              TMS=X(I)
```

```
0087            1 CONTINUE
0088              DO 2 I=1,N
0089            2 X(I)=(TMS-X(I))*(EXP*9)/(ST*60)
0090              RETURN
0091              END


0092              SUBROUTINE SPRINT(I1,I2)
0093              REAL LIM1,LIM2,LIM3,LIM4,LIM5,LIM6,LIM7
0094              COMMON/B/  IY(3000),X(3000),Z(3000)
0095              COMMON/C/AREA(100),PEAK(100),K
0096              S=0
0097              DS=0
0098              D=0
0099              DD=0
0100              K=0
0101              IEND=1
0102              DO 1 I=I1,I1+60
0103              S=S+IY(I)
0104            1 DS=DS+Z(I)
0105              AM=S/60
0106              DM=0.0
0107              DO 2 I=I1,I1+60
0108              D=D+(IY(I)-AM)**2
0109            2 DD=DD+Z(I)*Z(I)
0110              ERA=SQRT(D/59)
0111              ERD=SQRT(DD/59)
0112              LIM1=AM+5*ERA
0113              LIM2=DM+5*ERD
0114              LIM3=DM-5*ERD
0115              LIM4=DM+ERD
0116              LIM7=DM-ERD
0117              I=I1+5
0118              IEND=I
0119           20 IF(Z(I-2).GT.LIM2.AND.Z(I+2).LT.LIM3.AND.IY(I).GT.LIM1)GO TO 4
0120              GO TO 3
0121            4 K=K+1
0122              PEAK(K)=X(I)
0123              M=I-2
0124              ISTART=IEND
0125              DO 5 L=IEND,I
0126              IF(M.EQ.0)GO TO 6
0127              M=M-1
0128              IF(Z(M).GE.Z(M+1))GO TO 5
0129              IF(Z(M).LT.LIM6)ISTART=M
0130              IF(Z(M).LT.LIM6)GO TO 6
0131            5 CONTINUE
0132            6 DO 7 M=(I+2),I2
0133              IF(Z(M).LE.Z(M-1))GO TO 7
0134              IF(Z(M).GT.LIM7)IEND=M
0135              IF(Z(M).GT.LIM7)GO TO 8
0136            7 CONTINUE
0137            8 AREA(K)=0
0138              AP=AM
0139              DO 9 IN=ISTART,IEND
0140            9 AREA(K)=AREA(K)+(IY(IN)-AP)
0141              I=IEND
0142            3 I=I+1
0143              IF(K.EQ.0)GO TO 11
0144              IF(AREA(K).LT.1.0)K=K-1
0145           11 IF(I.LT.(I2-I2/20))GO TO 20
0146              RETURN
0147              END
```

ANALYSIS OF DIGITIZED SPECTRAL DATA BY THE SAVITZKY AND GOLAY METHOD

TI=   0.200 SEC   SCALE EXPANSION 1.00   SCAN TIME   5.00 MINS
             SPECTRUM TO BE SMOOTHED   1 TIMES
M= 0 S/N=   107.01 SIGMA=   5.1864 PEAK HEIGHT   1110 AT   7.09   MEAN VALUE   0
M= 1 S/N=   162.79 SIGMA=   3.3141 PEAK HEIGHT   1079 AT   7.09   MEAN VALUE   0

        PEAK   NUMBER 1   POSITION =   7.10   PEAK   AREA   =   6629.4 UNITS
        PEAK   NUMBER 2   POSITION =   6.96   PEAK   AREA   =   160.1 UNITS
        PEAK   NUMBER 3   POSITION =   2.80   PEAK   AREA   =   259.0 UNITS
        PEAK   NUMBER 4   POSITION =   2.68   PEAK   AREA   =   873.1 UNITS
        PEAK   NUMBER 5   POSITION =   2.56   PEAK   AREA   =   1119.6 UNITS
        PEAK   NUMBER 6   POSITION =   2.43   PEAK   AREA   =   410.3 UNITS
        PEAK   NUMBER 7   POSITION =   1.36   PEAK   AREA   =   1160.5 UNITS
        PEAK   NUMBER 8   POSITION =   1.23   PEAK   AREA   =   1899.0 UNITS
        PEAK   NUMBER 9   POSITION =   1.10   PEAK   AREA   =   812.5 UNITS

instrumental settings because it is easier to replace an erroneous card than to modify a paper tape.

In program NMRD (Table 6.4), the data card is read in line 13. The tape is read in line 23 until (line 25) — 9999 is found, denoting the end of the tape. The signal amplitudes are stored in the array IY(3000) and the corresponding times are computed and stored in X(3000). The subroutine REFR(ST, EXP, N) is then called and this examines the last $^1/_{18}$ high-field portion of the spectrum in order to find the maximum signal amplitude of the spectrum. The corresponding time element is taken as a zero reference point in order to calculate the corresponding chemical shift of every data point on the tape by the formula:

$$\delta_i = (x_0 - x_i) \cdot \text{EXP} \cdot 9/(\text{ST} \cdot 60)$$

This formula assumes that a complete scan with a scale expansion of unity corresponds to 9ppm and that the spectrum is scanned from the low-field end. (If your machine scans from 0 to 10 ppm, replace 9 by 10 in the formula.) On returning from this subroutine the IY array is unchanged but the X array now consists of precise chemical shifts (ppm) values referred to the internal standard.

The next step is to call the subroutine SAVAY (J,N,NS) which accepts the array of N data points and smooths them NS times by a 5-point smoothing formula (line 63). Prior to this an estimate of the signal-to-noise ratio is calculated (line 58) from the largest peak amplitude and the standard deviation of the 'noise' in the spectrum for the first 60 data points (this presumes that no signal is present in this region). Notice that, in order to conserve core store, we store the smoothed values of IY(I) in the locations IY(I−2), bearing in mind that the convolution operation uses the array elements IY(I−1) to IY(I+3) in the next cycle. The absolute positions of the elements in the array are restored in lines 64—66, following which we note (lines 67 and 68) that two data points are lost from the beginning and end of the array during each smoothing cycle. After the NS smoothing operations, we compute the first derivatives, Z(I) and call the integration routine, SPRINT(J,N). The purpose of this routine is to identify peak positions and to perform integration of the individual peaks. Referring to the subroutine SPRINT, the reader will find that certain limits (LIM1 to LIM3, LIM6 and LIM7) are calculated; these are used in the peak detection statement (line 119) which, essentially, tests for a change in sign of the first derivative, and for the peak amplitude to be simultaneously greater than LIM1; this very effectively eliminates the detection of minor 'noise' peaks. When a peak has been found, the program steps backwards (lines 128—131) and forwards (lines 133—135) to find the beginning and end of the peak repectively. The area of the peak is then calculated (lines 139—140) and the process is repeated until 95% of the spectrum has been processed. The reason for *not* processing the final 5% is that the reference peak is always to be found in this region (scanning from low- to high-field) and we do not wish to integrate this peak. Control is then returned to subroutine SAVAY and the peak positions and areas are listed (line 75).

An example of the output from the program is shown in Table 6.4, for the spectrum of ethyl benzene which was illustrated in Figure 6.7(a). The assignments are listed, below, together with their computed and theoretical peak areas expressed as a ratio (obtained from the peak areas in Table 6.4).

	Phenyl (7.10 p.p.m.)	CH_2 (2.43—2.80 p.p.m.)	CH_3 (1.10—1.36 p.p.m.)
Computed area	5.0	2.0	2.9
Theoretical area	5.0	2.0	3.0

Note that there is a small peak at 6.96 p.p.m. due to a small amount of impurity.

This spectrum was smoothed by SAVAY prior to peak detection; observe that the signal-to-noise ratio (S/N) increases and that the standard deviation of the noise (SIGMA) decreases.

There are other methods of treating NMR spectra and these are described in the literature. For example, Rondeau and Donlan[10] have described an off-line method in which spectra are scanned firstly in one direction and then in the other, this process being repeated several times. The digitized spectra were accumulated prior to analysis in order to remove spurious peaks. An on-line method is described in the article by Hoffmann et al.[4] who use a large computer, but their methods are of general applicability.

6.3.2 Data Acquisition in Other Areas in Chemistry

In this section, a representative method of data acquisition and analysis for each area will be described and reference will be made to other methods for those wishing to pursue the subject further. No computer programs are listed but the interested reader, on consulting the references and using the techniques already described, should be able to write suitable programs for himself.

Mass spectrometry In this technique,[11] a sample is introduced into the spectrometer and converted, by one of several techniques, into a positive ion. This ion and its possible fragments are then accelerated by an electric field, separated (usually by a magnetic field), and collected as an ion current which may be detected electrically or photographically. A diagrammatic view of a typical apparatus is shown in Figure 6.8. This relatively simple type of spectrometer is called a single focusing type.

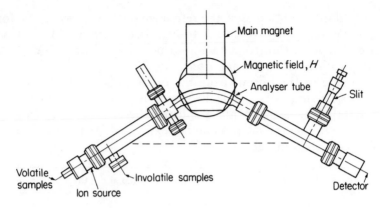

Figure 6.8   Simplified diagram of a single-focusing mass spectrometer. The accelerating voltage, V, is applied within the ion source.
(Reproduced by permission of Perkin–Elmer Limited)

The fundamental equation for such an instrument is

$$m/e = H^2 R^2 / 2V$$

in which

> m/e = ratio of the mass of the ion to the magnitude of the
>   electronic charge
> H = magnetic field intensity
> R = radius of the ion path in the magnetic field
> V = accelerating voltage

A mass spectrum consists of a series of peaks of differing intensities
arranged in order of m/e values.

Single focusing spectrometers have a relatively low resolution or
resolving power, this being the ability to separate two ions of different
mass. It is defined by the ratio $m_a/\Delta m$ where m_a and m_b are the m/e
values of two ions and Δm is the difference $(m_b - m_a)$. Peaks are
assumed to be resolved when the overlap between them is less than a set
percentage (commonly 2% or 10%). The resolving power of a single
focusing instrument is commonly a few hundred (600 for an AEI model
M.S.2) but this rises to several thousand with the more sophisticated
double-focusing instruments. This magnitude of resolving power is
essential   for   accurate   determination   of   molar   masses   (see
Section 3.2.3).

The most striking feature of a mass spectrum is its complexity and
high information density, as can be seen in Figure 6.9. In most practical
applications it is necessary to scan the spectrum relatively quickly (e.g.
typically 1 minute for a scan of m/e from 10 to 600) and slow data
acquisition units as described previously are of little use. Therefore

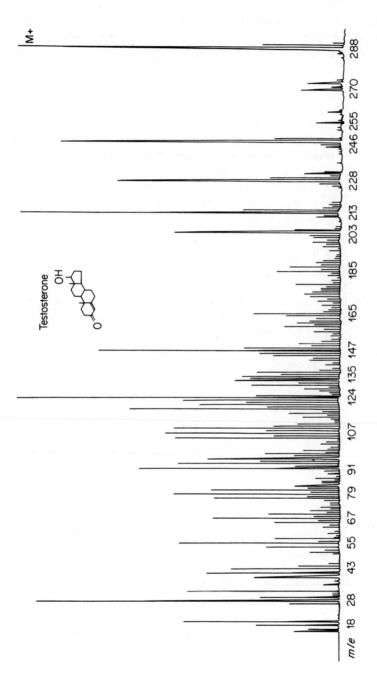

Figure 6.9 High information density in the mass spectrum of testosterone. (Reproduced by permission of Perkin–Elmer Limited)

252

faster systems should be considered such as:

(i) Magnetic tape or cassette.
(ii) Magnetic disk.
(iii) A ferrite-core store buffer plus a slow (e.g. punched tape) output device.

The last mentioned technique is used in the well-known 'Carrick' interface[12,12a] and this is one of the more economic means of acquiring mass spectrometric data.

As an example of data acquisition and processing in this area, we will describe an approach adopted by Christie, Smith and McKown[13] who used a small non-dedicated IBM 1130 computer. Although their programs are written for real-time processing, the reader should have little difficulty in devising suitable batch programs. A block diagram of their system is shown in Figure 6.10. On starting a scan, the timer and oscillograph begin to operate simultaneously. Data are taken at a rate of 5000 per second for a scan rate of 10 mass units per second until the spectrum is completed. In accordance with common practice, a small amount of reference substance, for which accurate m/e values are known, is introduced with the sample in order that mass reference peaks can be obtained. The substance used was perfluorokerosene (PFK) which has an abundance of suitable peaks. Data is collected by an assembly language program which stores, on disk, the values of all signals above a pre-set mass value; molecular formulae are then computed with a similar program to MASS (Section 3.2.3). Confirma-

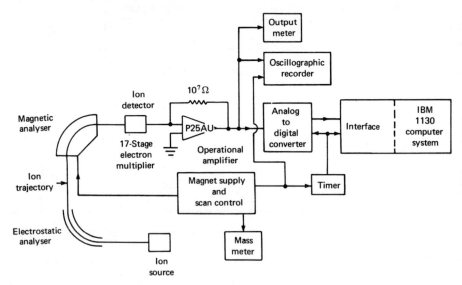

Figure 6.10   Block diagram of a high resolution mass-spectrometer/computer system (Reprinted from Reference 13, p. 56 by courtesy of Marcel Dekker, Inc.)

tion that the computed molecular formula is correct is obtained by comparison of the proposed formula with that of the fragment ions of lower mass. If the fragment could have originated from the proposed molecular ion then its formula, together with that of the neutral fragment lost, is printed for visual inspection (Table 6.5).

Another FORTRAN program of interest in this area, is listed in Reference 14. This calculates the predicted mass numbers and relative intensities of the lines in a cluster from any fragment ion containing one or more polyisotopic elements such as B, C, S, Cl and Br. Descriptions of many other programs for MS data acquisition and analysis have been published and the interested reader should consult the appropriate references,[15,16] Reference 15 being a general review of the subject.

Gas chromatography (GC) The data acquisition equipment for this type of instrumentation can be relatively slow, e.g. paper tape recording as described previously. Sampling rates of up to 10 readings per second can be achieved with paper tape punches and this is adequate for most GC purposes. Faster rates can be achieved with magnetic tape (at somewhat higher cost) and this permits the sequential sampling of many chromatographs working simultaneously.

There are two main techniques of area apportionment generally adopted in the analysis of GC data. The more sophisticated approach, as exemplified by the work of Littlewood *et al.*,[6] attempts to construct theoretical chromatograms on the basis of calculated heights, widths and positions. These parameters are then used to construct a peak shape which may, for example, be Gaussian and this is refined iteratively until the best fit is obtained to the experimental results (see Section 4.3). By this means, a complex pattern of overlapping peaks can be resolved into component Gaussian peaks and their areas can then be apportioned. A simpler method is described by Fozard *et al.*[7] who have devised a real time area apportionment routine, which could easily be adapted to a batch-processing FORTRAN procedure. Their method is capable of analysing both simple and multiple peaks, as illustrated in Figure 6.11. The areas of well-resolved overlapping peaks (b in Figure 6.11) are apportioned by dropping a perpendicular from the valley to the baseline. Curves (c), (d) and (e) are related in that they are poorly resolved and area allocation is generally more difficult in these cases. One method is to use a tangential skim procedure (Figure 6.12), taking the valley as one point and successive points along the curve until a maximum area is skimmed. This area is then subtracted from that of the major peak. A simple procedure based on the above and using absolute signal values only is illustrated in Figure 6.13 in which a decision as to whether a skimming or 'perpendicular drop' approach is chosen is made on the basis of relative peak heights. The program normally operates on

Table 6.5

Computer output showing possible ion sequence for the accurate masses measured for 3-indole acetonitrile. Ion sequence No. 1 is correct for this material. (Reprinted from Reference 13, p. 46 by courtesy of Marcel Dekker, Inc.)

3-Indole acetonitrile ion sequence				Exact mass	Error	Analysis of sequential fragmentation						
						Neutral lost			Composition			
Empirical formula						Meas.	Calc.	Error				
C	H	O	N						C	H	O	N
No. 1												
10	8	0	2	156.0687	0.0							
9	8	0	1	130.0656	−0.0	26.0031	26.0030	0.1	1	0	0	1
9	7	0	1	129.0578	−0.8	27.0118	27.0108	1.0	1	1	0	1
9	6	0	1	128.0500	−0.1	28.0189	28.0187	0.2	1	2	0	1
9	5	0	1	127.0421	0.1	29.0264	29.0265	−0.0	1	3	0	1
8	7	0	1	117.0578	1.2	39.0097	39.0108	−1.0	2	1	0	1
No. 2												
7	10	3	1	156.0660	2.6							
6	10	3	0	130.0629	2.6	26.0031	26.0030	0.1	1	0	0	1
6	9	3	0	129.0551	1.7	27.0118	27.0108	1.0	1	1	0	1
6	8	3	0	128.0473	2.5	28.0189	28.0187	0.2	1	2	0	1
6	7	3	0	127.0395	2.8	29.0264	29.0265	−0.0	1	3	0	1
5	9	3	0	117.0551	3.8	39.0097	39.0108	−1.0	2	1	0	1
No. 3												
5	8	2	4	156.0647	4.0							
4	8	2	3	130.0616	3.9	26.0031	26.0030	0.1	1	0	0	1
4	7	2	3	129.0538	3.1	27.0118	27.0108	1.0	1	1	0	1
4	6	2	3	128.0460	3.8	28.0189	28.0187	0.2	1	2	0	1
4	5	2	3	127.0381	4.1	29.0264	29.0265	−0.0	1	3	0	1
No fit found												

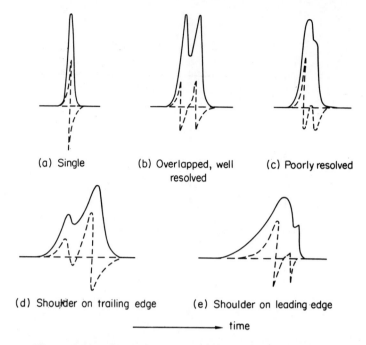

Figure 6.11   Single and overlapping GC peaks—dashed line
is the derivative curve

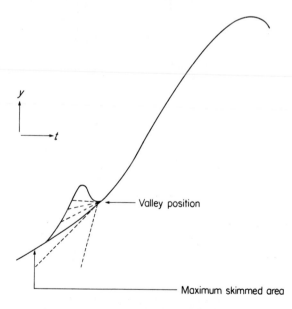

Figure 6.12   Tangential skim procedure for obtaining the area of a shoulder peak (d and e in
Figure 6.11)

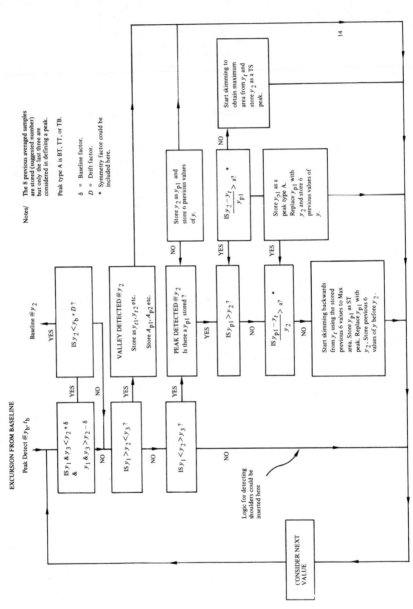

Figure 6.13 A real-time area apportionment routine. (Reproduced from *Applications of Computer Techniques in Chemical Research*, Applied Science Publishers, 1972, p. 47, by permission of Applied Science Publishers Limited)

the last three smoothed values of the signal but, for skimming purposes, at least 8 points are stored.

In Figure 6.13, the following notation is used:

B = baseline,
T = trough,
S = skimming point,
BB = fully resolved peak,
TS = trailing shoulder peak,
D = Drift factor (linear rate of change of baseline),
δ = baseline factor proportional to random noise.

As can be seen the routine will handle peaks of types BT, TT, TB, TS or ST but does not fully accommodate shoulders. This can however, easily be accomplished by a small extension of the skimming method. It is worth mentioning that although the procedure in Figure 6.13 is not used in a commercial instrument, it is a practical suggestion. The reader is referred to the literature for other references to GC data treatment.[17,18]

Gamma-ray spectrometry   The γ-ray spectrum of radionuclides can be recorded by a detector consisting of a crystal in contact with a photomultiplier or with a solid state detector. For high-resolution work, it is preferable to use lithium-drifted germanium or silicon detectors and to pre-process the signal with a multichannel analyser. The theory and practice of this type of spectrometry is discussed in detail by Nielson.[19] If we consider the emission of a γ-ray of a specific energy then the corresponding spectrum would resemble that in Figure 6.14. The peak arises from the total energy absorption of the γ-ray energy by the crystal and it is called a 'photopeak'. We note that the count rate is higher on the low-energy side of the peak so that, in order to find the photopeak area, we must subtract that area subtended by a sloping baseline. Observe also, the discontinuous appearance of the spectrum; this is due to the fact that the spectrum is stored in a multichannel analyser (up to 4096 channels) prior to being output either onto a display device, such as a recorder, or onto computer-readable media. This can be paper tape or in some cases the computer memory itself. Single channel analysers can also be used. These give a continuous analogue signal but have poorer resulution.

McDermott[20] has described a simple real-time FORTRAN approach which could easily be used for batch processing of paper tape data. A flow-chart based on McDermott's program is illustrated in Figure 6.15, from which a FORTRAN procedure could be devised. Peak postions are located by the use of first derivatives, as described in previous sections. The program then examines the highest energy peak and determines if this is resolved from the next energy peak. If it is not, successive peaks

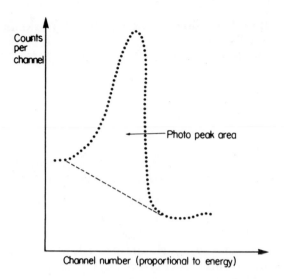

Figure 6.14   A photopeak in a gamma spectrum

are examined until a multiplet it is successfully isolated. The limits of the peak, or multiplet, are found by looking for three successive increases in the smoothed channel contents such that the difference between the first and third channels exceeds twice the square root of the first. The first channel in the group is then used as one limit of the peak. When both limits have been found, a straight line is joined between them and the nett photopeak is calculated. If the area corresponds to an unresolved multiplet, it is resolved into its n component peaks by using an iterative least-squares fit to the function.

$$y = a + bx + \sum_1^n c_j \exp \left(-s^2 \left(x - d_j\right)^2\right) \qquad (6.3)$$

where

$\qquad y$ = contents of channel x
$\qquad x$ = channel number
$\qquad a, b, s$ = constants for all component distributions
$\qquad c_j, d_j$ = empirical constants in the jth Gaussian distribution

The component areas can then be obtained from the relative values of the contributions to the summation in equation (6.3) as a proportion of the total area. It is interesting to note that the use of such a fitting function is more justified for multiple peaks in γ-spectometry than in chromatographic data analysis because photopeaks tend more frequently to approximate to the true Gaussian shape, whereas adjacent GC peaks can have different shapes.

Gamma-ray spectrometry is an important field of computer applica-

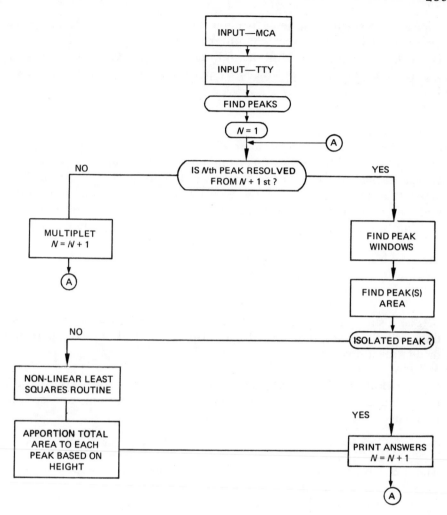

Figure 6.15 Proposal by McDermott[20] for real-time FORTRAN processing of gamma spectra

tions, not only in data processing but also in identification of the individual radionuclides responsible for the spectrum.[21]

Complete commercial systems have been developed for the collection and processing of nucleonics data, including gamma spectrometry. The CAMAC modular data transfer system is one of the most sophisticated and flexible collections of equipment available in this field. Bidirectional communication between instrument and computer is provided and individual systems are easily assembled from modules, which communicate via a multi-wire data highway ('dataway'). Considerable computer independence is thereby achieved since a CAMAC system

260

appears as a single peripheral to a computer and only the interface need be altered when a different computer is used. An example of a possible CAMAC configuration for a multichannel analyser is shown in Figure 6.16 which is reproduced by permission of Nuclear Enterprises Ltd.

6.4 Problems

6.4.1 For this problem, you need to obtain a paper tape recording from an atomic absorption spectrometer or any other instrument that gives a displacement from the baseline in the presence of a sample. The paper tape should contain characters corresponding to signal amplitude, with the correct sign. Record a baseline (no sample) for a set time, say one minute, and terminate with -9999; introduce the sample to obtain a fairly steady reading and record this in the same way. Finally, record the new baseline. Write a program that will allow for linear slope in the baseline (Section 6.2.1.) and will calculate (a) the mean displacement, (b) the standard deviation due to the sample.

6.4.2 Using any digital recording write a subroutine to smooth the data using (a) a 5-point and (b) an 11-point smooth and compare the difference in performance (Section 6.2.2.).

6.4.3 Modify program NMRD so that second derivatives are used for peak location. For a 5-point convolution operation, the second derivative at the centre point can be calculated from:

$$d^2y/dx^2 = (2y_{-2} - y_{-1} - 2y_0 - y_{+1} + 2y_{+2})/7\Delta x^2$$

where Δx is the timing interval. If possible, compare the performance of this method with that using first derivatives. Under what conditions do you expect 2nd derivative peak-picking to give inferior results?

6.4.4 Try to reduce the core store required by NMRD by at least 10,000 words (assume one real variable occupies 2 words) without reducing the size of spectrum that can be processed.

6.4.5 Improve NMRD so that small peaks (less than say 1% of the total spectrum area) are rejected and that any two peaks with an apparent spacing less than the resolution of the instrument are merged together.

6.4.6 Write a program that will carry out a cross-correlation of two digital records (Section 6.2.4). One record should be relatively noise-free, whilst the other may represent a more noisy spectrum of, for example, a more dilute solution of the same compound. Investigate the

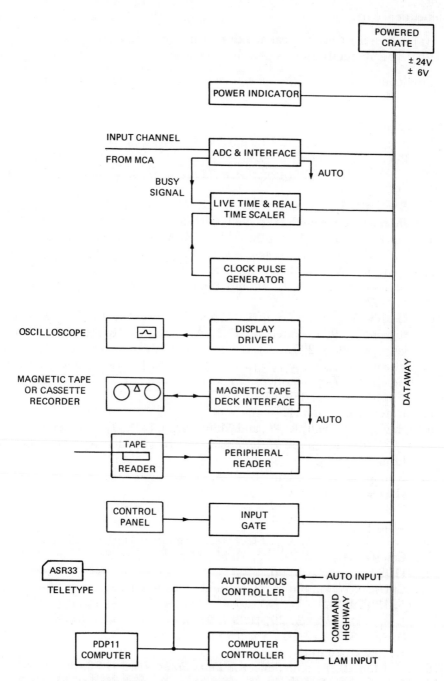

Figure 6.16 A possible CAMAC—MCA system configuration. (Reproduced by permission of Nuclear Enterprises Ltd.)

dependence of:

(a) The height of the peak maximum in the correlogram,
(b) The integrated cross-correlogram area,

on the concentration of the substance giving rise to the noisier signal.

References

1. Shepherd, T. M., and Vincent, C. A., *Chem. in Brit.*, 9, 66 (1973).
2. Beech G., *American Laboratory*, 53 (September 1973), (in identical form in *International Laboratory*, 35 (January/February 1974).
3. Bendat, J. S., and Piersol, A. G., *Random Data*, Wiley-Interscience, New York, 1971, p. 288.
4. Hoffman, E. G., Stempfle, W., Schroth, G., Wiemann, B., Ziegler, E., and Brandt, J., *Angew. Chem. Internat. Edit.*, 11, 375 (1972).
5. Savitzky, A. and Golay, M. J. E., *Anal. Chem.*, 36, 1627 (1964).
5a Steiner, J., Termonia, Y. and Deltour, J., *Anal. Chem.*, 44, 1906 (1972).
6. Littlewood, A. B., Gibb, T. C., and Anderson, A. H., *Gas Chromatography 1968*, edited by C. L. A. Harbourn, Institute of Petroleum, 1969, p. 297.
7. Fozard, A., Franses, J. J., and Wyatt, A. J., *The Applications of Computer Techniques in Chemical Research*, edited by P. Hepple, Applied Science Publishers Ltd., London, 1972.
8. Reference 3, p. 331.
9. Haws, E. J., Hill, R. R., and Mowthorpe, D. J., *The Interpretation of Proton Magnetic Resonance Spectra*, Heyden and Sons Ltd., London, 1973.
10. Rondeau, R. E., and Donlan, V. L., *Anal. Chem.*, 43, 1699 (1971).
11. Hill, H. C., *Introduction to Mass Spectrometry*, Heyden and Son Ltd., London, 1966.
12. Carrick, A., *Radio and Electronic Engineer*, 41, 453 (1971).
12a. Carrick, A., and Paisley, H. M., *Organic Mass Spectrom*, 8, 229 (1974).
13. Christie, W. H., Smith, D. H., and McKown, H. S., *Chem. Intrum.*, 5, 43 (1973).
14. Lee, J. D., *Talanta*, 20, 1029 (1973).
15. Ward, S. D., *Mass Spectrom.* (Chemical Society Specialist Reviews), 264 (1973).
16. Gordon, B. J., and Self, R., *Lab. Pract. (G.B.)*, 22, 267 (1973).
17. Schomburg, G., Weeke, F., Wirmann, B., and Ziegler, E., *Angew. Chem. (Internat. Edit.)*, 5, 366 (1972).
18. Kizer, K. L., *Amer. Lab.*, 5, 40 (1973).

19. Nielson, J. M., in *Physical Methods of Chemistry*, Wiley-Interscience, 1972, p. 661.
20. McDermott, W. E., *NASA Tech. Memo. 1972* (CNASA TM X–2440), p. 744.
21. Reference 17, p. 683 and references therein.

Chapter 7

Further Areas of Study

In the preceding chapters, an attempt has been made to describe some typical applications of a medium-size computer in the general area of chemistry. Hopefully, many of the programs will be used in their entirety or with some slight modifications.

Many readers, however, will make extensive modifications and improvements to the programs described and also they may wish to write completely new programs. It is for these persons that this chapter has been written as it is intended mainly as a guide to the literature. This consists mainly of brief articles or notes and it is unusual to find a complete program listing. For the simpler programs, listings are probably not required since many experienced computer users will prefer to do their own programming based on the ideas presented in the papers. Listings of the more advanced programs can, however, usually be obtained from the authors.

7.1 The Literature of Computing in Chemical Education
In such a rapidly expanding field, any guide to sources of literature is certain to become quickly out of date. This section, therefore, only relates to the time of writing but should be useful to those wishing to access relevant literature. Fortunately, it is possible to subdivide this area.

7.1.1 Textbooks, Original Articles and Abstracting Services
Many of these have been cited in this text and more references follow in Section 7.2, so that only an indication will be given here of the more important primary sources of information.

These include:

(i) *Journal of Chemical Education* (The majority of computing in chemistry articles are published here).

(ii) *Education in Chemistry* (Published by the Royal Institute of Chemistry).

(iii) *Chemical and Engineering News* — although not predominantly concerned with chemical education at least two articles are very relevant as background material: 'Computers — no longer a classroom novelty' by S. Smith, 47 (25), 48 (1969); 'Computers in Chemical Education' by F. D. Tubbutt, 48 (3), 44 (1970).

(iv) *International Journal of Mathematical Education in Science and Technology.*

Other journals which occasionally carry articles on educational aspects of computing in chemistry are:

(v) *School Science and Mathematics*
(vi) *Journal of Research in Science Teaching.*
(vii) *SIGCSE Bulletin.*
(viii) *Science Teacher.*
(ix) *Journal of College Science Teaching.*
(x) *Newsletter of the Association for Development of Instructional Systems.*

The conventional information services can be valuable, particularly those such as UKCIS which search the Chemical Abstracts data bases. For those having access to UKCIS services, an SDI (selective dissemination of information) profile used by the author in collecting information on computing in the physical sciences is given in Appendix B. From January 1975, a macro-profile on this subject will be available from UKCIS. Another abstracting service which is complementary to the above is the INSPEC TOPICS scheme which offers a low-cost standard service on the subject of computer applications in physics and chemistry. The address of INSPEC is:

Institute of Electrical Engineers,
Savoy Place,
London WC2R OBL.

A general guide to commercial information retrieval systems is to be found in *A Guide to Selected Computer Based Information Services* by R. Finer, published by ASLIB, London, May 1972.

7.1.2 Bibliographies and Other Reference Texts
A few bibliographies on computer assisted methodology make reference to the application of computers in chemistry. These include:

(i) *Computer Assisted Instruction — A Selected Bibliography,* edited by O. D. Barnes and D. B. Schreiber, Association for Educational Communications and Technology (1201 Sixteenth Street, N. W. Washington, D.C. 20036), March 1972. This contains 15 entries on the subject of chemistry.
(ii) 'A Comprehensive Annotated Bibliography on Computer

Assisted Instruction — Part II', by J. D. Testerman and J. Jackson in *Computing Reviews,* November 1973, p. 543. Of the 79 entries in 'Scientific Applications', at least 15 are relevant to chemical education.

(iii) *Index to Computer Assisted Instruction* 3rd Edition, Edited by H. A. Lekan, Instructional Media Laboratory, University of Wisconsin, Milwaukee. Almost 80 detailed descriptions of programs and packages appear in the 'Chemistry' section which are mainly of an interactive nature.

(iv) *International Computer Bibliography,* Vols. 1 and 2, published by National Computing Centre Ltd., Manchester, England, 1968 and 1971. All aspects of computer technology are covered in this compilation of over 7000 abstracts.

The National Computing Centre (U.K.) publishes a wide range of books; their most useful publications for scientists are their authorative guides on programming standards.

From time to time the Science Research Council publishes informative books on computing facilities in the U.K. For example, the recent Computing Science Committee report of June 1974, *Science Research Council Computer Networks,* includes a section on Universities and Research Institutions networks in the U.K. The SRC also published, in 1974, *Computational Chemistry and Physics,* which is a useful state-of-the-art summary for selected areas of research.

The Computer Users Year Book (published by the Computer Users Year Book, 18 Queens Road, Brighton, Sussex), is a valuable source of information, particularly regarding commercial companies and the services that they offer. Of some interest is the compilation of Software Houses and Programming Services which includes a tabular summary of the services offered, language availability, method of operation and application areas.

Complementary to the Year Book is the *International Directory of Computer and Information System Services (1974),* published for The Intergovernmental Bureau for Informatics, Rome, Italy, by Europa Publications Ltd., London. This lists, for each country, the types of establishments and the services available on specified equipment. Such information may be valuable when contemplating the acquisition of programs from other institutions.

Conference proceedings tend to be not so useful as original articles but can serve as fruitful sources of ideas. Particularly relevant are:

(i) *Conference on Computers in the Undergraduate Curricula, June 16—18, 1970, University of Iowa.* Published by the Centre for Conferences and Institutions, University of Iowa.

(ii) Title as (i), *June 23—25, 1971, Dartmouth College, New Hampshire.*

(iii) Title as (i) *June 12—14, 1972, Atlanta, Georgia.* Published through Southern Regional Education Board, Atlanta, Georgia 30313.

(iv) Title as (i) *June 18—20, 1973, The Claremont Colleges, Claremont, California.*

(v) *Computers in Higher Education, 25—27 June, 1973, Wolverhampton, U.K.* (Abstracts only from The Polytechnic Library, Wolverhampton).

(vi) *Computers in Higher Education, Lancaster University, March 1974*; proceedings published in Issues 3 and 4 of *International Journal of Mathematical Education in Science and Technology,* 5 (1974).

(vii) *Computer Assisted Learning in Science and Engineering (CAL 75), Oxford Polytechnic, March 1975.* (Abstracts only).

7.2 Programs I'd Like to Have written
The title to this section is not strictly accurate but the implication is that, in the following annotated bibliography, the selection from the literature is a personal one. It is not intended to be complete but, rather a cross-section of the literature relevant to the preceding chapters. The references cover the period 1967—1974 and, as will be seen, the rate of proliferation has increased rapidly. For convenience the field has been broken down into five subsections. A good general introduction to the literature can be found in the February 1970 issue of the *Journal of Chemical Education* which contains 21 articles and 5 notes in a special issue concerned with computers in chemistry.

7.2.1 The Computer as a General Teaching Aid
In chemical education, there is increasing emphasis on the use of computers as teaching aids to enrich the classical lecture/tutorial/practical situation. Many educators have expressed concern about this trend on the grounds of 'dehumanization' of the learning experience and, also, plain economics; there is no doubt that the provision of fifty or more terminals is an expensive portion of a normal educational budget, especially when the effectiveness of some computer aided learning (CAL) packages is unproven. When sensibly used, however, the computer can complement conventional methods by the provision of novel teaching situations. Furthermore, many of the newer teaching packages can be implemented on minicomputers with very modest core store requirements.

An example is provided by the package described by Clark *et al.*[1] which provides drill and practice in one-step organic reaction syntheses. The package is written in FORTRAN IV and is interactive. A larger collection of programs is described by Castleberry *et al.*[2] and Rodewald *et al.*[2a] This collection includes programs for tutorial work and for simulated experiments and is a fruitful source of ideas.

The element of fun is not lacking in instructional packages. The article by Breneman[3] describes a FOCAL package implemented on a minicomputer. Gambling games and witticisms abound in such serious matters as mole to gram conversions and experimental simulations! The package appears to be useful but it is worth noting that teletypes are slow output devices and are best used for essential output.

A glance at any recent *Journal of Chemical Education* annual index will reveal a number of programs designed for automatic record keeping and student grading. A sophisticated example implemented on an IBM 370/165 for 700 students, is described by Macmillan and Epstein.[4] This system is of a very general type, intended to produce records using data from various sources. A more specialized use of a computer terminal for interactive laboratory report grading was reported by Johnson.[5] This package provides immediate evaluation of student results, in addition to creation of a record for grading purposes. It is likely that there will be a considerable increase in the use of computers for the evaluation of student performance and, as a result, the generation of personalized laboratory and tutorial work.

7.2.2 General Laboratory Computing

Although the uses of a computer as a teaching/learning aid will proliferate in undergraduate education, there is little doubt that the analysis of laboratory data will remain an important activity. Wilkins and Klopfenstein[6] have written two articles which serve as a good introduction to a variety of applications using the BASIC language with a minicomputer. The applications include kinetics, spectroscopy, flash photolysis, potentiometric titrations and nuclear magnetic resonance simulations. Similarly, Wise[7] has made available a collection of FORTRAN programs for use in the undergraduate physical chemical laboratory. Applications include calculation of:

(i) The molar refraction of compounds from measurements of refractive indices.
(ii) Molar mass of benzene from the Dumas method.
(iii) Rate of inversion of sucrose (first-order reaction).
(iv) Distribution coefficients for a solute between two immiscible liquids.

Beech has also published[8] a description of the student programs used at Wolverhampton Polytechnic.

There are many more publications which describe programs intended to enrich the laboratory experience. A recent example is the 'totally integrated approach' described by Davis *et al.*[9] This approach, though oriented by the authors towards the gas laws, could be applied in other fields. The student performs some experimental work in the laboratory and then reinforces his understanding of the theoretical principles by simulating the same experiment with parameters that may not be available in the laboratory.

An area which might not be thought to be amenable to computer assistance is that of qualitative organic analysis. We have found, however, that by using the program published by Luteri and Denham[10] we are able to add some interest to an otherwise routine student activity. The basis of the program is that the student records his observations (such as melting point, boiling point, solubilities in various solvents, presence or absence of elements) and submits them for computer analysis. The program matches the input data against that of more than 6000 compounds (the data being stored in a magnetic tape file). Two lists are output — firstly, the compounds that match (within experimental error) in every respect and, secondly, those that match except for solubilities. A similar, slightly simplified program was described by Gasser and Emmons.[11]

Spectroscopy receives continued attention as can be seen from Chapters 4 and 6. To take just one field of experimentation we find that a recent textbook[12] is devoted solely to computer techniques in NMR and serves as a good introduction to this field. The review by Yamamoto and Someno,[13] in Japanese only at the time of writing, covers aspects of both NMR and ESR. Rafalski and Barciszewski[14] discuss the application of computers to the analysis of data from NMR experiments with lanthanide shift reagents, a class of compound in which there is considerable topical interest from the structural aspect.[15]

Mass spectrometry has been dealt with in Section 6.3.2 and references therein. The article by Robertson[16] includes a general review of mass spectrometry computer applications. Much effort in this area is concerned with the elucidation of molecular structures and also the storage, interpretation and identification of mass spectra. A BASIC program for the numerical identification of mass spectrum peaks has been described by Mantei and Hunter.[17] This program was written specifically for spectrometers which scan logarithmically and accepts as input, pairs of known M (mass) and X (distance of peak from zero position) values. These are fitted to an equation.

$$X = C_1 \log M + C_2 C_3^M + C_4$$

which permits the identification of unknown M-values from a series of input X. An interactive mass spectral search system which would be valuable in many undergraduate courses was described by Heller et al.[18] Their system uses an 'abbreviated spectrum' file consisting of the two most intense peaks in every interval of 14 a.m.u. for 8782 compounds.

Computer storage and subsequent file searching has also been applied to infrared spectroscopy in commercial systems and techniques have been described by Penski[19] and by Drobyshev.[20] The latter in Russian, uses the Sadtler collection of spectra as does a commercial system.[21]

Another area which is certain to increase in importance in the undergraduate laboratory is that of data acquisition and processing. The first task of data collection is that of interfacing between the analytical instrument and data collection device. This is likely to be facilitated by the newer solid state devices described by Dessy and Titus.[22] Their review is extremely readable and is strongly recommended as an introduction for those contemplating construction of their own interfaces. For those intending to run courses for students on the subject of data acquisition and related subjects, the article by Perone[23] is relevant since it describes the 3-week summer course 'Digital Computers in Chemical Instrumentation' at Purdue University. The course deals with both on-line and off-line computing and provides experience in digital logic, programming fundamentals and interfacing to laboratory experiments.

Large computer programs have been used for a considerable time in the elucidation of thermodynamic parameters, particularly equilibrium constants, from spectroscopic and potentiometric data.[24] Programs continue to be published such as that by Likussar[25] for the continuous variations method, and that by Robrecht et al.[26] The article by Robrecht (in English) makes the valid point that such general programs as LETAGROP[27,28] require a large computer and also a familiarization period by the user. This may be unacceptable in undergraduate chemical education and the simplified approaches exemplified by Robrecht facilitate the calculation of equilibrium constants by undergraduate students using laboratory computers. Magnell[29] has introduced a sophisticated program (SPECTRO 1130) at the Central Michigan University for undergraduate laboratory work. The program is designed to refine estimates of formation constants and spectral information for systems with up to three complex species, including binuclear, at up to five wavelengths. Although the program by Sabatini, Vacca and Gans[30] was developed for research, there is no reason why it should not be applied at undergraduate level provided that the principles of program MINIQUAD (listing provided in Reference 30) are understood. This program is the most recent of its type to be developed at the time of writing, and is said to be the most powerful program for the calculation of formation constants from potentiometric data.

A useful annotated bibliography of programs for chemical kinetics has been published recently by Hogg.[31] In line with the philosophy of this section, the references selected for the bibliography refer to programs which are either easily obtainable from the authors or reproduced in full. The information base was *Chemical Abstracts* and this was searched up to January 29, 1973. The bibliography is comprised of 47 references and it is not necessary to reiterate them here. Three references worth adding to Hogg's list are the papers by Seyse and Rose[32] (APL program for gas phase kinetics), Cummins and Wartell[33] (minicomputer-generated kinetics experiment) and Grinwald and Steinberg[34] (fluorescence decay kinetics).

Nuclear chemistry is also well suited to computer assistance because, in part, of the necessity for statistical calculations and also because of the electronic instrumentation which, having inherent digital circuitry, is easily interfaced to output devices. The two techniques which have received the greatest attention in this area are liquid scintillation counting and gamma spectrometry. In the former, quench correction curves using the channels ratio or external standard methods are used to determine absolute disintigration rates. Some instruments have small hard-wired computer circuitry but more flexibility is obtained from user-written programs. Spratt discusses[35] the acquisition and handling of liquid scintillation counting data and his review is useful as an introduction to the field, sufficient references are cited to enable the interested reader to pursue the subject in more detail. Gamma spectrometry was discussed in Section 6.3.2; FORTRAN programs have been published for the analysis of lithium-drifted germanium detector data[36, 37] and also for thallium activated sodium iodide detectors.[37] Other computer procedures are described in References 38–41. Reference 41 pertains to undergraduate work, using a BASIC program package and describes applications in the teaching of statistics and treatment of nuclear data and 'spectrum stripping' in gamma spectrometry.

A useful and unusual application has been reported by Seim and Prydz[42] who have devised a computerized method for two-dimensional mapping of radiochromatograms. The program is based on the contour mapping procedures used for the preparation of meteorological weather maps, although other contour generating programs, such as that in Appendix A, could be used.

7.2.3 Experimental Design and Optimization

This subject is of such great industrial importance that it is to be expected soon to make a considerable impact at the undergraduate level. A most readable account of the principles and practice of experimental design (including a thorough treatment of the computing aspects) has been written by Szonyi,[43] whose article includes 50 references to the subject. To quote from Szonyi, the questions to be asked when considering, for example, the feasibility of preparing a product, are:

What factors influence product yield?
What factors determine product purity?

What variables are responsible for the outcome of a given test?
What is their precision?

Statistically designed experiments can provide data to answer these
questions so long as the problem is correctly analysed and the correct
variables are chosen. For example we may know that the yield of a
reaction is influenced by temperature, T, pressure, p, and reaction time,
t. Mid-points for these variables are chosen, with steps ΔT, Δp and Δt
chosen such that the upper and lower levels cover what is regarded as
the significant range of each variable. For example, a mid-point value of
T might be 200 °C with ΔT equal to 100 °C; this would give lower and
upper levels of 100 °C and 300 °C for T. The yield, y, might be given
by an equation of the form:

$$y = b_0 + b_1 T + b_2 p + b_3 t + b_1 Tp + b_{12} Tt + b_{23} pt + b_{123} Tpt$$

$$(7.1)$$

Table 7.1
Selection of variables for
experimental design

Experiment	T	p	t
1	$T_1{}^a$	p_1	t_1
2	T_2	p_1	t_1
3	T_1	p_2	t_1
4	T_2	p_2	t_1
5	T_1	p_1	t_2
6	T_2	p_1	t_2
7	T_1	p_2	t_2
8	T_2	p_2	t_2

[a]The subscripts 1 and 2 denote the
lower and upper levels of each
variable.

The values and the statistical significance of each parameter b_0, b_1, b_2
etc. can be calculated by performing just 8 experiments (generally 2^n
where n is the number of variables), as in Table 7.1. The eight equations
of the form (7.1) can then be solved to find the values of the
parameters and, by further statistical tests, the significance of each
parameter can be assessed. The final step is to find the 'best' values (i.e.
to maximize the yield) and this is done via a steepest ascent method.
The programs used by Szonyi are written in BASIC although
FORTRAN versions are also available.

7.2.4 Quantum Chemistry
This intrinsically numerical subject is well suited to computer assistance
and considerable progress has been made in using computer techniques

to enhance students' understanding of this subject. The latter is exemplified by the considerable use of graphical output for the representation of atomic and molecular orbitals (see Chapter 5) and the test by Streitweiser and Owens[44] is the logical starting point. Parrett and Peterson[45] have described a FORTRAN program for producing hydrogen atom electron density plots which are similar to those in Chapter 5.

An interactive BASIC program has been described[46] for the production of accurate sp^{α} hybrid orbitals which is intended to assist in the teaching of such topics as orbital hybridization and chemical bonding.

A computer-aided instructional program has been described by Cox and Elton[47] in their paper 'Solving the Schrodinger equation with a desk-calculator plotter' which introduces such concepts as eigenvales and eigenfunctions. In a similar vein an interactive BASIC package for teaching the Huckel theory of bonding has been described by Janis and Petersen[48] who use a modular approach which allows for such experiments as (i) changes in bond angles and/or bond lengths, (ii) selection of the most stable geometry and (iii) effects of substituents on charge density distributions. An interesting complementary approach by Stoklosa[49] describes a simple program to calculate electronic distributions using the Sanderson electronegativity scale. The results are very similar to those obtained from Extended Huckel Theory (EHT) calculations. Another relatively simple program, useful for obtaining numerical values for the energy levels of the H_2^+ molecule has been published;[50] the program makes use of the LCAO method with no further refinements. In contrast, Frenkel and Davis[51] have successfully introduced Hartree—Fock calculations into undergraduate curricula as examples of 'state of the art' theoretical chemistry. Their package is intended for above average students who are encouraged to write part of the program package. At a similar level, Alderdice and Watts[52] have published programs for the precise calculation of atomic energy levels of the isoelectronic series Li^+, Be^{2+}, etc.

Quantitative molecular orbital calculations on transition metal complexes has proven difficult to implement into the undergraduate curriculum and the majority of publications are, to date, at research level. The book by Ballhausen and Gray[53] provides an introduction to the popular Wolfsberg—Helmholz methods and further details are given in the review by Davies and Webb.[54] Details of extended Wolfsberg—Helmholz calculations have been described by Litinski and Rakuaskas.[55]

7.2.5 Simulation of Chemical and Physical Processes
Several examples in preceding sections overlap with the general area of simulation and the following references should be regarded as being

complementary. The great advantage of simulated experiments is that they permit the student to investigate systems or conditions that may be available in the undergraduate laboratory. On the other hand, they can suffer from the disadvantage of being 'too ideal' in that it is unlikely that new discoveries will be made although new relationships may often be devised.

A useful example of simulation techniques in the undergraduate laboratory is the simulated Rutherford scattering experiment.[56] The classical Coulomb scattering equations, together with a random number generator, are used to predict the scattering data which is presented diagrammatically for angles from 0 to 180 degrees at 10 degree intervals. A more sophisticated program BACKS,[57] for the simulation of back scattering takes into account energy losses and thermal motion and efficiently generates backscattering data for particles with energies from 1 KeV to 2 MeV.

The liquid state lends itself well to simulation but most reports to date are at the research level. A good review by McDonald and Singer,[58] in addition to providing background material on liquid structure, includes references to computer applications in calculations of viscosity, diffusion, intermolecular forces and the thermodynamics of liquid mixtures. Scholfield[59] deals with simulated atomic motion in liquids, interparticle potentials and the method of molecular dynamics.

Magnetic resonance experiments, as indicated in Section 4.4.3, can be simulated and, in addition to the references on NMR already quoted, many-spin NMR spectra (e.g. of polymers) have now been simulated.[60] Electron spin resonance is also a suitable area as shown by Ling[61] who has published a FORTRAN IV program for simulation of first-order ESR spectra. The program generates up to nine individual spectra, with up to 1000 lines per spectrum, as absorption or derivative curves or as 'stick' spectra. ESR powder spectra have also been simulated, allowing for ligand hyperfine interactions.[62]

Simulation of electrode processes has reached an advanced level[63] and applications in various types of potentiometry have been cited.[64]

7.2.6 Representation and Manipulation of Chemical Information
This subject is an interdisciplinary one between the fields of chemistry and information science and has already achieved considerable industrial importance. The areas of concern to this section are:

Representation and storage of chemical information (formulae, structures) in a computer.
Retrieval of the stored information.
Analysis of spectroscopic data, mainly by pattern recognition, to ultimately yield chemical structures (which can then be stored in a computer!).

These three areas are not independent and there are others which we will not discuss here such as molecular modelling and computer-assisted design of organic syntheses.[65] Storage and retrieval of chemical information — the first two categories above — are grouped here for convenience.

One of the most advanced methods of chemical information handling currently in use is the Wiswesser Line-Formula Notation (WLN). Chemical structures can be encoded into WLN in an unambiguous manner, are intelligible to chemists with only a basic knowledge of WLN, and are computer-searchable. A good introduction to WLN is given by Palmer[66] and a fuller definition by Smith.[67] WLN uses letters, numerals and certain characters to represent the chemical elements simple groups, and their environments. The character set is:

space & $-/0123 - - - 9\ 10\ 11 - - - -$ ABCD $- - 0 -$ XYZ

Chemical elements whose symbols are 2 lettered are retained almost unchanged, e.g. NA for Na and many single letter symbols remain, e.g. F, I, S and P. Some examples of WLN symbols are given in Table 7.2.

Table 7.2
Examples of WLN symbols and characters

Symbol	Meaning	Symbol	Meaning
1, 2, 3, . . . , etc.	Number of C atoms in unbranched internally saturated chain e.g. 1 for $-CH_3$, $-CH_2-$, $-CH=$, 2 for $-CH_3CH_3$, $-CH_2CH_2-$,	U UU V	Double bond $=$ Triple bond $\equiv$ $\overset{\diagdown \diagup}{\underset{O}{\overset{C}{\|}}}$
E G	$\left.\begin{array}{l} Br \\ Cl \end{array}\right\}$ other halogens unaltered	W	Dioxo group e.g. $-SO_2-$
K	$+N\!\!<$	X	$-\overset{\|}{\underset{\|}{C}}-$
M	$\overset{}{\supset}NH$ or $=NH$	Y	$-\overset{\|}{C}H$
N	Hydrogen-free nitrogen	Z	$-NH_2$
O	Hydrogen-free oxygen		
Q	$-OH$ group		
R	Unfused benzene ring		

276

Some examples based on Table 7.2 are:

Chemical formula	WLN
$CH_3 . CO . CH_3$	1V1
$CH_3 . CO . O . CO . CH_3$	1V O V 1
$ClC{=}C{-}CH{=}CH{=}C{=}CCl$	G 1 U 1 1 U 1 U 1 U 1 G
$C_2 H_5 CN$	NC2
$CH_3 \quad CO . CH{=}C(OH)CH_3$	Q 1 Y U 1 V 1

Note in the last two examples that the *latest position* principle is used which selects, from the alternative notations, that which would begin the notation with the latest letter in the list (above) of allowed characters. Therefore, $C_2 H_5 CN$ would not be called 2CN since N falls later than 2 in the character list. Rules exist regarding branched chains, aromatic compounds and ring structures which can also be un-ambiguously coded. An average notation is about 20 characters in length and it is rare for any notation to exceed 50 characters. Therefore, 80 column cards are ideal for WLN. Having coded a collection of compounds in WLN, there are many subsequent uses, as shown in Figure 7.1. The 'connection table' in Figure 7.1 refers to a

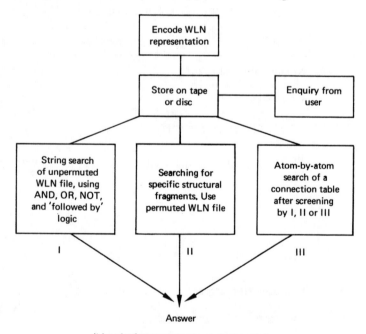

Figure 7.1   A possible information-retrieval system based on WLN representations

method of converting a WLN representation into a record which explicitly defines all atom-to-atom connections within a molecule. It is a more detailed description of the molecule and of course, uses more storage than the simple WLN strings. The best known procedure for this is the CROSSBOW system.[68]

Analysis of spectroscopic data　This is an area well suited to FORTRAN, being intrinsically numerical. There are at least two approaches to this problem:

  (i)　Learning machine (pattern recognition) method.[69, 70]
  (ii)　Imitation of a human analyst.[71]

Method (i) is better known and has been applied to a wide range of problems, both chemical and otherwise. A pattern-recognition method is normally composed of the following steps:

Computer-readable　　　　Feature　　　　　　Pattern
data from instrument　———→　extraction　———→　classifier

The data could be, for example, infrared spectrum intensities encoded every 0.1 nm from 2.0 to 15.0 nm. This data could be collected manually or, for large numbers of spectra, by data acquisition equipment. The 'Feature extractor' could ensure that noise or spurious peaks are removed, and that only data in the correct range and at correct intervals are stored.

The purpose of the pattern classifier is to attempt to isolate types of compounds or molecular fragments, into separate categories, dependent on physical and/or chemical properties. For example, we could record the solubility in a solvent and the melting point of two types of compounds and display the results on a graph (Figure 7.2a). Hopefully, one type of compound, e.g. carboxylic acids, will form one cluster of data points while the other type, e.g. primary amines, will form another cluster as illustrated. These two clusters will be separated by a decision line; if more than two properties are used, the clusters will be separated by a plane (three dimensions) or a hyperplane. Rather than attempt to construct such planes, it is easier to add one more arbitrary dimension to the n-dimensional system and hence to allow the decision plane always to pass through the origin (Figure 7.2b). The way in which this is done is to add a component of unit magnitude in the (arbitrary) $n + 1$ dimension to each data point. In our example of an originally 2-dimensional pattern, the effect is to 'lift' the data points out of the x, y plane along the z-direction. As can be seen, a plane can then be drawn from the origin to separate the two clusters.

In general, we would have more than two dimensions to start with. In the example quoted earlier, an infrared spectrum at 0.1 nm intervals

278

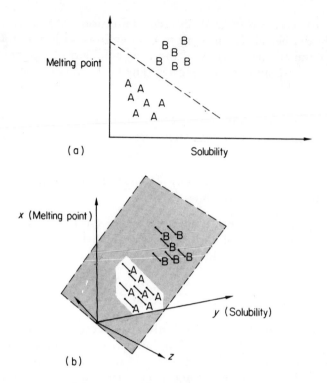

(a)

Melting point

Solubility

x (Melting point)

y (Solubility)

z

(b)

Figure 7.2   (a) Two clusters separated by a decision line.
(b) The same clusters, separated by a decision plane, by the
addition of a third component of unit magnitude

from 2.0 to 15.0 nm would yield an original 130-dimensional vector, X:

$$X = (x_1, x_2, x_3, \ldots, x_{130})$$

In order to draw a hyperplane passing through the origin, we would
redefine the data points by vectors Y:

$$Y = (x_1, x_2, x_3, \ldots, x_{130}, 1)$$

where the magnitude of component x_{131} is assigned as unity. Both the
3-dimensional plane in Figure 7.2 (b) and the 131-dimensional hyperplane needed to separate infrared spectra into clusters have the
common feature of a normal vector W at the origin, perpendicular to
the decision plane. This can be seen in Figure 7.2 (b). Therefore, by
finding the direction of this vector we automatically fix the orientation
of the plane. It is easy to calculate the components of this vector since
the dot product, s, of W with any vector, Y (corresponding to a data
point) will be positive or negative:

$$s = W \cdot Y = W Y \cos \theta$$

where θ is the angle between the vectors **W** and **Y**. If θ is less than $90°$, s is positive, and the data point is on one side of the plane (set A in Figure 7.2b). If θ is greater than $90°$ (s negative) **Y** is on the other side of the plane (set B in Figure 7.2b). Arbitrarily we can divide our data points into categories 1 and 2 with positive and negative dot-products respectively. If the first item in the training set gives the correct sign of the dot product on the basis of its category, no action is taken. If the dot product is of the incorrect sign, the normal vector **W** can be corrected by:

$$W' = W - \frac{2W \cdot Y_i}{Y_i \cdot Y_i} Y_i$$

This causes the decision plane to move to the same distance on the correct side of the point as it was previously on the incorrect side.[69] The remainder of the training set is examined in the same way until all points have been classified. The separation will be 100% if the points are well-clustered. Next, an unknown set of data points are examined, using the dot-product approach, in order to test the predictive efficiency of the decision hyperplane.

This type of technique has been applied to the classification of various types of spectroscopic information: NMR;[72] mass spectrometry;[73] infrared spectrometry[74] and combined patterns.[75] There are also applications in spectrum simulation[76] and conventional analytical chemistry.[77] Reference 77 should be of interest as it provides an introduction to a different method of pattern classification — 'non-linear mapping' — which uses graphical output to display the extent of clustering.

7.3 Organizations That Try to be Helpful
The more important organizations within the U.K. that are likely to be helpful to readers of this book are listed in Table 7.3. (The information for this Table was contributed by The National Development Programme in Computer Assisted Learning.)

The National Computing Centre (Table 7.3) is the authoritative source of information and advisory services in the U.K. Other activities include a full range of publications, a regular journal and an annual programme of conferences and courses.

For educational users seeking information about computing facilities in the U.K., the March/April 1972 issue of Computer Survey lists local authorities, universities, educational establishments and public bodies giving the type of computer and an indication of use.

Many local authorities make computer time available through Polytechnics and Universities. Information can be obtained from:

(i) Local Authority Management Services and Computer Committee, 35 Belgrave Square, London, SW1.

Table 7.3
Computer services available to U.K. physicists and chemists in education

Category	Name	Address	Nature of service
General Information Sources	National Computing Centre Limited	Quay House, Quay Street, Manchester, M3 3HU	Computing Software index
	British Computer Society	29 Portland Place, London, W1N 4AP	Courses and literature on computer science with some education aspects for schools
	National Development Programme in Computer Assisted Learning	37/41 Mortimer St., London, W1N 7RJ	Details of specific projects within NDPCAL sphere
	International Information Centre for Computing in Secondary School Education	Moray House College of Education, Holyrood Road, Edinburgh, EH8 8AO	Inquiry and information services. List of packages available
Manufacturers' Education Services	ICL	Computer House, 322 Euston Rd., London, NW1 3BD	
	IBM	389 Chiswick High Road, London W4	
	Honeywell	Honeywell House, Great West Rd., Brentford, Middx.	
	Hewlett Packard	Southern Office, Slough	
	Digital Equipment Ltd.	3, Arkwright Rd., Reading	
Program Exchanges and Resource Centres	Education Resources Information Centre (ERIC)	Centre for Research and Development in Teaching School of Education, Stanford University, Stanford, California, U.S.A.	Clearing house for work being done by schools, colleges, research institutes, industry and government in U.S.A.
	College & University Systems Exchange (CAUSE)	737, Twenty Ninth St., Boulder, Colorado 80303, U.S.A.	Cataloguing, storing, copying and distribution of systems and documentation

Table 7.3 (continued)

Category	Name	Address	Nature of service
Program Exchanges and Resource Centres (continued)	ENTELEK	Newburyport, Massachusetts 01950, U.S.A.	Updated directory of abstracts of the CAI, CMI, and PI literature; specifications of CAI and PI programs
	Quantum Chemistry Program Exchange (QCPE)	Indiana University, Indiana, U.S.A.	Storage and distribution of programs relevant to theoretical chemistry
	Physical Sciences Program Exchange (PSPE)	Department of Physical Sciences, The Polytechnic, Wolverhampton, WV1 1LY, U.K.	Storing, testing and distributing programs for the physical sciences
	Unite de Coordination, de la Documentation, D'Incitation a la Recherche, (UCODI)	UCODI, Celestijnenlaan, 200c, 3030- Heverlee, Belgium	Compiling a directory of CAL departments and CAL keywords

(ii) Advisory Unit for Computer Based Education, 19, St. Albans Rd., Hatfield, Herts.

(iii) Nuffield Foundation Computing Unit, Elliott School, Pullman Gardens, Putney, London, SW15.

The latter have a list of 30 programs useful at school level and the number is likely to grow considerably.

The Computer Services Association (CSA) at

109 Kingsway,
London, WC2B 6PU.

has over 100 members, some of which could offer valuable services at the inception of a large project.

References
Teaching Aids

1. Clark, H. A., Marshall, J. C., and Isenhour, T. L., *J. Chem. Educ.*, 50, 645 (1973).
2. Castleberry, S. J., Culp, G. H., and Lagowski, J. J., *J. Chem. Educ.*, 50, 469 (1973).
2a. Rodewald, L. B., Culp, G. H., and Lagowski, J. J., *J. Chem. Educ.*, 47, 134 (1970).
3. Breneman, G. L., *J. Chem. Educ.*, 50, 473 (1973).

4. Macmillan, J. G., and Epstein, M., *J. Chem. Educ.*, **50**, 459 (1973).
5. Johnson, R. C., *J. Chem. Educ.*, **50**, 223 (1973).

General Laboratory Computing
6. Wilkins, C. L., and Klopfenstein, C. E., *Chem. Technol. (USA)*, 564 (September 1972) and 681 (November 1972).
7. Wise, G., *J. Chem. Educ.*, **49**, 559 (1972).
8. Beech, G., *Int. J. Math. Educ. in Sci. and Tech.*, **5**, 259 (1974).
9. Davis, L. N., Coffey, C. E., and Macero, D. J., *J. Chem. Educ.*, **50**, 711 (1973).
10. Luteri, G. F., and Denham, J. M., *J. Chem. Educ.*, **48**, 670 (1971).
11. Gasser, W. L., and Emmons, J. L., *J. Chem. Educ.*, **47**, (1970).
12. Diehl, P., Kellerhals, H., and Lustig, E., *Computer Assistance in the Analysis of High Resolution NMR Spectra (NMR Basic Principles and Progress, Vol. 6)*, Springer, New York, 1972.
13. Yamamoto, L., and Someno, Z., *Kagaku No Ryoiki, Zokan*, No. 98, 27 (1972) (*Chem. Abs.*, **78**, 10, Abstract 64319, Sect. 73)
14. Rafalski, A. J., and Barciszewski, J., *J. Mol. Struct.*, **19**, 223, (1973).
15. Mayo, B. C., *Chem. Soc. Rev.*, **1**, 49 (1973).
16. Robertson, A. J. B., *MTP (Med. Tech. Publ. Co.) Int. Rev. Sci.: Phys. Chem. Ser. One*, **13**, (Analytical Chemistry, Part 2), 127–151 (1973).
17. Mantei, K., and Hunter, R. L., *J. Chem. Educ.*, **51**, 213 (1974).
18. Heller, S. R., Fales, H. M., and Milne, G. W. A., *J. Chem. Educ.*, **49**, 725 (1972).
19. Penski, E. C., Padowski, D. A., and Bouck, J. B., *Anal. Chem.*, **46**, 955 (1974).
20. Drobyshev, Yu. P., Nigmatullin, R. S., and Lovanov, V. I., *Izv. Sib. Otd. Akad. Nauk, SSSR, Ser. Khim. Nauk*, 108 (1972).
21. 'Iris' System, Heyden and Son Ltd., Spectrum House, Alderton Crescent, London NW4.
22. Dessy, R. E., and Titus, J., *Anal. Chem.*, **46**, *294A (1974)*.
23. Perone S. P., *J. Chem. Educ.*, **47**, 105 (1970).
24. Rossotti, F. J. C., Rossotti, H. S., and Whewell, R. J., *J. Inorg. Nucl. Chem.*, **33**, 2051 (1971).
25. Likussar, W., *Anal. Chem.*, **45**, 1926 (1972).
26. Robrecht, R., Steyaert, H., and Thun, H. P., *Bull. Soc. Chem. Belg.*, **82**, 505 (1973).
27. Sillen, L. G., *Acta. Chem. Scand.*, **16**, 159 (1962).
28. Dyrssen, D., Jagner, D., and Wengelin, F., *Computer Calculation of Ionic Equilibria and Titration Procedures*, John Wiley and Sons Ltd., London, 1968.

29. Magnell, K. R., *J. Chem. Educ.*, 50, 619 (1973).
30. Sabatini, A., Vacca, A., and Gans, P., *Talanta*, 21, 53 (1974).
31. Hogg, J. L., *J. Chem. Educ.*, 51, 109 (1974).
32. Seyse, R. J., and Rose, T. L., *J. Chem. Educ.*, 51, 112 (1974).
33. Cummins, J. D., and Wartell, M. A., *J. Chem. Educ.*, 50, 544 (1973).
34. Grinwald, A., and Steinberg, I. Z., *Anal. Biochem.*, 59, 593 (1974).
35. Spratt, J. L., *Liquid Scintillation Counting, Volume 2,* edited by M. A. Crook, P. Johnson and B. Scales, Heyden Ltd., London, 1971 (2 volumes).
36. Larson, R. E., and Repace, J. L., *U.S. Nat. Tech. Inform. Serv., Ad. Rep.,* 1973, No. 769219/7GA, 53 pp., *NTIS/Govt. Rep Announce (U.S.)* 74 (1), 152 (1974).
37. Baba, H., Sekine, T., Baba, S., and Okashita, H., *Nucl. Sci. Abstr.*, 29, 18473 (1974).
38. Lux, F., and Gierl, A., *Nucl. Sci. Abstr.*, 28, 17846 (1973).
39. Hertogen, J., Dedonder J., and Gijbels, R., *Nucl. Instrum. Methods,* 115, 197 (1974).
40. Yules, H. P., *J. Radioanal. Chem.*, 15, 695 (1973).
41. Galzan, R. H., and Ryan, V. A., *J. Chem. Educ.*, 49, 591 (1972).
42. Seim, T. O., and Prydz, S., *J. Chromatog.*, 73, 183 (1972).

Experimental Design
43. Szonyi, G., *Chem. Technol.*, 36 (January 1973).

Quantum Chemistry
44. Streitweiser, A., Jr., and Owens, P. H., *Orbital and Electron Density Diagrams. An Application of Computer Graphics,* Macmillan, Riverside, New Jersey, 1973.
45. Parrett, F. W., and Peterson, E., *J. Chem. Educ.*, 50, 122 (1973).
46. Holmgren, S. L., and Evans, J. S., *J. Chem. Educ.*, 51, 189 (1974).
47. Cox, M., and Elton, L. R. B., *Am. J. Phys.*, 42, 340 (1974).
47a. Cox, M., Elton, L. R. B., and Gray, R. G., *Int. J. Math. Educ. in Sci. and Technol.*, 5, 157 (1974).
48. Janis, F. T., and Peterson, E., *J. Chem. Educ.*, 50, 622 (1973).
49. Stoklosa, H. J., *J. Chem. Educ.*, 50, 290 (1973).
50. Castro, E. A., Ferro, C., and Amorebieta, V., *Amer. J. Phys.*, 42, 612 (1974).
51. Frenkel, E. C., and Davis, D. D., *J. Chem. Educ.*, 50, 80 (1973).
52. Alderdice, D. S., and Watts, R. S., *J. Chem. Educ.*, 47, 123 (1970).
53. Ballhausen, C. J., and Gray, H. B., *Molecular Orbital Theory,* Benjamin, New York, 1965.

284

54. Davies, D. R., and Webb, G. A., *Coord. Chem. Rev. (C)*, **6**, 95 (1971).
55. Litinski, A. O., and Rakuaskas, R. I., *Theor. and Exp. Chem.*, **4**, 2 (April/March 1971).

Simulation
56. Garbarino, J. R., and Wartell, M. A., *J. Chem. Educ.*, **50**, 792 (1973).
57. Hutchence, D. K., and Hontzeas, S., *Nucl. Instrum. Methods*, **116**, 217 (1974).
58. McDonald, I. R., and Singer, K., *Chem. Brit.*, **9** 54 (1973).
59. Schofield, P., *Comput. Phys. Commun. (Netherlands)*, **5**, 17 (1973).
60. Ferguson, R. C., *J. Magn. Resonance*, **12**, 296 (1973).
61. Ling, A. C., *J. Chem. Educ.*, **51**, 174 (1974).
62. Tennant, W. C., *N. Z. Dept. Sci. Ind. Res., Chem. Div. Rep.* (Technical Report), 1972, No. C.D. 2153. (*Chem. Abs.*, **78**, 20, Abstract 130297, section 73).
63. Joslin, T., and Pletcher, D., *J. Electroanal. Chem. and Interfacial Electrochem.*, **49**, 171 (1974).
64. Sandifer, J. R., and Buck, R. P., *J. Electroanal. Chem. and Interfacial Electrochem.*, **49**, 161 (1974).

Chemical Information
65. *Computer Representation and Manipulation of Chemical Information*, edited by W. Todd Wipke, S. R. Heller, R. J. Feldmann and E. Hyde, John Wiley and Sons, New York, 1974.
66. Palmer, G., *Chem. Brit.*, **6**, 422 (1970).
67. Smith, E. G., *The Wiswesser Line-Formula Chemical Notation*, McGraw-Hill, New York, 1968.
68. Campey, L. H., Hyde, E., and Jackson, A. R. H., *Chem. Brit.*, **6**, 427 (1970).
69. Jurs, P. C., Reference 65, p. 265.
70. Isenhour, T. L. and Jurs, P. C., *Anal. Chem.*, **43**, 20A (1971).
71. Beech, G., Jones, R. T., and Miller, K., *Anal. Chem.*, **46**, 714 (1974).
72. Kowalski, B. R., and Reilly, C. A., *J. Phys. Chem.*, **75**, 1402 (1971).
73. Isenhour, T. L., and Jurs, P. C., *The Applications of Computer Techniques in Chemical Research*, edited by P. Hepple, Institute of Petroleum, London, 1972, (Overseas sales: Applied Science Publishers Ltd., Barking, Essex, England).
74. Kowalski, R., Jurs, P. C., Isenhour, T. L., and Reilly, C. N., *Anal. Chem.*, **41**, 1945 (1969).

75. Jurs, P. C., Kowalski, B. R., and Isenhour, T. L., *Anal. Chem.*, 41, (1969).
76. Schechter, J., and Jurs, P. C., *Appl. Spectrosc.*, 27, 30 (1973).
77. Kowalski, B. R., *Chem. Technol.*, 300 (May 1974).

Contour Plotting Program for a CALCOMP Plotter

(Program written by Mr. J. P. H. Burden, Department of Mathematics
and Computing, Wolverhampton Polytechnic.)

Two programmes are used for contour plotting on the ICL1903A at
this Polytechnic, namely:

(1) ZCMT — this copies data from punched cards to magnetic tape.

(2) CONT — which generates and plots contours using data supplied
on magnetic tape. A listing of this program is given in Table A1.

For each contour map, input to CONT, whether from ZCMT or from
a magnetic tape file created by the user for his own program, consists of
the following series of unformatted records in a file named CONTOUR
DATA:

(i) Size of largest side of a map in plotter units ($\geqslant$ 15).

(ii) The number of rows of data (negative if curved contours
are required) and the number of columns of data (negative if marking
of data points is required). Neither must exceed 60.

(iii) The number of contour levels (negative if data points are to be
annotated).

(iv) The values of the contour levels.

(v) A title up to 32 characters in length.

(vi) The data, row by row (the number of items of data per card
must equal the number of columns in record (ii)). One row of data is
contained in one record.

Records (ii) and (iii) are integer, the others (except (v)) are real. The
file block size used was 128 words. Program CONT produces, in
addition to graphical output, line-printer listings of the data together
with certain error messages.

Table A.1
Listing of the CONT contour plotting routine

```
      MASTER CONT                                                      EJ000100
      LOGICAL FK,FF,FT,LP(500),CSW,ISW,INDIC1,CONTYPE,SPIN             EJ000110
      DIMENSION Z(60,60),XP(500),YP(500),TITLE(4),CC(50)              EJ000120
      INTEGER ER1(15),ER2(15),ER3(15),ER4(15),ER5(15)                EJ000130
      DATA  ER1(1)/' LARGEST SIDE > 30,0,  SET TO 30,0'/,            EJ000140
    C       ER2(1)/' TOO MANY ROWS OR COLUMNS OF DATA, RUN ABANDONED,'/,EJ000150
    C       ER3(1)/' TOO MANY CONTOUR LEVELS, RUN ABANDONED,'/,       EJ000160
    C       ER4(1)/' ERROR IN INPUT DATA, RUN ABANDONED,'/,           EJ000170
    C       ER5(1)/' TOO MANY POINTS AT THIS CONTOUR LEVEL, CONTOUR ABANEJ000180
    -DONED'/                                                          EJ000190
      COMMON/CONDATA/XP,YP,HP,SCF,CONTYPE                             EJ000192
      CALL STARTPLOT                                                  EJ000200
      CALL HGPLOT (0,0,34,0,0,4)                                      EJ000210
      YSOFAR,YLAST,XTC=0,0                                            EJ000215
  254 READ (4,END=450) SLS                                            EJ000220
      SPIN,CONTYPE,CSW,INDIC1=,FALSE,                                 EJ000230
      IF(SLS,LE,30,0) GO TO 255                                       EJ000240
      SLS=30,0                                                        EJ000250
      WRITE (9,500) ER1                                               EJ000260
  255 READ (4) IMAX,JMAX                                              EJ000270
      IF(IMAX,GT,0) GO TO 2555                                        EJ000271
      IMAX=-IMAX                                                      EJ000272
      CONTYPE=,TRUE,                                                  EJ000273
 2555 IF(JMAX,GT,0) GO TO 2556                                        EJ000274
      JMAX=-JMAX                                                      EJ000275
      SPIN=,TRUE,                                                     EJ000276
 2556 IF(IMAX,LE,60,AND,JMAX,LE,60) GO TO 256                         EJ000277
      IF(IMAX,LE,60,AND,JMAX,LE,60) GO TO 256                         EJ000280
      WRITE (9,500) ER2                                               EJ000290
      STOP                                                            EJ000300
  256 FJ=JMAX-1                                                       EJ000310
      FI=IMAX-1                                                       EJ000320
      SCF=SLS/AMAX1(FI,FJ)                                            EJ000330
  257 READ (4) KN                                                     EJ000340
      IF(KN,GT,0) GO TO 258                                           EJ000350
      CSW=,TRUE,                                                      EJ000360
      KN=-KN                                                          EJ000370
  258 IF(KN,LE,50) GO TO 259                                          EJ000380
      WRITE (9,500) ER3                                               EJ000390
      STOP                                                            EJ000400
  259 READ (4) (CC(J),J=1,KN)                                         EJ000410
      READ (4) TITLE                                                  EJ000420
      WRITE (9,501) TITLE                                             EJ000430
      WRITE (9,504) IMAX,JMAX                                         EJ000435
      SPIN=SPIN,OR,CSW                                                EJ000437
  260 DO 261 I=1,IMAX                                                 EJ000440
  261 READ (4,ERR=262,END=263) (Z(I,J),J=1,JMAX)                      EJ000450
      GO TO 264                                                       EJ000460
  262 WRITE (9,500) ER4                                               EJ000470
      STOP                                                            EJ000480
  263 ISW=,TRUE,                                                      EJ000490
  264 DO 265 I=1,IMAX                                                 EJ000500
  265 WRITE (9,502) I,(Z(I,J),J=1,JMAX)                               EJ000510
C                                                                     EJ000520
C                                                                     EJ000530
C              ALL DATA NOW INPUT                                     EJ000540
  500 WP=FJ*SCF                                                       EJ000550
      HP=FI*SCF                                                       EJ000560
      CH=WP*7,0/192,0                                                 EJ000570
      SCS=HP+0,1                                                      EJ000580
      YNEXT=1,1+CH+HP                                                 EJ000584
      CALL HGPWHERE (XPEN,YPEN)                                       EJ000586
      IF(YSOFAR+YNEXT,LT,54,0) GO TO 3001                            EJ000588
      XPEN=XPEN-XTC                                                   EJ000590
      YPEN=YPEN+YSOFAR-YLAST                                          EJ000592
      YSOFAR,YLAST=0                                                  EJ000594
      XTC=WP+2,0                                                      EJ000596
      GO TO 3002                                                      EJ000598
 3001 YPEN=YPEN-YLAST                                                 EJ000600
 3002 CALL HGPLOT (XPEN,YPEN,0,4)                                     EJ000602
      CALL HGPSYMBL (0,0,SCS,CH,TITLE,0,0,32)                         EJ000604
      CALL HGPDASHLN (WP,0,0,WP,HP,0,0)                               EJ000606
      CALL HGPDASHLN (WP,HP,0,0,HP,0,0)                               EJ000610
      CALL HGPDASHLN (0,0,HP,0,0,0,0,0)                               EJ000620
      CALL HGPDASHLN (0,0,0,0,WP,0,0,0,0)                             EJ000630
      IF(SPIN) GO TO 303                                              EJ000640
      TS=AMIN1(0,5,5,0/AMAX1(FI,FJ))                                 EJ000650
      DO 301 J=1,JMAX                                                 EJ000660
      X=(JMAX-J)*SCF                                                  EJ000670
  301 CALL HGPDASHLN (X,0,0,X,-TS,0,0)                                EJ000680
```

```
      DO 302 I=1,IMAX                                              EJ000690
      Y=(I-1)*SCF                                                  EJ000700
  302 CALL HGPDASHLN (0,0,Y,-TS,Y,0,0)                             EJ000710
      GO TO 100                                                    EJ000720
  303 CHC=SCF*0.075                                                EJ000730
      ASSIGN 304 TO IST                                            EJ000734
      IF(CSW) ASSIGN 3035 TO IST                                   EJ000736
      DX=2.0*CHC/7.0                                               EJ000740
      DY3=9.0*CHC/7.0                                              EJ000750
      DX3=3.0*DX                                                   EJ000762
      SAN=3.0*CHC                                                  EJ000764
      DO 304 J=1,JMAX                                              EJ000770
      X=(J-1)*SCF                                                  EJ000780
      DO 304 I=1,IMAX                                              EJ000790
      Y=(IMAX-I)*SCF                                               EJ000800
      XMK=X-DX3                                                    EJ000810
      YMK=Y-DY3                                                    EJ000815
      IF(.NOT.SPIN) GO TO 3035                                     EJ000820
      CALL HGPSYMBL (XMK,YMK,SAN,'+',0,0,1)                        EJ000822
      GO TO IST                                                    EJ000824
 3035 IF(Z(I,J).GE.999.95) GO TO 304                              EJ000840
      IF(Z(I,J).LE.-99.95)GO TO 304                               EJ000850
      XAN=X+DX                                                     EJ000860
      YAN=Y-DY3                                                    EJ000870
      CALL HGPNUMBER (XAN,YAN,CHC,Z(I,J),0,0,0,3,1)               EJ000880
  304 CONTINUE                                                     EJ000890
  100 DO 77 KKC=1,KN                                               EJ000900
      HC=CC(KKC)                                                   EJ000910
C                                                                  EJ000920
C                                                                  EJ000930
C             ANALYSIS AND PLOTTING FOR CONTOUR LEVEL HC.          EJ000930
C                                                                  EJ000940
      INDIC1=.FALSE.                                               EJ000950
  101 N=1                                                          EJ000960
      DO 107 J=1,JMAX                                              EJ000970
      DP=Z(1,J)-HC                                                 EJ000980
      DO 107 I=1,IMAX-1                                            EJ000990
      DN=Z(I+1,J)-HC                                               EJ001000
      P=DP*DN                                                      EJ001010
      IF(P) 105,102,107                                            EJ001020
  102 IF(DP) 104,103,104                                           EJ001030
  103 YP(N)=I-1                                                    EJ001040
      XP(N)=J-1                                                    EJ001050
      LP(N)=.FALSE.                                                EJ001060
      N=N+1                                                        EJ001070
      IF(DN.NE.0) GO TO 107                                        EJ001075
  104 IF(I.NE.IMAX-1) GO TO 107                                    EJ001080
      YP(N)=IMAX-1                                                 EJ001090
      GO TO 106                                                    EJ001100
  105 YP(N)=FLOAT(I-1)+DP/(DP-DN)                                  EJ001110
  106 XP(N)=J-1                                                    EJ001120
      LP(N)=.FALSE.                                                EJ001130
      N=N+1                                                        EJ001140
 1075 IF(N.LE.500) GO TO 107                                       EJ001143
 1076 WRITE (9,500) ER5                                            EJ001144
      GO TO 77                                                     EJ001145
  107 DP=DN                                                        EJ001150
      DO 109 I=1,IMAX                                              EJ001160
      DP=Z(I,1)-HC                                                 EJ001170
      DO 109 J=1,JMAX-1                                            EJ001180
      DN=Z(I,J+1)-HC                                               EJ001190
      P=DP*DN                                                      EJ001200
      IF(P) 108,109,109                                            EJ001210
  108 YP(N)=I-1                                                    EJ001220
      XP(N)=FLOAT(J-1)+DP/(DP-DN)                                  EJ001230
      LP(N)=.FALSE.                                                EJ001240
      N=N+1                                                        EJ001250
 1085 IF(N.GT.500) GO TO 1076                                      EJ001255
  109 DP=DN                                                        EJ001260
      MM=N-1                                                       EJ001260
      WRITE (9,550) MM,HC                                          EJ001270
  550 FORMAT (1H ,I4,' CONTOUR POINTS AT HEIGHT ',G14.4,'.')      EJ001280
      IF(MM.EQ.0) GO TO 77                                         EJ001290
      LCL=1                                                        EJ001320
      IPL=0                                                        EJ001330
   27 L=LCL                                                        EJ001340
      LP(L)=.TRUE.                                                 EJ001350
   28 DMIN=20.0                                                    EJ001360
      FF=.FALSE.                                                   EJ001370
      FK=.FALSE.                                                   EJ001380
      IXL=XP(L)                                                    EJ001390
      IYL=YP(L)                                                    EJ001400
      FIXL=FLOAT(IXL)                                              EJ001402
      FIYL=FLOAT(IYL)                                              EJ001403
      FT=XP(L).EQ.FIXL.AND.YP(L).EQ.FIYL                          EJ001404
```

```
      DO 36 K=1,MM                                                        EJ001410
      IF(LP(K)) GO TO 36                                                  EJ001420
      FF=.TRUE.                                                           EJ001430
C                                                                         EJ001440
C             THERE IS AT LEAST ONE POINT WITHOUT A LINE STARTING         EJ001450
C             ON IT                                                       EJ001460
C                                                                         EJ001470
      AY=ABS(YP(K)-YP(L))                                                 EJ001480
      IF(AY.GT.1.0) GO TO 36                                              EJ001490
      AX=ABS(XP(K)-XP(L))                                                 EJ001500
      IF(AX.GT.1.0) GO TO 36                                              EJ001510
      IX=XP(K)                                                            EJ001520
      IY=YP(K)                                                            EJ001530
      FIX=FLOAT(IX)                                                       EJ001532
      FIY=FLOAT(IY)                                                       EJ001533
      IF(FT) GO TO 35                                                     EJ001535
      IF((XP(K)-FIX)*(XP(L)-FIX)) 36,295,295                              EJ001590
  295 IF((XP(K)-FIXL)*(XP(L)-FIXL)) 36,296,296                            EJ001600
  296 IF((YP(K)-FIY)*(YP(L)-FIY)) 36,297,297                              EJ001610
  297 IF((YP(K)-FIYL)*(YP(L)-FIYL)) 36,298,298                            EJ001620
  298 IF(IPL) 0,35,0                                                      EJ001622
      IF(FIXL.NE.XP(L)) GO TO 299                                         EJ001625
      IF((XP(IPL)-XP(L))*(XP(K)-XP(L))) 299,299,36                        EJ001627
  299 IF(FIYL.NE.YP(L)) GO TO 55                                          EJ001629
      IF((YP(IPL)-YP(L))*(YP(K)-YP(L))) 35,35,36                          EJ001630
   35 DS=AX**2+AY**2                                                      EJ001650
      IF(DS.GE.DMIN) GO TO 36                                            EJ001660
      DMIN=DS                                                             EJ001670
      LN=K                                                                EJ001680
      FK=.TRUE.                                                           EJ001690
   36 CONTINUE                                                            EJ001700
C                                                                         EJ001710
C             XP(LN),YP(LN) NEAREST ALLOWABLE POINT.                      EJ001720
C                                                                         EJ001730
      IF(FK) GO TO 59                                                     EJ001740
C                                                                         EJ001750
C             TO 59 FOR 'NORMAL' JOINING UP.                              EJ001760
C                                                                         EJ001770
      IF(L.EQ.LCL) GO TO 55                                               EJ001775
      IF(ABS(XP(L)-XP(LCL)).GT.1.0) GO TO 365                             EJ001781
      IF(ABS(YP(L)-YP(LCL)).GT.1.0) GO TO 365                             EJ001782
      FXCL=FLOAT(IFIX(XP(LCL)))                                           EJ001783
      FYCL=FLOAT(IFIX(YP(LCL)))                                           EJ001784
      IF((XP(L)-FIXL)*(XP(LCL)-FIXL)) 365,801,801                         EJ001785
  801 IF((YP(L)-FIYL)*(YP(LCL)-FIYL)) 365,802,802                         EJ001786
  802 IF((XP(L)-FXCL)*(XP(LCL)-FXCL)) 365,803,803                         EJ001787
  803 IF((YP(L)-FYCL)*(YP(LCL)-FYCL)) 365,50,50                           EJ001788
  365 IF(XP(L).EQ.0.0.OR.XP(L).EQ.(JMAX-1)) GO TO 51                      EJ001790
      IF(YP(L).EQ.0.0.OR.YP(L).EQ.(IMAX-1)) GO TO 51                      EJ001800
      DO 52 JJ=1,MM                                                       EJ001810
      IF(XP(JJ).NE.AINT(XP(JJ))) GO TO 52                                 EJ001820
      IF(YP(JJ).NE.AINT(YP(JJ))) GO TO 52                                 EJ001830
      IF(((XP(JJ)-XP(L))**2+(YP(JJ)-YP(L))**2).GT.2.0) GO TO 52           EJ001840
      IF(JJ.EQ.IPL) GO TO 52                                              EJ001845
      LP(JJ)=.FALSE.                                                      EJ001850
      GO TO 28                                                            EJ001860
   52 CONTINUE                                                            EJ001870
      GO TO 53                                                            EJ001880
   50 IF(INDIC1) GO TO 51                                                 EJ001885
      IF(LCL.EQ.IPL)GO TO 365                                             EJ001888
      CALL LINKUP (L,LCL)                                                 EJ001890
      CALL CONTOUPPLOT                                                    EJ001900
      IF(.NOT.FF) GO TO 77                                                EJ001940
      GO TO 53                                                            EJ001950
   51 IF(.NOT.INDIC1) GO TO 515                                           EJ001960
  511 CALL CONTOURPLOT                                                    EJ001962
      IF(.NOT.FF) GO TO 77                                                EJ001964
      GO TO 53                                                            EJ001966
  515 INDIC1=.NOT.INDIC1                                                  EJ001970
      IF(XP(LCL).EQ.0.0.OR.XP(LCL).EQ.(IMAX-1)) GO TO 511                 EJ001972
      IF(YP(LCL).EQ.0.0.OR.YP(LCL).EQ.(IMAX-1)) GO TO 511                 EJ001973
      L=LCL                                                               EJ001975
      IPL=ISL                                                             EJ001977
      GO TO 28                                                            EJ001980
   53 DO 55 K=1,MM                                                        EJ001990
      IF(LP(K)) GO TO 55                                                  EJ002000
      LCL=K                                                               EJ002010
      INDIC1=.FALSE.                                                      EJ002013
      ISL,IPL=0                                                           EJ002015
      CALL CONTOURPLOT                                                    EJ002017
      GO TO 27                                                            EJ002020
   55 CONTINUE                                                            EJ002030
      GO TO 77                                                            EJ002031
   59 CALL LINKUP (L,LN)                                                  EJ002032
```

```
      IF(IPL.EQ.0) ISL=LN                                          EJ002037
      IPL=L                                                         EJ002039
      LP(LN)=.TRUE.                                                EJ002041
      L=LN                                                          EJ002043
      GO TO 28                                                      EJ002045
 77 CONTINUE                                                        EJ002047
      CALL CONTOURPLOT                                              EJ002048
      IF (ISW) GO TO 450                                            EJ002050
      YSOFAR=YSOFAR+YNEXT                                           EJ002060
      YLAST=YNEXT                                                   EJ002070
      XTC=AMAX1 (XTC,WP+2.0)                                        EJ002075
      GO TO 254                                                     EJ002080
450 CALL FINISHPLOT                                                 EJ002090
      STOP                                                          EJ002100
500 FORMAT (15A4)                                                   EJ002110
501 FORMAT ('1CONTOUR DATA FOR := ',4A8)                            EJ002120
502 FORMAT ('  ROW ',I2,9(G10.3,2X)/(8X,9(G10.3,2X)))             EJ002130
504 FORMAT (1H ,I3,' ROWS OF DATA,'/1H ,I3,' COLUMNS OF DATA.')    EJ002135
      END                                                          EJ002140
      SUBROUTINE LINKUP(K,L)                                        EJ003000
      LOGICAL OPEN                                                  EJ003010
      COMMON/PTARRY/LINK(500)                                       EJ003020
      COMMON/DATA/OPEN,JEND,J                                       EJ003030
      IF(J.EQ.0) GO TO 10                                           EJ003040
      IF(LINK(J).EQ.K) GO TO 11                                     EJ003050
      IF(K.EQ.LCL) GO TO 12                                         EJ003060
      IF(LINK(J).NE.LINK(1)) OPEN=.TRUE.                           EJ003065
      CALL CONTOURPLOT                                              EJ003070
 10 LCL.LINK(1)=K                                                   EJ003080
      OPEN=.TRUE.                                                   EJ003085
      J=1                                                           EJ003090
 11 J=J+1                                                           EJ003100
      LINK(J)=L                                                     EJ003110
      IF(L.EQ.LCL) OPEN=.FALSE.                                     EJ003115
      RETURN                                                        EJ003120
 12 OPEN=.TRUE.                                                     EJ003130
      JEND=J                                                        EJ003140
      GO TO 11                                                      EJ003150
      END                                                           EJ003160
      SUBROUTINE CONTOURPLOT                                        EJ004000
      LOGICAL OPEN,CONTYPE                                          EJ004010
      COMMON/CONDATA/XP(500),YP(500),HP,SCF,CONTYPE                EJ004020
      COMMON/PTARRY/LINK(500)                                       EJ004030
      COMMON/DATA/OPEN,JEND,J                                       EJ004040
      DATA TWOPI/6.28318/                                           EJ004042
      DIMENSION CVX(500),CVY(500)                                   EJ004050
      IF(J.EQ.2.AND.LINK(1).EQ.LINK(2)) GO TO 56                   EJ004055
      IF(OPEN) GO TO 3                                              EJ004070
      DO 1 K=1,J                                                    EJ004080
      LK=LINK(K)                                                    EJ004090
      CVX(K)=XP(LK)*SCF                                             EJ004100
  1 CVY(K)=HP-YP(LK)*SCF                                            EJ004110
      N=J                                                           EJ004140
      IF(J.LT.4) GO TO 14                                           EJ004151
      IF(CONTYPE) GO TO 15                                          EJ004152
 14 CALL HGPLINE (CVX,CVY,N,1)                                     EJ004154
      GO TO 2                                                       EJ004156
 15 TH1=ATAN2(CVY(1)-CVY(J-1),CVX(1)-CVX(J-1))                     EJ004158
      TH2=ATAN2(CVY(2)-CVY(1),CVX(2)-CVX(1))                       EJ004160
      IF(TH1.LT.0) TH1=TWOPI+TH1                                    EJ004162
      IF(TH2.LT.0) TH2=TWOPI+TH2                                    EJ004164
      TH=0.5*(TH1+TH2)                                              EJ004166
      CALL HGPSCURVE (CVX,CVY,N,3,TH,TH)                           EJ004168
  2 JEND,J=0                                                        EJ004170
      RETURN                                                        EJ004180
  3 IF(JEND.EQ.0) GO TO 45                                          EJ004185
      DO 4 K=1,JEND                                                 EJ004190
      LK=LINK(JEND+1-K)                                             EJ004200
      CVX(K)=XP(LK)*SCF                                             EJ004210
  4 CVY(K)=HP-YP(LK)*SCF                                            EJ004220
 45 DO 5 K=JEND+1,J                                                 EJ004230
      LK=LINK(K)                                                    EJ004240
      CVX(K)=XP(LK)*SCF                                             EJ004250
  5 CVY(K)=HP-YP(LK)*SCF                                            EJ004260
      IF(J.LT.4) GO TO 54                                           EJ004262
      IF(CONTYPE) GO TO 55                                          EJ004264
 54 CALL HGPLINE (CVX,CVY,J,1)                                     EJ004266
      GO TO 56                                                      EJ004268
 55 CALL HGPSCURVE (CVX,CVY,J,0,0,0,0,0)                           EJ004270
 56 JEND,J=0                                                        EJ004272
      OPEN=.FALSE.                                                  EJ004274
      RETURN                                                        EJ004300
      END                                                           EJ004310
      FINISH                                                        EJ009990
```

The contours are formed by:

(i) Linear interpolation between the data points, which are assumed to form a regular lattice.

(ii) Joining all the points found in (i) by straight line segments.

Two forms of annotation are available depending on the sign of record (iii) as shown in Figures A.1(a) and A.1(b).

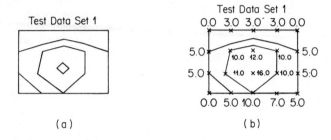

(a)   (b)

Figure A.1   The two forms of annotation by CONT: record (iii) is (a) positive (b) negative

Specimen data for ZCMT are as follows:

```
1
5.00
4   5
3
5.0   10.0   15.0
TEST DATA SET 1
    0    5    3    3    0
    5   10   12   10    5
    5   11   16   10    5
    0    5   10    7    5
```

This data would produce the first of the illustrated plots. Another example of the use of this program is given in Section 5.2.2. In order to use program CONT on other computers, the plotting subroutines used on the ICL1903A will be discussed. Before any plotting subroutines are called, initialization is achieved by the statement CALL STARTPLOT which leaves the pen in the extreme $+y$ position and initializes a magnetic tape file for the plotter. This is followed by a call to HGPLOT which assigns an origin relative to the initialized pen position. The remainder of the ICL plotting subroutines used in CONT on the ICL1900 series are defined as follows (in order of appearance in

CONT):

Name	Purpose	Result
HGPWHERE(X,Y)	Provide pen position relative to assigned origin	X and Y are current pen positions relative to the origin
HGPLOT(X,Y,IC,L)	To move the pen to a new location (or to assign an origin — see above)	With IC = 2 and L = 0 the pen moves to a new X, Y position in the pen down position. With IC = 3, the pen remains up; L = 1 is used for initialization
HGPSYMBL(X,Y,H,S,T,N)	To draw alphanumeric characters on the plotter	X and Y are the coordinates of the lower left-hand corner of the first character in the string S of N characters. H is the character height. T is the angle between the X-axis and the base of the symbol
HGPDASHLN(X0,T0,X1, Y1,D)	To draw a solid or dashed line	The line is drawn from point X0, Y0 to point X1, Y1. With D = 0.0 a solid line is drawn, with D = 0.5, a dashed line
HGPNUMBER(X,Y,H,F,T, I,IP,IQ)	To draw numbers on the plot	The number F is drawn at position X, Y. H and T are defined as for HGPSYMBL. If I is 0 the number is output in fixed-point form and with I equal to 1 the number will be in floating point. IP and IQ are the number of decimal places before and after the point respectively
HGPLINE(X,Y,N,K)	To draw straight lines between successive pairs of points	Lines connect the N points specified in the X and Y coordinate array (every successive point is connected with K equal to 1)
HGPSCURVE(X,Y,N,L, YOP,YFD)	To draw a smooth curve through a set of points	The curve is drawn through the array of N points in the X and Y arrays. With L equal to 3, YOP and YFP are the initial and final slopes of the curve specified in radians relative to the X-axis

The final plotter statement in CONT is CALL FINISHPLOT which closes the plotter tape file and leaves the pen in the extreme $+y$ position. Both STARTPLOT and FINISHPLOT make use of the ICL plotter routines HGPLOT and HGPTAPE.

The rest of the program is fairly conventional FORTRAN though making use of some special features. In particular note line numbers EJ000734 and EJ000736 which use the ASSIGN statement. The general form of this statement is

ASSIGN label$_i$ TO i

i is a variable to be used in a GO TO statement (line EJ000824) and label$_i$ is the statement number of an executable statement in the same program segment. In program CONT, if CSW is false, IST is assigned the value 304; line EJ000824 will then cause control to be tranferred to the statement

304 CONTINUE.

UKCIS SDI Profile

Explanation of UKCIS Terminology

Each line of the profile, which is matched against the data base being searched, consists of a *term* or *logic* line.

A term line consists of a term type (e.g. TEXT, KEYWORD, AUTHOR, JOURNAL CODEN, SECTION NUMBER (in *Chemical Abstracts*) and a term related to the particular type. Lines numbered 32–40, 42–46, 48–58, 60 and 61 are examples.

A logic line, such as 41, relates preceding lines by Boolean operators. For example, line 41 requires any entrant from 32 to 40 whereas line 47 requires entrant 42 to occur with any entrant 43–46.

Truncated terms such as line 56 allow retrieval of any word which begins with the sequence of letters preceding the asterisk.

Line 59 informs us that any items satisfying logic lines 41 or 47–56 *and* which is in *Chem. Abs.* sections 65–75 (line 57) will be retrieved. Line 62 is an amendment line which rejects any such items which contain the words ALGORITHM or PROGRAMMED.

The final line defines the end of the search profile. This also limits the numbers of retrievals per search (100 in this case).

Error code	Symbolic name	Line number	Search term type or operation	Display rank	Operand
		32 TEXT			COMPUT*
		33 TEXT			PROGRAM*
		34 TEXT			REAL-TIME
		35 TEXT			FORTRAN
		36 TEXT			ALGOL
		37 TEXT			INTERACTIVE
		38 TEXT			FITTING
		39 TEXT			BEST FIT
		40 TEXT			LEAST SQUARES
		41	LOGICAL		32–40 OR
		42 TEXT			DATA
		43 TEXT			PROCESSING
		44 TEXT			ACQUISITION
		45 TEXT			HANDLING
		46 TEXT			ANALYSIS
		47 TEXT	LOGICAL		42 AND 43–46
		48 TEXT			ONLINE
		49 TEXT			OFFLINE
		50 TEXT			ON-LINE
		51 TEXT			OFF-LINE
		52 JOURNAL CODEN			JCHDA
		53 TEXT			DIGIT
		54 TEXT			DIGITS
		55 TEXT			DIGITAL
		56 TEXT			DIGITIZ*
		57 R. SECTION NUMBER			66–75
		58 R. SECTION NUMBER			77–80
		59	LOGICAL		41, 47–56 and 57
		60 TEXT			ALGORITHM*
		61 TEXT			PROGRAMMED
		62	LOGICAL		59 NOT 60–61
		63	ENDSUBQUEST		C 100

Hints and Answers to Problems

Chapter 2

2.8.1 After the standard deviation and mean have been calculated, set up a DO loop to check all of the values:

```
N1 = 0
N2 = 0
N3 = 0
DO 20 I = 1, J
A = ABS (DEV(I) )
IF(A. LT STD) N1 = N1 + 1
IF (A. LT. (2*STD) ) N2 = N2 + 1
IF (A. LT. 3* STD) N3 = N3 + 1
20 CONTINUE
```

We would then print N1, N2 and N3.

2.8.2 The COMMON statement must be altered, as must all statements in which IDATA (J) appears (including input and output).

2.8.3 This is straightforward — use the equations (2.15) to (2.19).

2.8.4 Using subroutine HYPB, χ^2 is 4.448×10^{-2} whereas with subroutine POLY, χ^2 is 6.131×10^{-2}. Therefore, in this case the use of HYPB is the preferred method.

2.8.7 The easiest procedure is to define a function such as:

$$F(Z) = (Z + 1)**2$$

Then we use equation (3.41), first calculating

$$C1 = F(X1) + F(X2)$$

where X1 and X2 are the integration limits (-1 to $+1$ in this case). We then decide on a small increment (Δx in (2.41)) which can be defined by DX = (X2 − X1)/N, where N is an integer.

The two other terms in (2.41), C2 and C3, are calculated in a DO loop and the value of the integral is then calculated. We then increase N (for example, by a factor of 2) and recalculate the value of the integral until convergence is obtained. Using N = 4096, Jurs and Isenhour obtained a value of 2.66395. (Reference 13, Chapter 2.)

Chapter 3
3.4.1 Fit the expression

$$\ln (k/T) = m(1/T) + b$$

where $m = \Delta H^{\ddagger}$ and $b = \Delta S^{\ddagger}$. The answers are:

$\Delta H^{\ddagger} = 73.0$ kJ mol^{-1} and $\Delta S^{\ddagger} = -56.3$ J mol^{-1} K^{-1} (see *J. Chem. Ed.*, **47**, 114 (1970)).

3.4.2 Standard thermodynamic equations are used, including

$$\Delta G^{\theta} = -RT \ln K$$

where K is the desired equilibrium constant. From the test data provided you should obtain 0.11397 for K.

3.4.3 You must express an equation for K in a logarithmic form and then use a least squares technique. The test data should result in

$n = 1.068$ and $K = 1.693, 1.780$,   and   1.693.

3.4.4 Replace the volume readings by any other extensive measurement that is a measure of the extent of reaction. Change the format statements accordingly.

3.4.5 Using program MULT (Section 3.2.1) as a model we set the 'path length' and 'concentrations' to unity; use m/e values in place of wavelengths and peak heights in place of absorbances. From the data provided, our modification of MULT provided 'concentrations' of A,B,C and D of 0.3859, 0.2065, 0.3869, 0.9686 respectively. From these relative values it is a simple matter to calculate partial pressures.

3.4.6 The most economical method, in terms of core store used by most graph plotters is to use one set of symbols for pH (e.g. *, +) and another set for the derivative values. The visual effect is improved by joining one set by a solid line and the other by a dashed line.

3.4.7 Sufficient details are to be found in Reference 21 (Chapter 3).

3.4.8   Use the polynomial fitting routine POLY (Section 2.4.2).

Chapter 4
4.5.1   Read in one character after each half-life to denote the time unit (e.g. S for second) in A1 format. Use a logical IF to compare this character with the time unit used for the total observation time and scale appropriately

4.5.2   Smaller relative values of N1, N2 and N3 should yield a more accurate and less 'lumpy' graph.

4.5.3   The full equation for $[H^+]$ should be used, not an approximation.

4.5.4   A useful addition would be to output the data graphically. Note that a similar random number generator to that used in program MONT (Table 4.2) could be used.

4.5.5   Consult the paper by Jorgensen for guidance here. (Reference 32, Chapter 4)

Chapter 5
5.5.1   Consider the way in which the energy levels split on reducing the symmetry from cubic through the flattened cube to the unsymmetrically shaped box. Examine the numbers of levels with identical energy, in each case.

5.5.2   This is a straightforward application.

5.5.3   The most probable distance corresponds to a maximum in the radial distribution function.

5.5.4   This, with other orbital maps, is a straightforward application.

5.5.5   Use equation (5.35).

5.5.6   Statements need to be included which use equation (5.35), with appropriate output formats.

5.5.7   Equation (5.39) must be used towards the end of a DO loop which encompasses the usual Huckel calculations in BEE2 to generate new α-values. The iteration should be terminated when an acceptably small change in, e.g., bond orders has occurred.

Chapter 6

6.4.1 The equations required for this problem are given in Section 6.2.1. It is important that the total recording time is kept sufficiently small so that the computer memory does not overflow.

6.4.2 The standard deviation should be calculated before and after smoothing. This should be reduced by factors of N5 and 3 respectively, for 5- and 9-point smoothing.

6.4.3 This will give similar results although second derivative methods are more noise-prone than others. Therefore smoothing should be carried out prior to peak location.

6.4.4 You can virtually eliminate the array $Z(3000)$ by calculating the first derivatives point by point, rather than for the whole spectrum, and carrying out this task in the subroutine SPRINT.

6.4.5 Checks are easily built into the program to allow for small peaks. The best procedure is to express the peak areas as percentages of the total and to reject those below, say, 1%. Further details are to be found in Reference 2.

6.4.6 It would be useful if your program lists the value of the cross-correlation function with respect to a variable lag between the two recordings. You should find a near-linear dependence of the peak height or correlation area on the concentration of the substance. Some form of graphic display is invaluable here.

Index